IntechOpen

Supporting open minds since 2005

Polyimide for Electronic and Electrical Engineering Applications
http://dx.doi.org/10.5772/intechopen.77597
Edited by Sombel Diaham

Contributors
Daomin Min, Xiaoping Wang, Shengtao Li, Yan Yang, Guangning Wu, Jin Li, Boxue Du, Ranran Xu, Jiwen Xing, Janet Ho, Marshall Schroeder, Sang Youl Kim, Kunihiko Tajiri, Takahiro Muto, Didier Marty-Dessus, Laurent Berquez, Gilbert Teyssedre, Marie-Laure Locatelli, Sombel Diaham, Virginie Griseri, Flora Carrasco, Jerome Castellon, Jean Pierre Habas, Serge Agnel, Shakeel Akram, Andrea Cavallini, Tao Han, Jun-Wei Zha, George Chen, Xue-Jie Liu, Yaya Tian, Zhi-Min Dang, Baoxing Chen, Zhongfu Zhou, Wenlin Chen, Bin Chen, Hui Ding, Ying Zhang, Rao Fu, Yanya Jin, Xuejiao Sun, Jiansu Yu, Hiroaki Miyake, Yasuhiro Tanaka, Marc Abadie, Radu Dan Rusu, Sun Dal Kim, Taejoon Byun, Jin Kim, Im Sik Chung

First published in London, United Kingdom, 2021 by IntechOpen
IntechOpen is the global imprint of INTECHOPEN LIMITED, registered in England and Wales, registration number: 11086078, 5 Princes Gate Court, London, SW7 2QJ, United Kingdom
Printed in Croatia

British Library Cataloguing-in-Publication Data
A catalogue record for this book is available from the British Library

Additional hard and PDF copies can be obtained from orders@intechopen.com

Polyimide for Electronic and Electrical Engineering Applications
Edited by Sombel Diaham
p. cm.
Print ISBN 978-1-83880-097-0
Online ISBN 978-1-83880-098-7
eBook (PDF) ISBN 978-1-83880-379-7

For EU product safety concerns:
IN TECH d.o.o., Prolaz Marije Krucifikse Kozulić 3, 51000 Rijeka, Croatia, info@intechopen.com or visit our website at intechopen.com.

Meet the editor

Sombel Diaham obtained an MSc and Ph.D. in Electrical Engineering from the University of Toulouse, France, in 2005 and 2007, respectively. In 2008, he joined the Laboratory on Plasma and Conversion of Energy (LAPLACE) as an associate professor. He is an expert in electrical isolation for HV power and integrated electronics. He works on polymer films, thin inorganic layers, and encapsulation resins. His interests concern the physics of dielectrics (charge injection, breakdown mechanisms), electrical characterizations, properties-process relationships, modeling and multi-stress aging of insulation, and the application to real devices. Dr. Diaham develops advanced composites and nanocomposites to enhance devices' and systems' reliability. In this role, he designs field grading composites tailored by (di)electrophoresis and applies them to power electronics isolation.

Contents

Preface

Polyimide is one of the most efficient polymeric materials in many industries due to its excellent thermal, electrical, mechanical, and chemical properties, as well as its easy processability. In particular, in the electronic and electrical engineering industries, polyimide has widely been used for decades thanks to its very good dielectric and insulating properties at the high electric field and up to temperatures of around 200°C in long-term service, and sometimes at even higher temperatures for short-term service.

Since polyimide's discovery in the mid-1950s, it has become a key material for designing reliable electrical devices and systems for use in applications ranging from low to high voltage. Polyimide is also essential for the development of new types of electronic devices where further considerations such as high power density, integration, higher temperature, thermal conduction management, energy storage, reliability or flexibility are required in order to sustain the growing electrical energy consumption needs of the global society.

Consequently, polyimide materials have faced and will face new and exciting fundamental, technological and environmental challenges, including:

- a better understanding of its intrinsic electrical and dielectric properties to identify current limitations and propose new, advanced device designs

- the development of innovative composites and nanocomposites structuration to tailor its physical properties (electrical, thermal, mechanical) by involving classical and original nanoparticles such as graphene, carbon nanotubes, metal, silicates, nitrides, and so on

- the development of polyimide-based composites for energy storage (high-k composites, conductive electrodes for lithium-ion batteries), thermal management (high thermal conductive composites), reinforced nanodielectrics, and corona-resistant nanocomposites (high electric field, partial discharge resistant)

- the development of new low dielectric constant (low-k and ultra-low-k) polyimide chemistries for microelectronics (fluorinated polyimides, nanoporous, mesoporous)

- the development of new higher temperature reliable polyimide (high glass transition, high degradation temperature)

- the processability, performance, and flexibility required for new digital isolators, organic substrates, or thin-film transistor applications

- the emergence of bio-based, eco-friendly polyimide process routes to sustain societal environmental challenges

With all these challenges addressed, polyimide should be well placed for technological breakthroughs in the field of electronic and electrical engineering applications in the coming decades. In addition, many challenges regarding the understanding of the aging mechanisms under single or multiple constraints and the prediction of accurate lifetimes using both powerful physical modeling and design of experiments is a ubiquitous question in most electronic industries.

This book reviews the main applications of polyimide in the field of electronic and high-voltage engineering. The volume is organized as follows:

- **Section 1: General Considerations and Technological Processes of Polyimide for Electronics and Electrical Systems**

Chapter 1: *Polyimide in Electronics: Applications and Processability Overview* (by S. Diaham)

- **Section 2: Polyimide in Microelectronic Applications**

Chapter 2: *Synthesis and Properties of Fluorinated Polyimides from Rigid and Twisted Bis(Trifluoromethyl)Benzidine for Flexible Electronics* (by S.D. Kim et al.)

Chapter 3: *Design, Fabrication, and Application of Colorless Polyimide Film for Transparent and Flexible Electronics* (by C. Wenlin, C. Bin, D. Hui, F. Rao, Z. Ying, S. Xuejiao, J. Yanya, Y. Jiansu, Z. Zhongfu)

- **Section 3: Polyimide in Space Applications**

Chapter 4: *Charging and Discharging Mechanism of Polyimide under Electron Irradiation and High Voltage* (by X. Wang, D. Min, S. Li)

Chapter 5: *Polyimide Used in Space Applications* (by V. Griseri)

- **Section 4: Polyimide in Power Conversion Applications**

Chapter 6: *Polyimide Films for Digital Isolators* (by B. Chen and S. Diaham)

Chapter 7: *Behavior of Space Charge in Polyimide and the Influence on Power Semiconductor Device Reliability* (by K. Tajiri, H. Muto, D. Marty-Dessus, L. Berquez, G. Teyssedre, M.-L. Locatelli, S. Diaham, V. Griseri, F. Carrasco)

- **Section 5: Polyimide for Energy Storage**

Chapter 8: *Polyimides as High Temperature Capacitor Dielectrics* (by J. Ho, M. Schroeder)

Chapter 9: *High-Temperature Polyimide Dielectric Materials for Energy Storage* (by J.-W. Zha, X.-J. Liu, Y. Tian, Z.-M. Dang, G. Chen)

- **Section 6: Breakdown, Aging and Lifetime of Polyimide**

Chapter 10: *Space Charge Accumulation Phenomena in PI Under Various Practicable Environment* (by H. Miyake, Y. Tanaka)

Chapter 11: *Lifetime of Polyimide under Repetitive Impulse Voltages* (by Y. Yang, G.N. Wu)

Chapter 12: *Electrical Endurance of Corona-Resistant Polyimide for Electrical Traction* (by T. Han, A. Cavallini)

- **Section 7: Future Trends in Polyimide Functionalization for High Voltage Applications: Nanocomposite Structuration and Surface Treatments**

Chapter 13: *Synthesis Process Optimization of Polyimide Nanocomposite Multilayer Films, Their Dielectric Properties, and Modeling* (by S. Akram, J. Castellon, S. Agnel, J.-P. Habas)

Chapter 14: *Effect of Molecular Structure Modification and Nano-Doping on Charge Transportation of Polyimide Films for Winding Insulation* (by B. Du, R. Xu, J. Xing, J. Li)

- **Section 8: New Sustainable and Eco-Friendly Polyimides**

Chapter 15: *New High-Performance Materials: Bio-Based, Eco-Friendly Polyimides* (by R.D. Rusu, M.J. Abadie)

Sombel Diaham
Associate Professor,
University of Toulouse,
LAPLACE Institute,
Toulouse, France

Section 1

General Considerations and Technological Processes of Polyimide for Electronics and Electrical Systems

Chapter 1

Polyimide in Electronics: Applications and Processability Overview

Sombel Diaham

Abstract

Polyimides are nowadays quite famous dielectrics and insulating materials widely used in electronics and electrical engineering applications from low voltage microelectronics up to high voltage engineering industry. They are well appreciated because of their excellent physical properties (i.e., thermal, electrical, and mechanical properties), as well as, their coating process ease either from a liquid or a gas phase. Consequently, polyimides appear in a various range of applications to efficiently separate metal levels or electrodes at different electrical potentials. This chapter intends to review the main chemical generalities of polyimides, the different monomer families, the coating and curing processes, and the main physical properties for electronic and high voltage industrial applications.

Keywords: polyimide, imidization, deposition techniques, curing process, physical properties, microelectronics, high voltage

1. Introduction

Polyimides are advanced polymeric materials that are well known for their excellent thermal, electrical, mechanical, and chemical properties [1]. Polyimides are particularly interesting for microelectronics as well as for high voltage electrical engineering industry, thanks to their high thermal stability, high glass transition temperature, low dielectric constant, and overall very good electrical insulation properties (high breakdown field, low dielectric loss factor, and low conductivity). Polyimides present also an ease of processability making them patternable for many types of integrated electronic devices [2]. In the last 20 years, with the emergence of a wide range of novel electronic applications, polyimides have regained an increasing interest from both fundamental researches and applicative research and development sides, as shown in **Figure 1** where the publication map across the scientific fields and the total number of publications in Electrical Engineering and Electronics edited journals between 1975 and 2019 are presented.

Thus, one can observe that two publication thresholds occurred in 1999 and 2011 leading now to an annual dissemination activity on polyimides close to 4,000 publications for only the Electrical Engineering and Electronics domains. This clearly proves the large interest that these polymeric materials arouse and this trend should still progress over the next coming years.

Polyimides are present in different ways (from substrates to thin coatings) and for different purposes (purely mechanical one to advanced electrical insulation) in

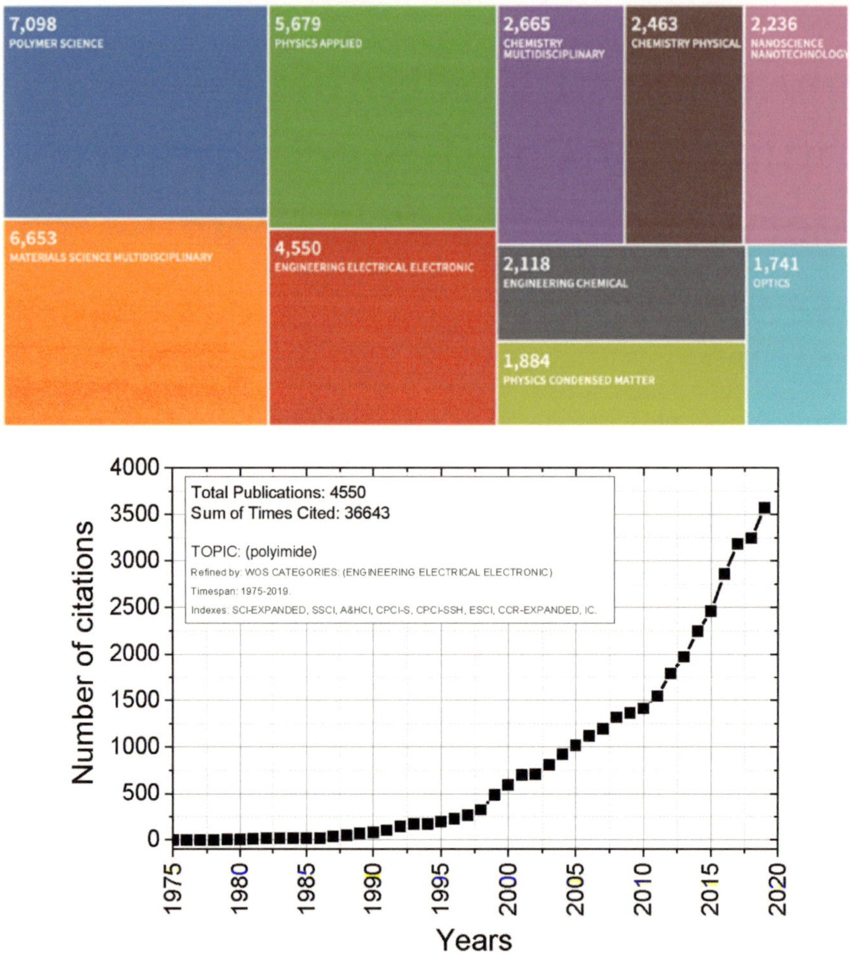

Figure 1.
All publication records map on 'polyimide' ranked by topics from 1975 to 2019 (a) and number of citations per year in the category 'Electrical Engineering Electronic' (source: *Web of science).*

electronics and high voltage applications. **Figure 2** shows an overview of the main electrical industrial applications of polyimides.

They can be listed as:

- low-*k* intermetallic layer for ultra-large scale integration in microelectronics,
- bond pad redistribution and buffer layers for ICs
- films for flexible substrate for the development of a wide range of sensors,
- rigid substrate for PCB technologies,
- protective coating for space applications,
- films for high temperature capacitors in new energy storage challenges,
- insulating layer in next generation of high temperature aeronautic cables,

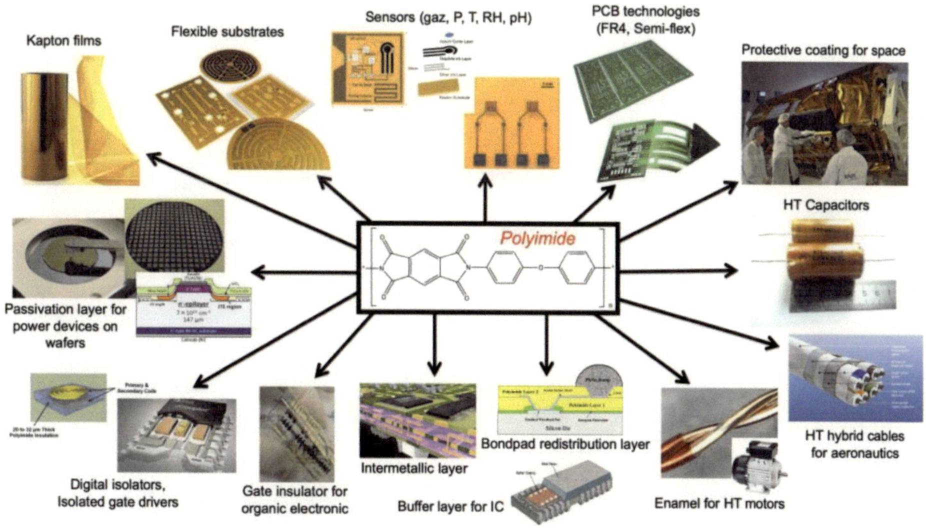

Figure 2.
Overview of the main electronic and HV engineering industrial applications of polyimide.

- high thermal class enamel for high temperature wires in electrical motor applications,
- passivation layer for surface insulation of high voltage and/or high temperature power electronic devices (Si, SiC, GaN, ...),
- insulating barrier for digital isolators in isolated gate drivers,
- gate insulator for new organic electronics, etc.

The present book overall intends to present most of these applications and the place of polyimides in their related context, the state-of-the-art as well as the next challenges to face to enable improving the efficiency of electrical devices and systems at different scales.

As a preliminary introduction, the present chapter will detail some generalities on polyimide chemistry, their thermal stability, the main chemical routes to synthesize them, their deposition and process techniques, their different curing processes, and a summary of their main physical properties.

2. Chemical generalities of polyimide

2.1 Thermal stability

The thermal stability of a polymer is defined as its ability to withstand high temperatures without initiating degradation processes such as thermolysis. It is usually evaluated by thermogravimetric analysis. This method consists in measuring the mass loss of a material either as a function of temperature or as a function of time at a constant temperature. By convention, a polymer is said to be thermostable if it can be used without losing its properties for 1000 hours at 300°C, 10 hours at 400°C, and a few minutes at 500°C. In the case of polyimides, the decomposition temperature generally appears between 500°C and 600°C. However, their typical

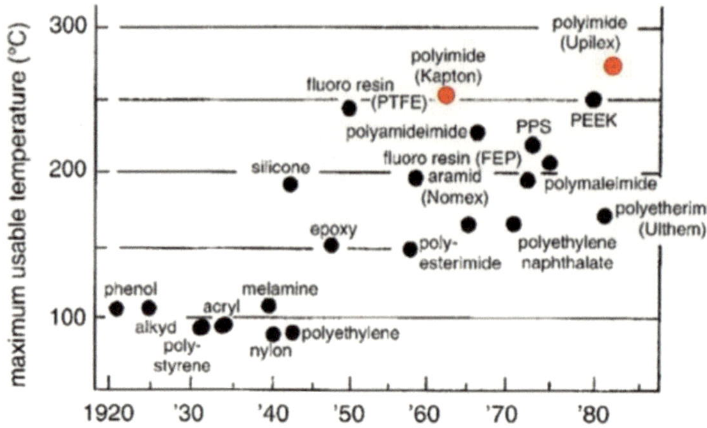

Figure 3.
Maximum working temperature for the main developed polymers over the last century including polyimides (reproduced and modified from *[1]*).

maximum usable temperature is usually between 250°C and 275°C, as shown in **Figure 3** where all the main thermostable polymers are also displayed for comparison. Thus, polyimides appear as the thermostable polymer having the highest maximum working temperature.

It has been shown that the increase in the number of benzene rings in polyimide monomer macromolecules contributes to increase their degradation temperature [1]. However, the degradation temperature can also be affected by the presence of low thermostable chemical bonds in the macromolecular structure like the C—O—C ether group [3]. Finally, a few studies even present thermal stability as high as 300°C that report on the potential use of polyimides in high temperature electronic applications for electrical insulation purposes [4, 5].

2.2 Synthesis and imidization reaction

The two-step synthesis method is the simplest and most commonly used method to obtain polyimides in industry. In 1955, Edwards et al. were the first to synthesize polyimides (PI) from polyamide salts [6]. Endrey was the first to successfully synthesize high molecular weight aromatic polyimides [7]. In the method described, the synthesis is carried out in two stages.

The first one is to prepare a polyamic acid (PAA) solution, which is the precursor of polyimide. The synthesis of PAA takes place via the reaction between two precursor monomers, a dianhydride and a diamine at room temperature and in polar aprotic solvents such as N-methyl-2-pyrrolidone (NMP), N,N-dimethylformamid (DMF) or N,N-dimethylacetamid (DMAc).

PAA is then cyclodehydrated using a thermal or chemical conversion process, called "imidization," to form the final, insoluble, and infusible polyimide. The steps for the synthesis and imidization of polyimides by this method are presented in **Figure 4**, where —R— and —R′— represent the radicals of the dianhydride and diamine monomers, respectively.

2.3 Main radicals for dianhydride and diamine precursor monomers

Many varieties of PAA can be synthesized leading to hundreds of different polyimide combinations. **Tables 1** and **2** list the main radicals —R— and —R′— of the dianhydride and diamine monomers, respectively, marketed for the synthesis of PAA.

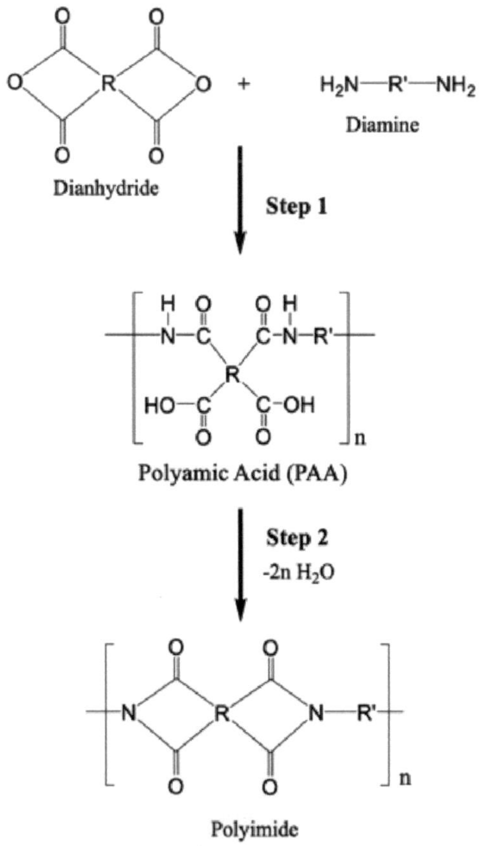

Figure 4.
Two-steps synthesis method of polyimide. Step 1: PAA synthesis. Step 2: PAA conversion into polyimide by imidization.

2.4 Polyimide deposition and process techniques

There are several routes to shape polyimides depending on the targeted application. Among them, one can cite the biaxial stretching for film production, the spin-coating and lithography process for wafer-level deposition and the vapor-deposition process. The following sections present all these.

2.4.1 Biaxial stretching for film production

Compared with the lab-scale preparation, the greatest difference for industrial manufacturing of polyimide films is the stretching process [9, 10]. Stretching process, either uniaxial or biaxial stretching of the gel-like PAA films, will result in the full orientation and extension for the polyimide molecular chains.

Figure 5 shows the main steps of the industrial scale production line of polyimide film by biaxial stretching from PAA precursors. In this procedure, the precursor monomers are firstly introduced into the polymerization reactor containing the solvent. After PAA synthesis, the obtained solution is deaerated and cast in the form of a continuous film onto the surface of a heated rotating stainless steel drum. The solvent is partially evaporated and a part of the imidization reaction takes place simultaneously. Thus, a self-supported PAA film is formed. The gel-like PAA film is peeled from the metal drum and first stretched in the machine direction (MD) while controlling the stretching rate. The PAA film is then stretched in the transverse direction (TD). The solvent is removed by evaporation, and the film is heat treated by means of hot air or radiant heat from an

Dianhydride		**Radical chemical structure (—R—)**
Pyromellitic dianhydride	PMDA	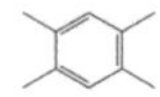
3,3′,4,4′-benzophenonetetracarboxylic dianhydride	BTDA	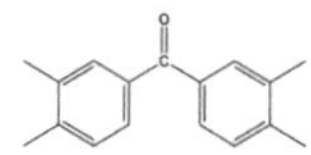
3,3′,4,4′-biphenyltetracarboxylic dianhydride	BPDA	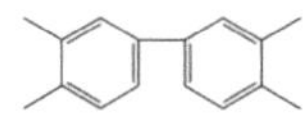
3,3′,4,4′-oxydiphthalic anhydride	ODPA	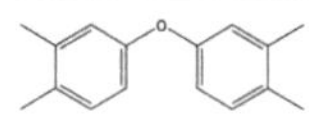
4,4′-hexafluoroisopropylidenebis(phthalic anhydride)	6FDA	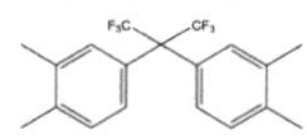
4,4′-bis(3,4-dicarboxyphenoxy) diphenylsulfide dianhydride	BDSDA	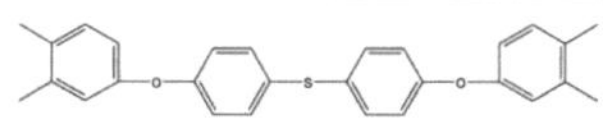
3,6-diphenylpyromellitic dianhydride	DPPMDA	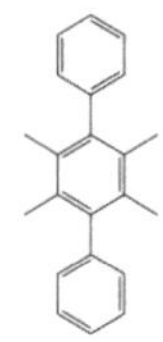
1,4-phenylenebis-(phenylmaleic anhydride)	1,4-P(PMA)	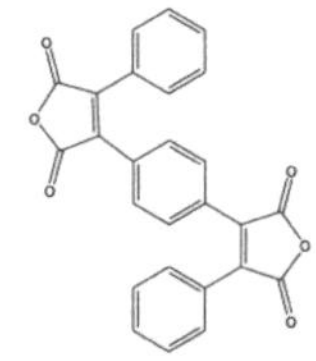
4,4′-(1-phenyl-2,2,2-trifluoroethylidene bis (phthalic anhydride)	3FDA	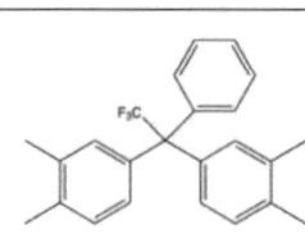
4,4′,5,5′-dioxydiphthalic anhydride	DODPA	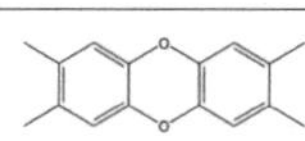
4,4′,5,5′-sulfonyldiphthalic anhydride	DSO_2DA or DSDA	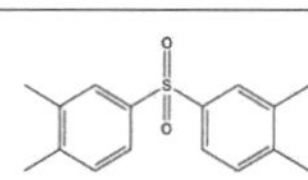

Table 1.
Main radicals for dianhydrides (reproduced from *[8]*).

electrical heater to give a biaxial oriented polyimide film. The transverse stretching is carried out at temperatures around 350°C to facilitate the imidization of PAA into polyimide. Such a procedure has been widely used for PI film production, and there has been significant patent activity in the past half century since the commercialization of PI films in 1960s. Up to now, most of the commercially available wholly aromatic PI films have been produced by such kind of procedure.

2.4.2 Spin coating and lithography for wafer-level deposition

Spin coating is a common technique widely used in microelectronic industry for applying resins in thin film form to silicon wafers. Its primary advantage over other methods is its ability to quickly and easily produce very uniform films.

Diamine		Radical chemical structure (—R′—)
p-phenylene diamine	*p*-PDA, PPD	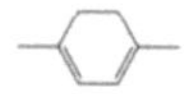
m-phenylene diamine	*m*-PDA, MPD	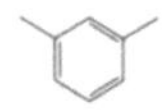
4,4′-methylene dianiline ou 4,4′-diaminodiphenyl methylene	MDA	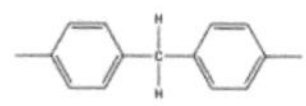
2-chlorophenylene diamine	C1PPD	–
benzidine	Bz or BDA	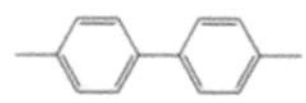
2,2′-dichlorobenzidine	2,2′-diC1Bz	–
3,3′-dimcthylbenzidine	3,3′-diMeBz or DBDA	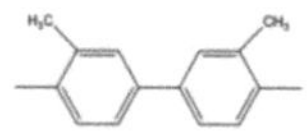
4,4′-oxydiphenylene ether or 4,4′-diaminodiphenyl ether	ODA or DDE	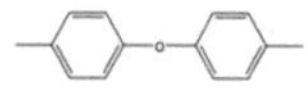
4,4′-diaminobenzophenone	4,4′-DABP	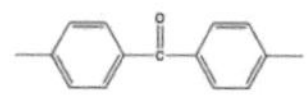
3,3′-diaminobenzophenone	3,3′-DABP	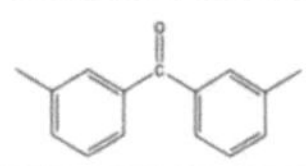
4,4′-diaminodiphenyl sulfone	4,4′-SO_2D or DDS	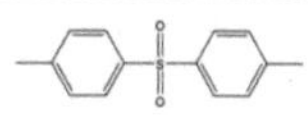
4,4′-diaminodiphenyl sulfide	SDA	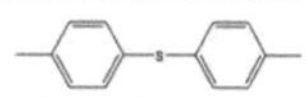
1,4-bis-(4-aminophenoxy)benzene	APB 4–1,4	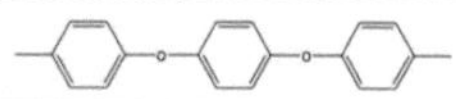
1,4-bis-(3-aminophenoxy)benzene	APB 3–1,4	–
1,3-bis-(4-aminophenoxy)benzene	APB 4–1,3	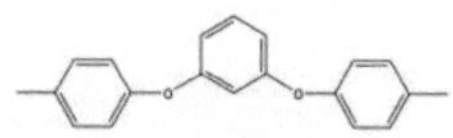
1,4-phenylindanc diamine	DAPI	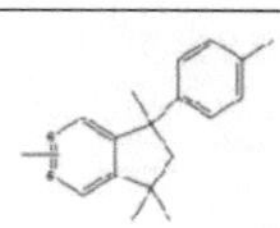
2,2-bis-[4-(4′-aminophenoxyphenyl)] hexafluoropropane	4-BDAF	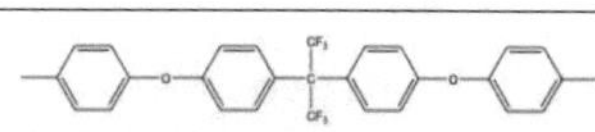
1,1-bis-(4-aminophenyl-1-phenyl-2,2,2) trifluoroethane	3FDAM	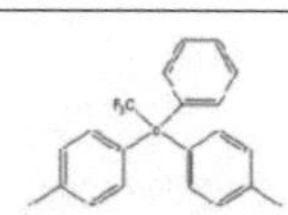
2,2-bis-(4-aminophenyl) hexafluoropropane	*p,p*′-6FDAM	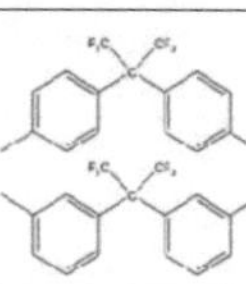
2,2′-di-bis-(trifluoromethyl) benzidine	PFMB, TFMB	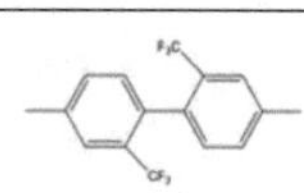
3,5-diaminobenzotrifluoride	DABTF	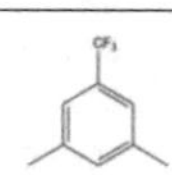

Table 2.
Main radicals for diamines (reproduced from *[8]*).

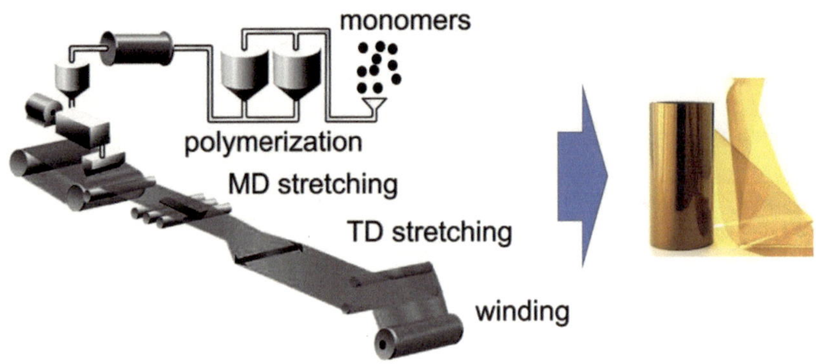

Figure 5.
*Industrial scale polyimide film stretching manufacturing from PAA precursors (*reproduced and modified from *[9]).*

When the solution of PAA containing NMP solvent is spun at high speeds, the centrifugal force enables to cover the substrate (see **Figure 6**). Spin coating results in a PAA thin film ranging from a few microns to a few tens of microns in thickness depending on coating parameters. The thickness of the final film is determined by the spinning speed, surface tension, and viscosity of the solution. The solvent is removed partly during the spinning process due to evaporation and partly by subsequent baking at elevated temperatures during imidization.

Many experiments have confirmed a mathematical model for photoresist deposition where the thickness variation versus the spin speed is like $\omega^{-0.5}$ [11]. In the

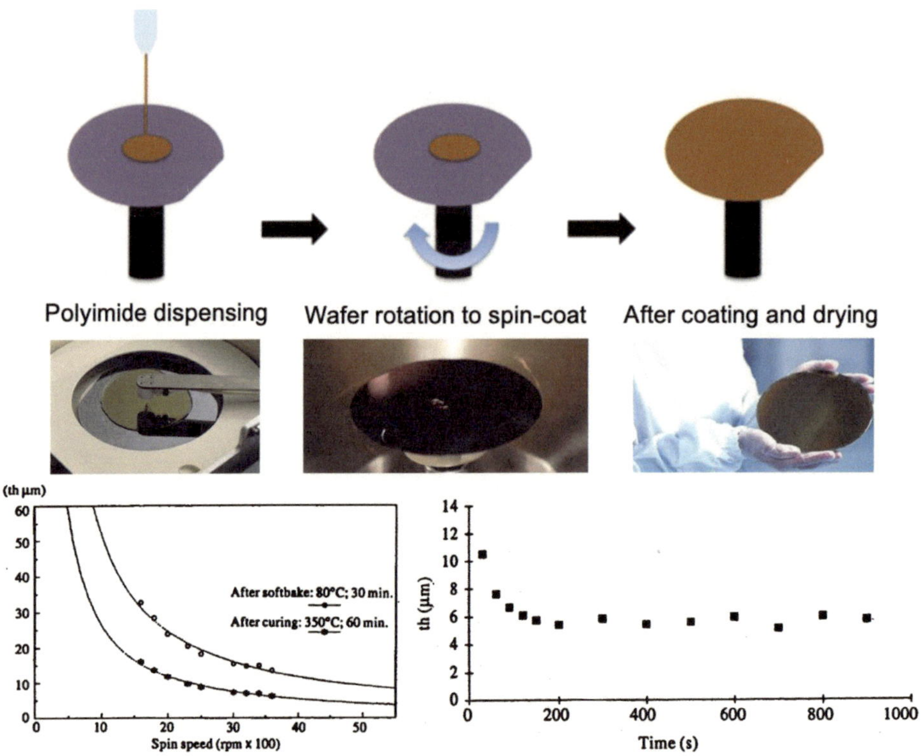

Figure 6.
*(a) Industrial polyimide spin-coating process on 8″-wafer scale from PAA precursor solution. (b) Polyimide thickness as a function of spin speed and time (*reproduced from *[11]).*

case of polyimide deposition, the thickness variation versus the spin speed shows experimentally different behavior ranging between $\omega^{-0.5}$ and ω^{-1}. It has been proved that the polyimide has a different behavior depending on the residual solvents at the end of the process. The double behavior is emphasized with two different power laws. This proves that the model is verified only if the solvent is completely removed at the end of the process.

Nowadays, the semiconductor manufacturing industry commonly process 300 mm wafers. Thus, polyimides have been developed to be spin-coated and patterned at this scale with very good thickness uniformity accuracy (around 2% of difference in thickness across the wafer) [12].

Conventional, or non-photosensitive, polyimide cannot be directly patterned on wafers due to the absence of photo-active agent (see **Figure 7a**). They require several process steps after the fabrication of the active device. To process non-photosensitive polyimide, a first thick polyimide film is spin coated on the wafer similarly to the photoresist process (see **Figure 7b**). Then, a thin layer of photoresist is applied and exposed using a photolithography tool. A standard development process of the photoresist using a mask is used to define the pattern. This pattern is transferred to the polyimide layer by wet etching through openings during the photoresist lithography step. The wet etch is an isotropic process that causes critical dimension and sidewall control issues. This technical difficulty, combined with the complexity of the process, has limited the non-photosensitive polyimide application.

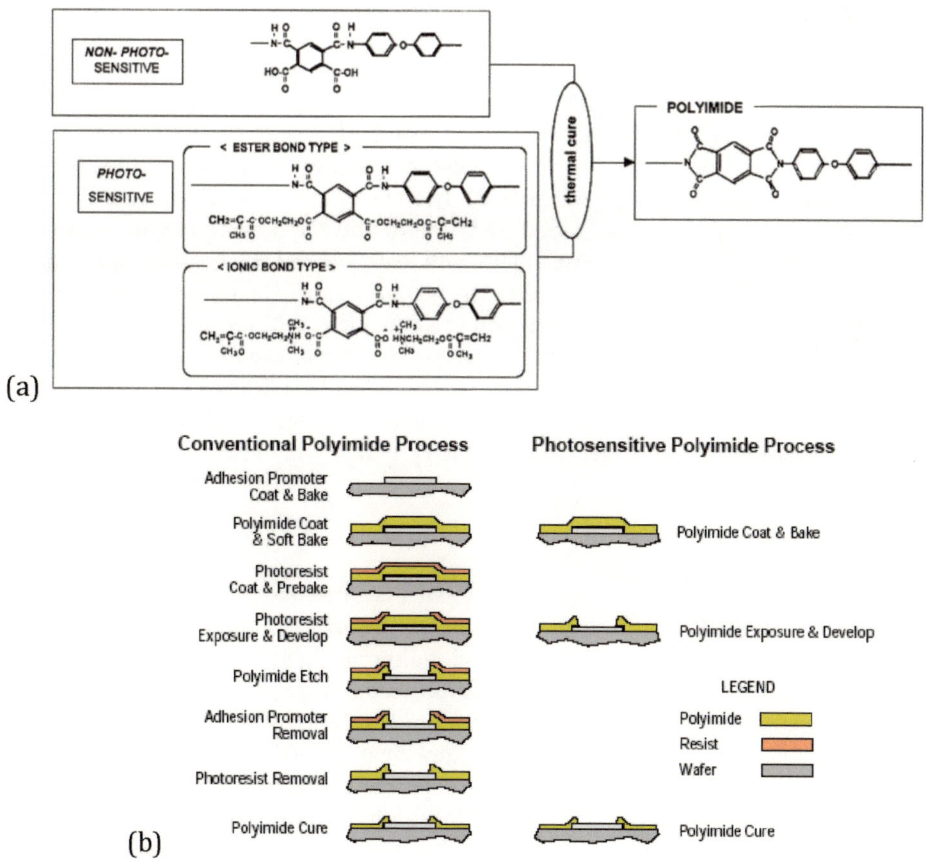

Figure 7.
Comparison of precursor PAA monomers between non-photosensitive and photosensitive polyimides (a). Lithography process steps comparison between conventional and photosensitive polyimides (reproduced from *[13]) (b).*

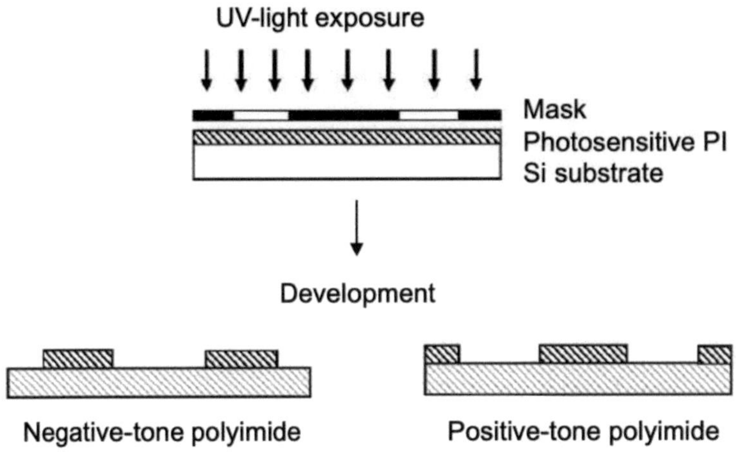

Figure 8.
Lithography process steps comparison between negative and positive photosensitive polyimide resins.

To overcome such difficulties, photosensitive polyimides have been developed to offer an alternative cost savings to the buffer coat polyimide application (see **Figure 7a**). Photosensitive polyimides can be processed similarly to standard resists using photolithography techniques, as shown in **Figure 7b**. Thus, the eight-step non-photosensitive polyimide process can be reduced into a three-step process using photosensitive polyimide. In addition to providing process simplification, this three-step process offers the significant advantages of superior resolution and improved sidewall profiles. As a consequence, cycle time and chemical consumption are reduced. All of these benefits translate into cost savings, ease of use and better quality.

Photosensitive polyimide, like photoresist, can be divided into two categories: positive and negative tones (see **Figure 8**). In the case of the positive tone, the photosensitive polyimide is degraded by UV light and the developer will dissolve away the regions that were exposed. That will leave behind the coating where the mask was initially placed. In the case of the negative tone, the photosensitive polyimide is cross-linked by UV light and the developer will remove only the unexposed regions, leaving behind the coating in areas where the mask was not placed. Application of positive photosensitive polyimide is limited because of the narrow film thickness range available. This makes the negative photosensitive one the most commonly used in electronic industry with a wide viscosity range.

2.4.3 Vapor-deposition process (VDP)

The vapor-deposition polymerization (VDP) is a method where polyimide is directly deposited and synthesized from its two precursor monomers (dianhydride and diamine), evaporated separately at high temperature in a vacuum chamber and collected on a heated substrate for imidization (see **Figure 9**).

The temperature of the transported vapors and of the substrate is usually between 100 and 200°C. A higher temperature post-annealing (≥300°C) is sometimes carried out to complete the imidization reaction. The synthesis of polyimides (with thickness between a few 1 to 10 μm) has been successfully demonstrated by this method for PMDA/ODA [16–19] and other variants of polyimides. Polyimides synthesized by VDP generally have a low oxygen permeability in comparison with conventional spin-coated versions and adhere relatively well to their substrate. However, this method remains difficult to fit with industrial manufacturing processes for electrical and electronic systems because of certain inhomogeneities in terms of thickness of the layers.

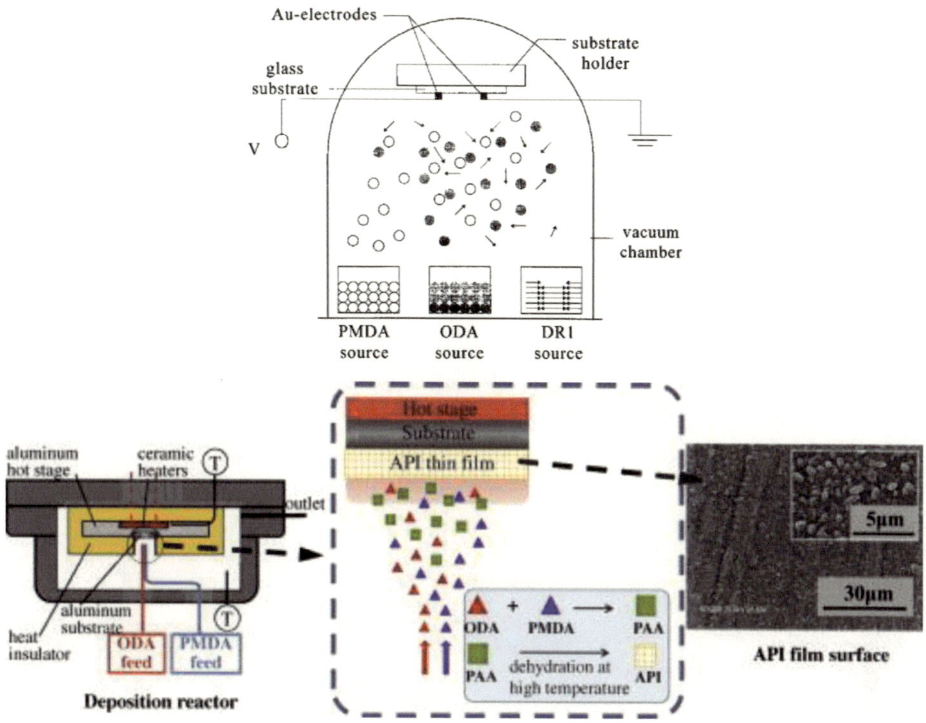

Figure 9.
Polyimide thin-films deposited by VDP from vapor phase (reproduced from *[14, 15]*).

2.5 Curing process techniques

2.5.1 Thermal curing

Different techniques for curing polyimides and complete the imidization reaction can be used and are reported in the literature. The curing enables to convert PAA into polyimide and so that to finalize the physical properties of the deposited layer. The most commonly used method is the thermal curing, which is carried out optimally at temperatures of at least 250°C under inert gas, like N_2. It is simple to implement and leads to good properties of the material.

During that critical process step, the PAA coating is slowly heated up until a first temperature plateau at 200°C corresponding to the NMP solvent boiling point. It is usually held on for at least 15 minutes in order to fully remove the solvent from the layer (see **Figure 10**). Then, a subsequent temperature rising is performed up to a second plateau at temperature from 250 to 400°C and between 30 minutes up to 2 hours to complete the imidization reaction and obtain the final polyimide. Of course, the final chemical structuration and physical properties strongly depend on the temperature and time duration [20]. Moreover, heating and cooling ramps also need to be controlled to avoid thermomechanical stress storage within the films [21].

2.5.2 Variable frequency microwave (VFM) curing

In addition, other annealing methods have started to emerge, leading to equivalent properties of the deposited polyimide layers. This is the case with variable frequency microwave (VFM) curing [22, 23].

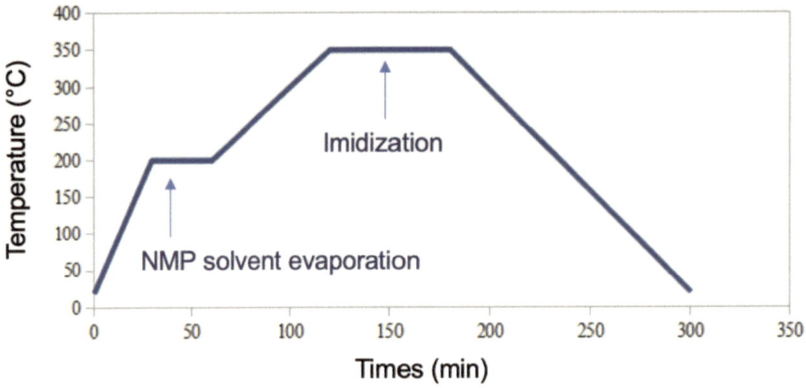

Figure 10.
Typical thermal curing profile for polyimide.

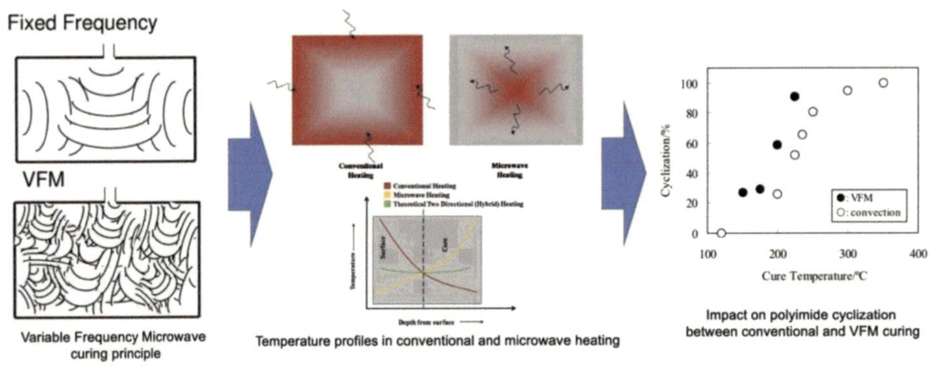

Figure 11.
Variable frequency microwave curing process for polyimide and impact on cyclization (reproduced from *[25]*).

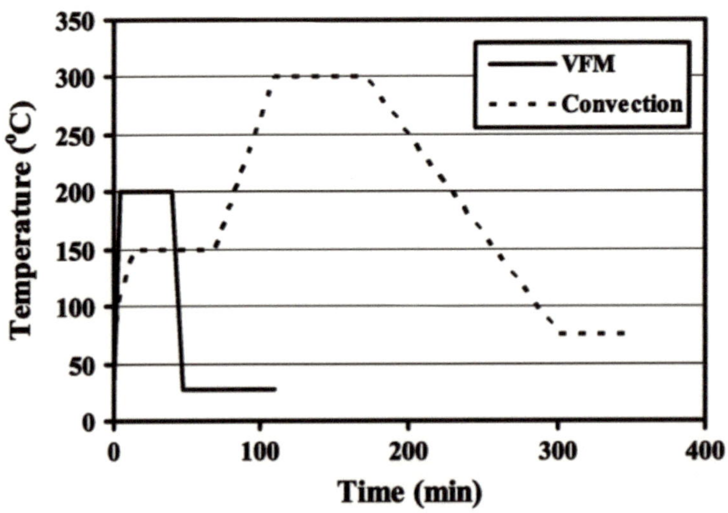

Figure 12.
Typical cure temperature–time profile of polymers using VFM and conventional thermal curing (reproduced from *[26]*).

Aromatic polyimide	T_g (°C)	Weigth loss (%)	σ (S/cm)	ε_r	tanδ	E_{BR} (MV/cm)	CTE (ppm/°C)	E′ (GPa)	E″ (MPa)	T_{BR} (MPa)	S_{BR} (%)	Moisture absorption (%)
PMDA/ODA	377–399	3%/100 h at 325°C	$< 10^{-18}$	3.1–3.5	10^{-3}–10^{-2}	4.5 (1 μm) 3.5 (10 μm)	22–40	3.0		170	40–110	1.3–3.5
BPDA/ODA							43	4.1	140	230	73–110	
BTDA/ODA	279				6×10^{-3}	1.8 (20 μm)	40	1.0		150	15	1.5
PMDA/PPD							2	12.2	400	296	5	
BPDA/PPD	477–500			2.9–3.1	2×10^{-3}		3–7.5	8.8–10.2	334–400	390–600	15–47	1.0
BTDA/PPD							30	7.1	232–400	248	18	
ODPA/PPD							35	8.1	251–350	263	18	
BTDA/ODA-MPD	320		10^{-20}				40–47	3.3				
PMDA-BPDA/ODA-PPD							19	4.72		260	44	
PMDA-BPDA/PPD (alternate)							2.3	9.6		254	7	
PMDA-BPDA/PPD (random)							5.9	9.1		248	10	
PMDA-BTDA/PPD (alternate)							9.3	8.4		189	4	
PMDA-BTDA/PPD (random)							15.2	7.4		216	9	
PMDA-ODPA/PPD (alternate)							6.8	8.1		223	10	
PMDA-ODPA/PPD (random)							20.8	5.3		184	17	
IPDA/MPD				2.8								
BTDA/DAPF	358			3.1–3.4								

Fluroninated polyimide	T_g (°C)	Weigth loss (%)	σ (S/cm)	ε_r	tanδ	E_{BR} (MV/cm)	CTE (ppm/°C)	E′ (GPa)	E″ (MPa)	T_{BR} (MPa)	S_{BR} (%)	Moisture absorption (%)
6FDA/ODA	290											
6FDA/MPD				3.0								
6FDA/PPD							48	3.8		108	6	
6FDA/DABTF	300			2.58								
6FDA/20FMDA	189	10% at 455°C		2.6								
6FDA/RfbMPD	257	5% at 472°C		2.7			86	1.7		72	6	0.5
6FDA-PMDA/TFMOB-PPD	>400			2.6–2.8			29	9.8			22	
3FXDA/ODA				2.8								
3FXDA/RfbMPD	394	5% at 465°C		2.5			67	1.9		115	25	1.1
6FXDA/ODA				2.8								
6FXDA/RfbMPD	347	5% at 458°C		2.3			70	2.0		116	28	0.6
6FXDA/TFMB	420			2.4			6					12
6FXDA/TFMOB	375			2.8			10					8
PMDA/3FDAM	420–430											

T_g: Glass transition temperature.
σ: Electrical conductivity; ε_r; Dielectric constant; tanδ: Dielectric loss factor; E_{BR}: Dielectric strength. Data given at 25°C.
CTE: coefficient of thermal expansion; E′: Young modulus; E″ Mechanical loss modulus; T_{BR}: Tensile at breakdown; S_{BR}: Elongation at mechanical breakdown.

Table 3.
Main physical properties of aromatic and fluorinated polyimides (data taken from *[8]*).

Microwave heating of polymers occurs because of dielectric loss mechanisms. When an external electric field is applied to a dielectric material, three types of polarization can occur [24]. These ones are the electronic, ionic or atomic, and orientational or dipolar polarization mechanisms. The main coupling mechanism between microwave radiation and polymer dielectrics is through dipole orientation under the applied electric field. The efficiency of coupling microwave energy into a polymer depends on different factors, which include the dipole strength, the dipole mobility and the dipole mass.

The VFM curing principle is to subject the deposited films to very high frequency waves (>1 GHz) causing the macromolecular structure to vibrate (see **Figure 11**). The vibrations then locally heat the coating which thus polymerizes. The advantages of the VFM curing method are the short annealing time. It is around 10 minutes only compared to a few hours for a conventional full thermal cycle. Moreover, the lower associated temperature (~200°C) during the application of microwaves enables shorter curing cycles (see **Figure 12**). The industrial interest could be to replace standard thermal curing techniques with faster and lower cost VFM method.

3. Polyimide physical properties ... at a glance

Polyimides are therefore materials which have very good thermal, electrical and mechanical properties. These depend essentially on their chemical structure generated by the choice and the reaction of the basic monomers constituting them.

This chapter ends by a summary of the physical properties of the main aromatic and fluorinated polyimides that have been developed over the last decades, as shown in **Table 3**.

4. Conclusion

This chapter has introduced polyimide materials which are now since few decades commonly used as dielectrics or insulating materials in the electronics and high voltage engineering industries for different purposes. It has been reviewed the state-of-the-art on the polyimide thermal stability. Moreover, the synthesis and imidization reactions, the main precursor monomers, the different deposition and process techniques, and the curing methods were presented. Finally, the main physical properties were summarized. This will offer a good overview as an introduction for the rest of this book.

Author details

Sombel Diaham
University of Toulouse, LAPLACE Research Institute, CNRS, Toulouse, France

*Address all correspondence to: diaham@laplace.univ-tlse.fr

References

[1] Sroog CE, Endrey AL, Abramo SV, Berr CE, Edwards WM, Olivier KL. Aromatic polypyromellitimides from aromatic polyamic acids. Journal of Polymer Science Part A: Polymer Chemistry. 1965;**3**:1373

[2] Gosh MK, Mittal KL. Polyimides, Fundamentals and Applications. New-York: Mercel-Dekker; 1996

[3] Diaham S, Locatelli ML, Khazaka R. BPDA-PDA polyimide: Synthesis, characterizations, aging and semiconductor device passivation. In: Abadie JM, editor. High Performance Polymers – Polyimides Based – From Chemistry to Applications. Rijeka: InTech. 2012. pp. 15-36. DOI: 10.5772/53994

[4] Khazaka R, Locatelli ML, Diaham S, Bidan P. Effects of mechanical stresses, thickness and atmosphere on aging of polyimide thin films at high temperature. Polymer Degradation and Stability. 2013;**98**:361-367

[5] Khazaka R, Locatelli ML, Diaham S, Bidan P. Endurance of thin insulation polyimide films for high-temperature power module applications. IEEE Transactions on Components, Packaging and Manufacturing Technology. 2013;**3**(5):811-817

[6] Edwards WM, Robinson IM. Polyimides of Pyromellitic Acid. E. I. Du Pont de Nemours and Company. US Patent No. 2710853; 1955

[7] Endrey AL. Process for Preparing Polyimides by Treating Polyamides-Acids with Lower Fatty Monocarboxylic Acid Anhydrides. E. I. Du Pont de Nemours and Company. US Patent No. 3179630; 1965

[8] Diaham S. Étude du Comportement sous Haute Température de Matériaux Polyimides en Vue de la Passivation de Composants de Puissance à Semi-Conducteur Grand Gap [PhD Thesis]. Toulouse: Université Paul Sabaiter; 2007

[9] Ni HJ, Liu JG, Wang ZH, Yang SY. A review on colorless and optically transparent polyimide films: Chemistry, process and engineering applications. Journal of Industrial and Engineering Chemistry. 2015;**28**:16-27

[10] Demeuse MT. Biaxial Stretching of Film: Principles and Applications. Cambridge: Woodhead; 2011

[11] Borsetto M, Carcano G, Ceriani M. Spin coating study for thick layers with high viscosity polyimide. MicroElectronics International. 1996; **13**(3):30-32

[12] Flack WW, Kulas S, Franklin C. Characterization study of an aqueous developable photosensitive polyimide on 300 mm wafers. In: Proceedings of SPIE; #4346–96. 2001. pp. 1-11

[13] Cheang P, Christensen L, Reynaga C. Optimization of photosensitive polyimide process for cost effective packaging. In: Proceedings of Surface Mount Technology Seminar. Vol. 1996, 1996. pp. 1-18

[14] Wu HS, Jou JH, Li YC, Huang JY. Real-time poling vapor co-deposition of dye-doped second-order nonlinear optical polymer thin films. Macromolecules. 1997;**30**:4410-4414

[15] Haruki M, Hasegawa Y, Fukui N, Kihara SI, Takishima S. Deposition of aromatic polyimide thin films in supercritical carbon dioxide. The Journal of Supercritical Fluids. 2014;**94**: 147-153

[16] Ito Y, Hikita M, Kimura T, Mizutani T. Effect of degree of imidization in polyimide thin films prepared by vapor deposition

polymerization on the electrical conduction. Japanese Journal of Applied Physics. 1990;**29**(6):1128-1131

[17] Iida K, Imamura Y, Liao C, Nakamura S, Sawa G. Evaluation of molecular orientation in aromatic polyimide films by FT-IR reflection absorption spectroscopy. Polymer Journal. 1996;**28**(4):352-356

[18] Karamancheva I, Stefov V, Soptrajanov B, Danev G, Spasova E, Assa J. FTIR spectroscopy and FTIR microscopy of vacuum-evaporated polyimide thin films. Vibrational Spectroscopy. 1999;**19**(2):369-374

[19] Anthamatten M, Letts SA, Day K, Cook RC, Gies AP, Hamilton TP, et al. Solid-state amidization and imidization reactions in vapor-deposited poly(amic acid). Journal of Polymer Science Part A: Polymer Chemistry. 2004;**42**(23): 5999-6010

[20] Diaham S, Locatelli ML, Lebey T, Malec D. Thermal imidization optimization of polyimide thin-films using fourier transform infrared spectroscopy and electrical measurements. Thin Solid Films. 2011; **519**(6):1851-1856

[21] Ree M, Kim K, Woo SH, Chang H. Structure, chain orientation, and properties in thin films of aromatic polyimides with various chain rigidities. Journal of Applied Physics. 1997;**81**(2): 698-708

[22] Farnsworth KD, Manépalli RN, Bidstrup-Allen SA, Kohl PA. Variable frequency microwave curing of photosensitive polyimide. IEEE Transactions on Components and Packaging Technologies. 2001;**24**(3): 474-481

[23] Tanikella RV, Bidstrup-Allen SA, Kohl PA. Novel low-temperature processing of polymer dielectrics on organic substrates by variable frequency microwave processing. In: Proceedings of the 8th International Symposium on Advanced Packaging Materials (IEEE ISAPM). 2002. pp. 254-259

[24] Tanikella RV, Bidstrup Allen SA, Kohl PA. Variable-frequency microwave curing of benzocyclobutene. Journal of Applied Polymer Science. 2002;**83**: 3055-3067

[25] Matsutani H, Hattori T, Ohe M, Ueno T, Hubbard RL, Fathi Z. Low temperature curing of polyimide precursors by variable frequency microwave. Journal of Photopolymer Science and Technology. 2005;**18**(2): 327-332

[26] Yota J, Ly H, Ramanathan R, Sun HC, Barone D, Nguyen T, et al. Variable frequency microwave and convection furnace curing of polybenzoxazole buffer layer for GaAs HBT technology. IEEE Transactions on Semiconductor Manufacturing. 2007; **20**(3):323-332

Section 2

Polyimide in Microelectronic Applications

Chapter 2

Synthesis and Properties of Fluorinated Polyimides from Rigid and Twisted Bis(Trifluoromethyl) Benzidine for Flexible Electronics

Sun Dal Kim, Taejoon Byun, Jin Kim, Im Sik Chung and Sang Youl Kim

Abstract

Fluorinated polyimides were prepared from the twisted benzidine monomer containing two trifluoromethyl (CF_3) groups on one aromatic ring. The diamine monomer having a rigid and nonplanar structure was polymerized with typical dianhydride monomers including BPDA, BTDA, ODPA, 6-FDA, and PMDA, to obtain the corresponding polyimides. Most polyimides are soluble in organic solvents due to their twisted chain structure and can be solution cast into flexible and tough films. These films have a UV-vis absorption cut-off wavelength at 354–398 nm and a light transparency of 34–90% at a wavelength of 550 nm. They also have tensile strengths of 92–145 MPa and coefficients of thermal expansion (CTE) of 6.8–63.1 ppm/°C. The polymers exhibited high thermal stability with 5% weight loss at temperatures ranging from 535 to 605°C in nitrogen and from 523 to 594°C in air, and high glass temperature (T_g) values in the range of 345–366°C. Interestingly, some of the soluble polyimides showed thermo-responsive behaviors in organic solvents presumably due to the multiple hydrogen bondings with unsymmetrically positioned two CF_3 groups along the polymer chains.

Keywords: fluorinated polyimides, rigid and nonplanar structure, flexible and tough films, thermal stability, high glass temperature, coefficients of thermal expansion

1. Introduction

Aromatic polyimides (PIs) are well known as high-performance polymeric materials having excellent thermal, mechanical, and electrical properties. As a result of these properties, many PIs have been commercialized and used widely in microelectronic and aerospace engineering [1, 2]. Recently, aromatic PIs are considered as a strong candidate for flexible plastic substrates applicable to flexible electronics, including flexible solar cell arrays and flexible organic light-emitting diode (OLED) displays [3]. Despite the outstanding results associated with the use of aromatic PIs, they also have a number of drawbacks, one of which is their poor processability caused by their limited degrees of solubility in organic solvents due to strong interchain interactions. Another shortcoming is the pale yellow or a deep brown color of PI films due to their highly conjugated aromatic structures and/or the formation of an intermolecular charge-transfer complex (CTC) between alternating electron-donor (diamine) and electron-acceptor (dianhydride) moieties, thus narrowing their applicability [4, 5].

To overcome these problems, much research effort has focused on the synthesis of soluble and transparent PIs in a fully imidized form without deterioration of their excellent properties [4, 6]. Several successful approaches to synthesize soluble and transparent PIs, including the insertion of flexible or unsymmetrical linkages or bulky substituents on the main chain and the use of noncoplanar or alicyclic monomers, have been introduced over the last few decades [4–8].

Among many approaches, the incorporation of trifluoromethyl (CF_3) groups onto polymer chains is considered as an effective means of realizing soluble and transparent PIs without deteriorating their excellent properties, not only because bulky CF_3 groups disturb the interactions and chain packing between the polymer chains, but also because the strength of the carbon-fluorine chemical bond is the one of the strongest single bonds [6, 9–30]. It is also possible to give the corresponding PIs have many attractive features, such as a low refractive index as well as low optical loss, dielectric constant, surface energy, and moisture absorption characteristics, due to the high electronegativity and low electric polarity of fluorine atoms [31–39].

Recently, we reported new soluble PIs which were prepared from 4-(4′-aminophenoxy)-3,5-bis(trifluoromethyl)aniline to introduce two CF_3 groups unsymmetrically onto the repeating units of the chain [40, 41]. Unsymmetrical incorporation of the substituents into the main chain of PIs can improve the solubility and optical transparency because increasing the irregularity of the microstructure of PIs disrupts the interchain interactions [42–47]. The PIs synthesized in earlier work showed good solubility while retaining their useful thermal and optical properties due to the unsymmetrical presence of CF_3 groups as substituents. Furthermore, the good solubility of the PIs led them to show lower critical solution temperature (LCST) behavior in organic solvents. This unprecedented phenomenon of the PIs may stem from a change of the interaction strength in the vicinity of CF_3 between the polymer chains and the acetyl-containing solvents [41].

Subsequently, we designed another monomer, 2,6-bis(trifluoromethyl) benzidine, which has two CF_3 groups at the 2,6-positions of the benzidine unit [48]. Although this monomer has more rigid structure compared to 4-(4′-aminophenoxy)-3,5-bis(trifluoromethyl)aniline, a series of poly(amide-imide)s synthesized from the monomer exhibited good solubility as well as good thermal and optical properties. Meanwhile, in terms of the structure, the new benzidine monomer has an isomeric relationship with 2,2′-bis(trifluoromethyl)benzidine, well known as a rigid/linear benzidine unit containing the CF_3 group and frequently employed in the synthesis of PIs having a high thermal resistance, a high T_g value, a low degree of thermal expansion, a low refractive index, and low water absorption capabilities [16–30]. Therefore, we envisioned that the PIs obtained from the new benzidine monomer would exhibit high thermal and mechanical properties while maintaining good solubility in organic solvents, as they have twisted structures while retaining the rigidity of the chains. The chemistry and the physical properties of the PIs prepared from the twisted benzidine monomer containing two trifluoromethyl (CF_3) groups on one aromatic ring are described herein.

2. Experiments

2.1 Materials

2-Bromo-5-nitro-1,3-bis(trifluoromethyl)benzene (**1**) and 2,6-bis (trifluoromethyl)benzidine (**3**) were synthesized as reported in our previous papers [31, 48]. The aromatic tetracarboxylic dianhydrides of pyromellitic dianhydride (PMDA), 3,3′,4,4′-biphenyltetracarboxylic dianhydride (BPDA), 3,3′,4,4′-benzophenone-tetracarboxylic dianhydride (BTDA), 4,4′-oxydiphthalic anhydride (ODPA), and 4,4′-hexafluoroisopropylidenediphthalic anhydride (6-FDA) were purified by vacuum sublimation. *m*-Cresol was stirred in the presence of P_2O_5 overnight and then distilled under reduced pressure. All other commercially available reagent-grade chemicals were used without further purification.

2.2 Measurements

The Fourier-transform infrared (FTIR) spectra of the compounds were obtained with a Bruker EQUINOX-55 spectrophotometer using a KBr pellet or film. The nuclear magnetic resonance (NMR) spectra of the synthesized compounds were recorded on a Bruker Fourier Transform Avance 400 spectrometer. The chemical shift of the NMR was reported in parts per million (ppm) using tetramethylsilane as an internal reference. Splitting patterns were designated as s (singlet), d (doublet), dd (doublet of doublets), dt (doublet of triplets), t (triplet), q (quartet), or m (multiplet). Elemental analyses (EA) of the synthesized compounds were carried out with a FLASH 2000 series device. The single-crystal diffraction data of the diimide model compound were collected on a Bruker SMART 1000 with graphite-monochromated Mo Kα radiation (λ = 0.71073 Å) at 120 K. The inherent viscosities of the polymers were measured using an Ubbelohde viscometer. Gel permeation chromatography (GPC) diagrams were obtained with a Viscotek TDA302 instrument equipped with a packing column (PLgel 10 μm MIXED-B) using tetrahydrofuran (THF) as an eluent at 35°C. The number and weight-average molecular weight of the polymers were calculated relative to linear polystyrene standards. Wide-angle X-ray diffraction (WAXD) measurements were performed at room temperature (ca. 25°C) on a Rigaku D/MAX-2500 X-ray diffractometer with a Cu Kα radiation under graphite monochromatic operation at 40 kV and 300 mA. The scanning rate was 1°/min over a range of 2θ = 2–45°. The mechanical properties of the films were measured with an Instron 5567 at a crosshead speed of 2 mm/min on strips approximately 40–50 μm thick and 11 mm wide with a 15 mm gauge length. The average of two individual determinations was used. Thermogravimetric analysis (TGA) and differential scanning calorimetry (DSC) were conducted on a TA Instruments TGA Q500 and a DSC Q100 instrument, respectively. The TGA measurements were conducted at a heating rate of 10°C/min in N_2 and air. The melting points (m.p.) of the synthesized compounds and the T_g values of the polymers were obtained with DSC instrument at a heating rate of 10°C/min in N_2. T_g values were taken from the second heating scan after cooling to 0°C from 400°C. The in-plane linear coefficients of thermal expansion (CTEs) of polymer films were measured by thermomechanical analysis (TMA) using a TA TMA-2940 thermomechanical analyzer. Specimens were 5 mm in width, 10 mm in length, and typically 70 μm thick. The measurements were carried out three times in a heating range up to 300°C at a heating rate of 5°C/min. After the first measurement (first run), the sample was cooled gradually to room temperature in a nitrogen atmosphere, after which the second measurement (second run) was taken. The same operation was carried out between the second run and the third run. The CTE values were determined as the mean at 50–250°C in the second and third heating runs. UV-visible spectra of the polymer films were recorded on an Optizen POP spectrophotometer in the transmittance mode. The refractive indices n_{TE} and n_{TM} for the transverse electric (TE) and transverse magnetic (TM) modes of the polymer films were measured with a Sairon SPA-4000 prism coupler equipped with a gadolinium gallium garnet (GGG) prism at a wavelength of 633 nm at room temperature. The birefringence values (Δn) were calculated as the difference between n_{TE} and n_{TM}. In order to confirm the thermal response behavior of the polymers in organic solvents, another UV-vis spectrophotometer (Shimadzu UV-3600) was used. The transmittance change was measured at a wavelength of 600 nm, and the heating and cooling rates were 0.5°C/min. The clouding point (T_{cp}) was determined as the temperature at which 90% transmittance was observed during the heating process. When T_{cp} was observed at a temperature above the boiling temperature of the solvent used, T_{cp} was measured by the naked eye, while the solutions were placed in screw-cap vials and gradually heated at 5°C intervals in a heating bath equipped with a mercury thermometer.

2.3 Polymerization

PI-1. Diamine monomer **3** (0.4012 g, 1.253 mmol) and BPDA (0.3689 g, 1.254 mmol) were initially dissolved in 4.8 mL of *m*-cresol to a concentration of 16 wt% in a 25 mL three-necked round-bottom flask equipped with a nitrogen inlet, a Dean-Stark trap, and a mechanical stirrer. After the mixture was stirred at room temperature for 30 min, isoquinoline (ca. 5 drops) was added, and further stirring was conducted at room temperature for 4 h. After the solution was diluted with 4.8 mL of *m*-cresol to a concentration of 8 wt%, the temperature was raised to 190°C slowly and the reaction mixture was stirred for 12 h at this temperature. During this time, the water released during the imidization process was removed by distillation as chlorobenzene/water azeotrope, and small amount of *m*-cresol (total additional volume = 4.8 mL) was added periodically to maintain proper viscosity of the reaction mixture. After cooling to room temperature, the solution was diluted with *m*-cresol and then slowly poured into an excess of vigorously stirred ethanol. The resulting polymer was collected by filtration, washed with ethanol, and then dried in vacuo at 180°C for 12 h (0.7444 g, 100% yield). FTIR (thin film, cm^{-1}): 1779 (asym C=O str); 1726 (sym C=O str); 1476 (aromatic C=C); 1373 (C—N str); 1123–1190 (C—F in CF_3); 738 (imide ring deformation). ^1H NMR (DMSO-d_6, 400 MHz, 100°C, ppm): 8.54–8.34 (m, 6H), 8.25–8.04 (m, 2H), 7.66 (d, J = 8.1 Hz, 2H), 7.56 (d, J = 8.5 Hz, 2H). ^1H NMR (THF-d_8, 400 MHz, 55°C, ppm): 8.49 (d, J = 18.6 Hz, 2H), 8.37 (s, 4H), 8.24–8.08 (m, 2H), 7.73 (d, J = 7.9 Hz, 2H), 7.51 (d, J = 8.6 Hz, 2H). Anal. calcd for $C_{30}H_{12}F_6N_2O_4$: C, 62.29; H, 2.09; N, 4.84. Found: C, 61.51; H, 2.06; N, 4.78.

PI-2. The same procedure used for **PI-1** was repeated with 0.4017 g (1.254 mmol) of **3**, 0.4043 g of (1.255 mmol) of BTDA, and 5 mL of *m*-cresol. Before heating the reaction mixture to 190°C, the solution was diluted with 5 mL of *m*-cresol and there was no injection of the additional solvent (0.7538 g, 99.1% yield). FTIR (thin film, cm^{-1}): 1783 (asym C=O str); 1732 (sym C=O str); 1679 (diaryl ketone of BTDA); 1476 (aromatic C=C); 1376 (C—N str); 1137–1191 (C—F in CF_3); 723 (imide ring deformation). ^1H NMR (DMSO-d_6, 400 MHz, 25°C, ppm): 8.40 (s, 2H), 8.35–8.14 (m, 6H), 7.60 (s, 4H). Anal. calcd for $C_{31}H_{12}F_6N_2O_5$: C, 61.40; H, 1.99; N, 4.62. Found: C, 61.20; H, 2.08; N, 4.44.

PI-3. The same procedure used for **PI-1** was repeated with 0.4030 g of (1.258 mmol) of **3**, 0.3907 g (1.259 mmol) of ODPA, and 4.9 mL of *m*-cresol. Before heating the reaction mixture to 190°C, the solution was diluted with 4.9 mL of *m*-cresol and there was no injection of the additional solvent (0.7548 g, 100% yield). FTIR (thin film, cm^{-1}): 1782 (asym C=O str); 1729 (sym C=O str); 1475 (aromatic C=C); 1375 (C—N str); 1276, 1239 (—O—); 1140–1190 (C—F in CF_3); 745 (imide ring deformation). ^1H NMR (DMSO-d_6, 400 MHz, 25°C, ppm): 8.36 (s, 2H), 8.14 (dt, J = 17.8, 6.7 Hz, 2H), 7.77–7.62 (m, 4H), 7.56 (s, 4H). Anal. calcd for $C_{30}H_{12}F_6N_2O_5$: C, 60.62; H, 2.03; N, 4.71. Found: C, 60.71; H, 2.01; N, 4.66.

PI-4. The same procedure used for **PI-1** was repeated with 0.4003 g (1.250 mmol) of **3**, 0.5560 g (1.252 mmol) of 6-FDA, and 6 mL of *m*-cresol. Before heating the reaction mixture to 190°C, the solution was diluted with 6 mL of *m*-cresol without injection of any additional solvent (0.8961 g, 98.4% yield). FTIR (thin film, cm^{-1}): 1789 (asym C=O str); 1733 (sym C=O str); 1476 (aromatic C=C); 1375 (C—N str); 1145–1193 (C—F in CF_3); 722 (imide ring deformation). ^1H NMR (DMSO-d_6, 400 MHz, 25°C, ppm): 8.34 (s, 2H), 8.28 (t, J = 7.8 Hz, 1H), 8.23 (t, J = 8.0 Hz, 1H), 8.04 (t, J = 6.6 Hz, 1H), 7.98 (t, J = 6.2 Hz, 1H), 7.80 (dd, J = 18.6, 15.8 Hz, 2H), 7.56 (s, 4H). Anal. calcd for $C_{33}H_{12}F_{12}N_2O_4$: C, 54.41; H, 1.66; N, 3.85. Found: C, 55.43; H, 1.56; N, 3.77.

PI-5. The same procedure used for **PI-1** was repeated with 0.4062 g (1.268 mmol) of **3**, 0.2768 g (1.269 mmol) of PMDA, and 4.2 mL of *m*-cresol. After the solution was diluted with 4.2 mL of *m*-cresol to a concentration of 8 wt%, the temperature was raised to 190°C slowly. The solution became turbid and heterogeneous as soon as the temperature reached 190°C. The heterogeneous reaction mixture was further stirred for 12 h at this temperature. After cooling to room temperature, the solution was poured into an excess of vigorously stirred ethanol. The solid polymer powder was collected by filtration, washed with ethanol, and then dried in vacuo at 180°C for 12 h (0.6392 g, 100% yield). FTIR (KBr, cm^{-1}): 1783 (asym C=O str); 1728 (sym C=O str); 1477 (aromatic C=C); 1367 (C—N str); 1125–1192 (C—F in CF_3); 724 (imide ring deformation). Anal. calcd for $C_{24}H_8F_6N_2O_4$: C, 57.38; H, 1.61; N, 5.58. Found: C, 56.81; H, 1.70; N, 5.46.

2.4 Preparation of polyimide films

An *N,N*-dimethylacetamide (DMAc) solution of the polymers (7.5 wt%) was prepared at room temperature. The DMAc solution was filtered and cast onto a glass plate. The solvent was evaporated in a vacuum oven at room temperature for 5 h and then heated to 180°C for 10 h to remove the residual solvent. To measure the refractive indices of the polyimides, the DMAc solution of the polymers (2.5 wt%) was filtered and cast onto a silicon substrate and then dried in the same manner described above.

3. Results and discussion

3.1 Monomer syntheses

The diamine monomer, 2,6-bis(trifluoromethyl)benzidine (**3**), was prepared in two steps, as reported previously (**Figure 1**) [48]. In the first step, 2-bromo-5-nitro-1,3-bis(trifluoromethyl)benzene (**1**) was reacted with 4-aminophenylboronic acid through a Suzuki coupling reaction in the presence of Pd as a catalyst to produce **2**. The mono-nitro compound was quantitatively converted to the corresponding diamine monomer **3** by hydrogenation with hydrogen in the presence of Pd/C catalyst.

The chemical structures of **2** and **3** were confirmed by ^1H and ^{13}C NMR, FTIR, and an elemental analysis. **Figure 2** shows the NMR spectra of **2** and **3**. Through a reduction reaction, the chemical shifts of proton H_1 and carbon C_1 moved upfield because they were more shielded by the change of the substituents from the NO_2

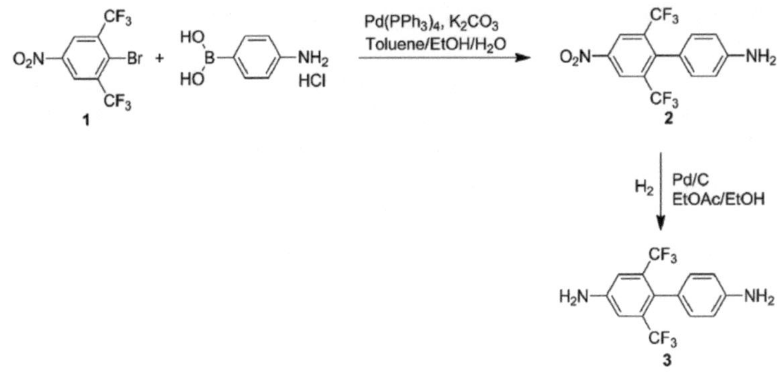

Figure 1.
Synthesis of the unsymmetrical diamine monomer.

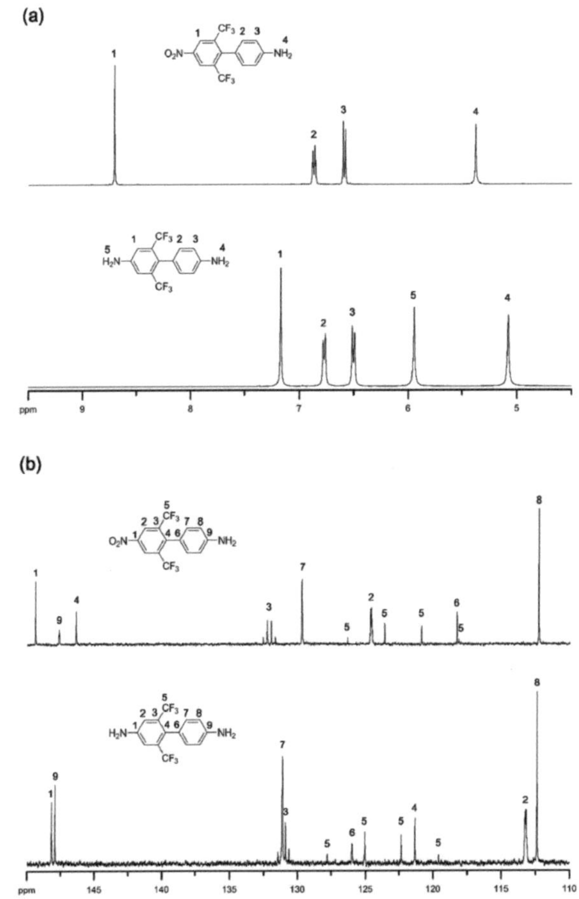

Figure 2.
(a) ^{1}H and (b) ^{13}C NMR spectra of **2** *and* **3** *(DMSO-d$_6$, 25°C) [48].*

group to the NH_2 group. Meanwhile, there were obvious differences in the chemical shifts between the two amine groups in monomer **3**. The chemical shifts of proton H_5 and carbon C_1 in **3** appear further downfield compared to those of H_4 and C_9 due to the deshielding effect of the electron withdrawing CF_3 groups, indicating that the amine group located far away from CF_3 groups in monomer **3** has a higher electron density and greater nucleophilic reactivity than in the opposite case. The FTIR spectra of **2** and **3** are shown in **Figure 3**. The compound **2** gave characteristic bands at 3499, 3401 cm^{-1} (N—H stretching), at 1620 cm^{-1} (N—H bending), and at 1333–1533 cm^{-1} (NO_2 asymmetric and symmetric stretching). After the reduction, the characteristic absorptions of the nitro group disappeared, and the amino group exhibited a pair of N—H stretching bands in the region of 3221–3486 cm^{-1} and an N—H bending band at 1640 cm^{-1}. All spectroscopic data obtained were in good agreement with the predicted structures.

3.2 Model reaction

A model reaction was conducted to investigate the suitability of a nucleophilic addition and cyclodehydration of the diamine monomer in the polymerization reaction condition as well as to obtain a model compound as a reference material for a structural analysis. The diamine monomer was reacted with two-equivalent of phthalic anhydride in *m*-cresol in the presence of a catalytic amount of isoquinoline

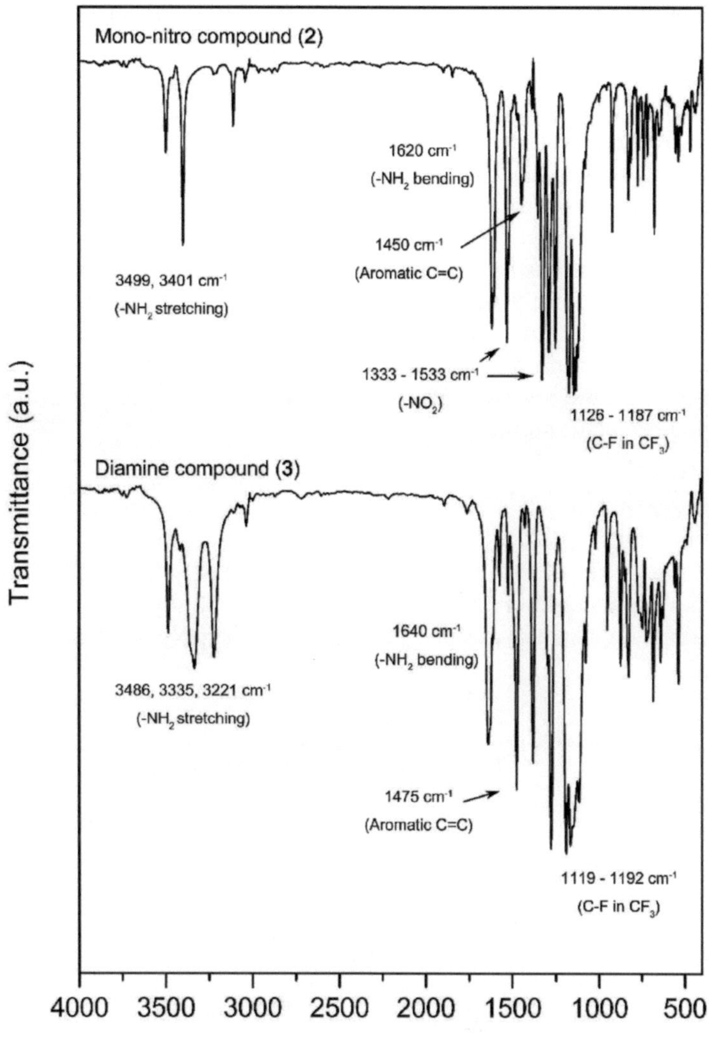

Figure 3.
FTIR spectra of mono-nitro (2) and diamine (3) compounds [48].

(**Figure 4**), and all reaction steps were examined by thin layer chromatography. Despite the difference in the electron density between the two amines resulting from the unsymmetrical structure, the two amines had sufficient nucleophilicity to react with the phthalic anhydride. As a result, the corresponding diamic acid form was generated quantitatively within 0.5 h at room temperature, and cyclodehydration was completed within 0.5 h at 190°C. Finally, diimide model compound (**4**) was obtained quantitatively. The structure of **4** was confirmed by FTIR, ^{1}H NMR and ^{13}C NMR spectroscopy (**Figure 5**). The FTIR spectrum of the model compound shows absorption bands at 1789, 1380, and 722 cm^{-1} corresponding to the C=O imide stretching, C—N imide stretching, and imide ring deformation, respectively, without the characteristic absorptions of the amino groups. The ^{1}H and ^{13}C NMR spectra also supported the formation of the diimide model compound. Owing to the unsymmetrical presence of CF_3 groups on the product, proton and carbon peaks on both phthalimide units appeared with different chemical shifts in the NMR spectra, in which the peaks of phthalimide connected to a trifluoromethylated phenyl ring appeared further downfield due to the electron-withdrawing characteristic of the CF_3 groups. All spectroscopic data obtained were in good agreement with the predicted structure.

The structure features of **4** were further detailed by single-crystal X-ray diffraction, and its X-ray quality crystals were obtained by the slow evaporation of saturated tetrahydrofuran/water solution at room temperature [48]. As shown in **Figure 6**, the solid-state structure of **4** also ensures the coupling reaction of **3** to the phthalic anhydrides. It is noteworthy that the plane of a phenyl moiety is almost orthogonally located relative to the adjacent *m*-$(CF_3)_2$Ph plane such that the biphenyl of **4** had a rigid but twisted structure with a dihedral angle (θ) of 76°. Compared with 2,2′-bis(trifluoromethyl)benzidine (θ = 59°) [27], the benzidine unit of **4** is more distorted. Therefore, it was expected that the obtained polyimides would exhibit good solubility and transparency with high mechanical and thermal properties, as they have a rigid but twisted structure with bulky CF_3 groups, thereby reducing the interchain interactions.

3.3 Polymer syntheses

Based on the results of the model reactions, several polyimides (PIs) were prepared from **3** and commercially available aromatic dianhydrides via a one-pot solution imidization method, as shown in **Figure 7**. The polymerizations of diamine

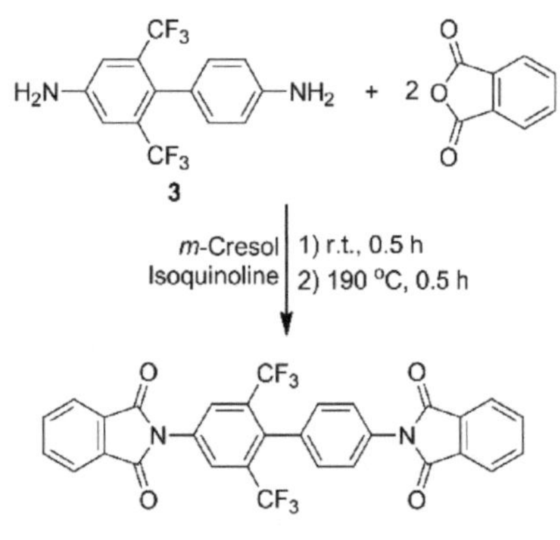

Figure 4.
Model reaction of 3 with phthalic anhydride.

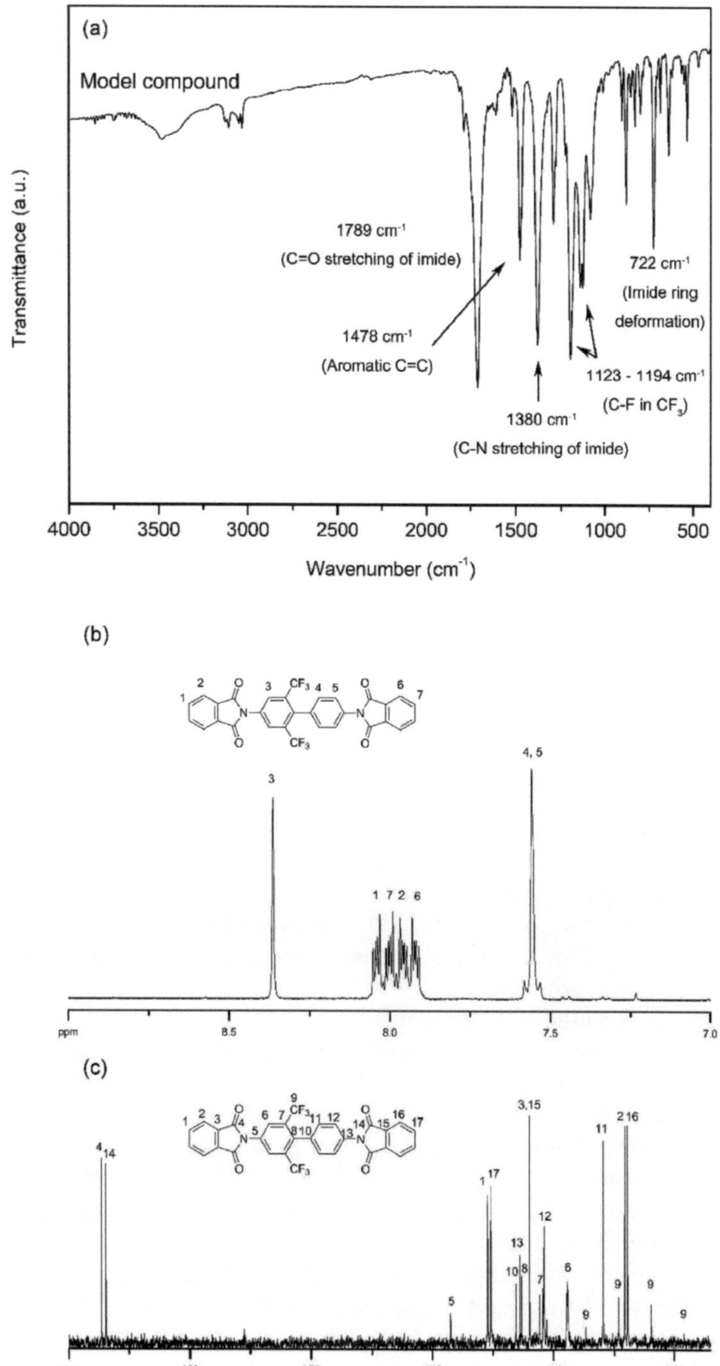

Figure 5.
*(a) FTIR, (b) ^{1}H, and (c) ^{13}C NMR spectra of model compound (**4**) (NMR: DMSO-d_6, 25°C) [48].*

monomer **3** with stoichiometric amounts of five different aromatic dianhydride monomers, BPDA (**PI-1**), BTDA (**PI-2**), ODPA (**PI-3**), 6-FDA (**PI-4**), and PMDA (**PI-5**), were carried out in *m*-cresol with catalytic amounts of isoquinoline at a solid content of about 16 wt%. The ring-opening polyaddition at room temperature for

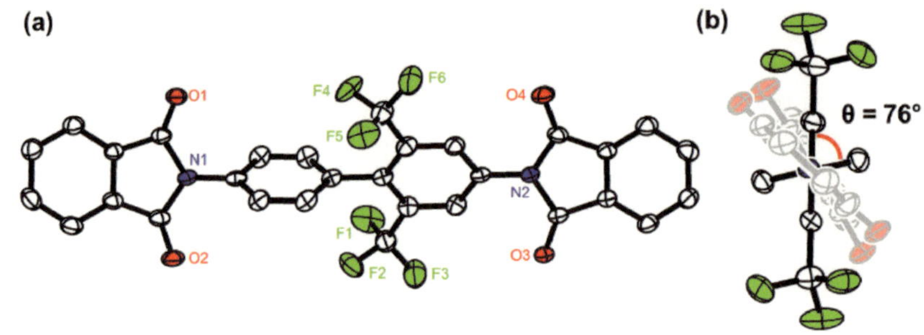

Figure 6.
*(a) Top view and (b) side view of displacement ellipsoid (50%) representations of **4**. All hydrogen atoms are omitted for clarity.*

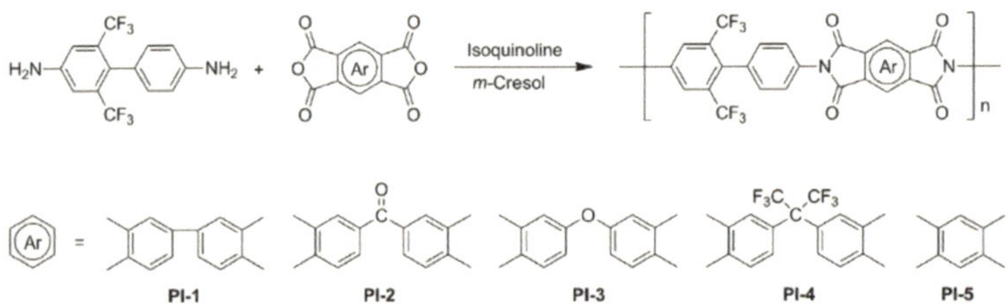

Figure 7.
*Polymerization of **3** with aromatic dianhydrides.*

4 h yielded poly(amic acid) solutions. After dilution of the solution to 8 wt%, subsequent cyclodehydration by heating at 190°C with the azeotropic distillation of chlorobenzene for 12 h gave fully imidized and homogeneous PI solutions except for that of PMDA (**PI-5**). When PMDA was used as the dianhydride, the solution became turbid with phase separation as soon as the temperature reached 190°C. This was likely due to the most rigid chain characteristic of **PI-5**. At the end of the reaction, pure solid polymers were obtained by precipitation of the corresponding polymer solutions into ethanol.

Table 1 shows the inherent viscosities and GPC data of the PIs. The inherent viscosities of the organosoluble PIs were in the range of 0.69–2.30 dL/g, as measured in DMAc at 30°C. Additionally, the PIs soluble in THF exhibited weight-average molecular weights (M_ws) in the range of 7.32–8.81 × 10^4 relative to the polystyrene standard. The molecular weights of the PIs were high enough to obtain flexible and tough polymer films by casting from their DMAc solutions.

The formation and the structures of the polymers were verified by elemental analyses, FTIR, and ^{1}H NMR spectroscopy. The elemental analysis values of the PIs (listed in Experiments) were in good agreement with the calculated values of the proposed structures. The typical FTIR spectrum of **PI-1** is shown in **Figure 8**. All PIs exhibited characteristic imide group absorptions around 1780 and 1730 (typical of imide carbonyl asymmetrical and symmetrical stretching), 1370 (C—N stretching), and 730 cm^{-1} (imide ring deformation), together with a number of strong absorption bands in the regions of 1120–1200 cm^{-1} due to C—F stretching. The absence of amide and carboxyl bands indicates the virtually complete conversion of the poly (amic acid) precursors into PIs. The ^{1}H NMR spectra of the PIs are illustrated in **Figure 9**. All proton peaks were also assigned to the predicted structures without amide and acid protons, demonstrating the successful preparation of the PIs. Additionally, the proton peaks of the dianhydride units in the polymer chains were divided into different chemical shifts due to the unsymmetrical structure of the diamine unit in the PI main chains. The chemical shifts of protons closer to the CF_3

Polymer code	η_{inh} (dL/g)[a]	M_w (kDa)[b]	M_n (kDa)[b]	M_w/M_n[b]
PI-1	2.30	—[c]	—[c]	—[c]
PI-2	0.78	73.2	28.2	2.60
PI-3	0.88	83.7	33.6	2.49
PI-4	0.69	88.1	32.9	2.68
PI-5	—[c]	—[c]	—[c]	—[c]

[a]*Measured in DMAc at a concentration of 0.5 g/dL at 30°C.*
[b]*Determined by GPC in THF at 35°C (relative to polystyrene standard).*
[c]*Insoluble.*

Table 1.
Inherent viscosities and elemental analyses results of the PIs.

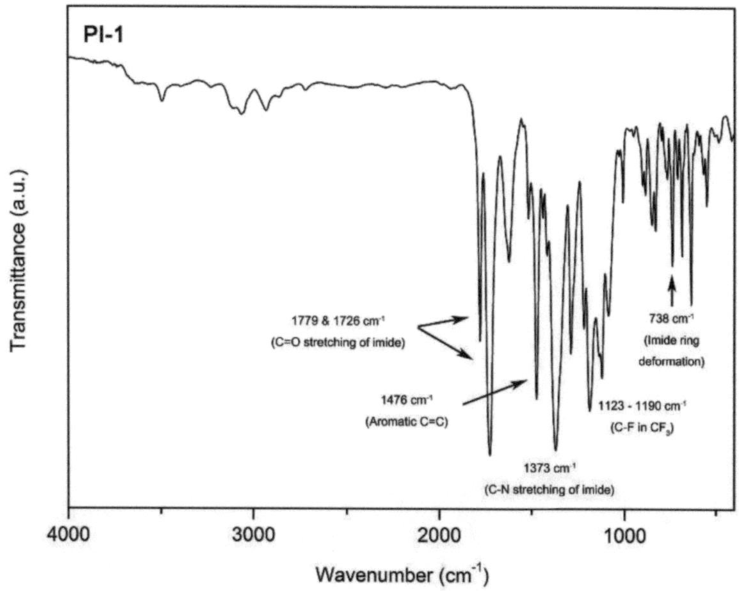

Figure 8.
*FTIR spectrum of **PI-1** (film).*

groups appeared further downfield in the NMR spectrum due to the deshielding effect of the electron-withdrawing CF_3 groups. This result was consistent with the model reaction result.

3.4 Polymer properties

The solubility of the synthesized PIs is summarized in **Table 2**. The synthesized polymers retained good solubility in organic solvents, although their rigidity increased compared to those synthesized from 4-(4′-aminophenoxy)-3,5-bis (trifluoromethyl)aniline [40]. All PIs except for **PI-5** exhibited good solubility in *N*-methyl-2-pyrrolidone (NMP), *N*,*N*-dimethylacetamide (DMAc), *m*-cresol, and anisole at room temperature. In addition, **PI-2**, **PI-3**, and **PI-4** showed good solubility in *N*,*N*-dimethylformamide, dimethyl sulfoxide, tetrahydrofuran, and ethyl acetate at room temperature. The good solubility of the PIs is attributed not only to the bulkiness of the two CF_3 groups on the polymer chains but also to the unsymmetrical structure resulting from the diamine monomer. Given the increased chain flexibility, **PI-3** and **PI-4** were also soluble in chloroform and **PI-4** was soluble even

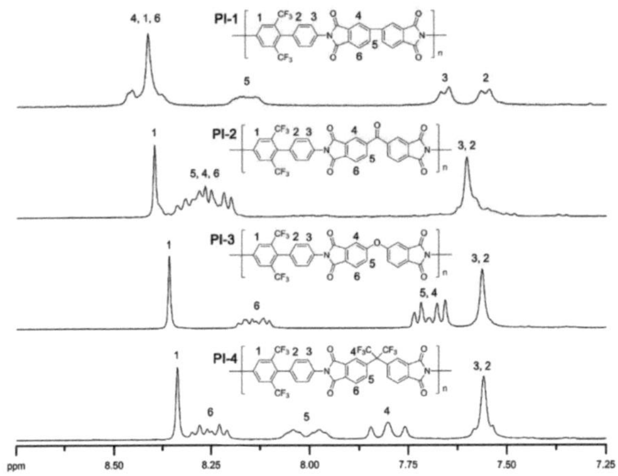

Figure 9.
^1H NMR spectrum of ***PI-1*** *(DMSO-d_6, 100°C); ^1H NMR spectra of* ***PI-2, PI-3,*** *and* ***PI-4*** *(DMSO-d_6, 25°C).*

Solvents	PI-1	PI-2	PI-3	PI-4	PI-5
NMP	++	++	++	++	+−
DMAc	++	++	++	++	+−
m-Cresol	++	++	++	++	−
Anisole	++	++	++	++	−
DMF	+	++	++	++	−
DMSO	+	++	++	++	−
THF	+	++	++	++	−
Ethyl acetate	−	++	++	++	−
Acetone	−	+−	+−	++	−
Chloroform	−	−	++	++	−
ODCB	+	+	+	+	−
Acetonitrile	−	−	−	−	−
Toluene	−	−	−	−	−
Diethyl ether	−	−	−	−	−
n-Hexane	−	−	−	−	−
Methanol	−	−	−	−	−

Solubility: ++, soluble at room temperature; +, soluble on heating; +−, partially soluble; −, insoluble. Abbreviations: NMP, N*-methyl-2-pyrrolidone; DMAc,* N,N*-dimethylacetamide; DMF,* N,N*-dimethylformamide; DMSO, dimethyl sulfoxide; THF, tetrahydrofuran; ODCB, 1,2-dichlorobenzene.*

Table 2.
Solubility of the PIs.

in acetone at room temperature. Upon a comparison of the solubility behavior with PIs prepared from symmetrical benzidine, 2,2′-bis(trifluoromethyl)benzidine [17–21], the PIs described here showed enhanced solubility. The improved solubility can be attributed to the unsymmetrical and more twisted chain structure, which inhibits close packing and reduces intermolecular interactions. Meanwhile, **PI-5** showed poor solubility in the organic solvents tested, although it was partially soluble in NMP and DMAc.

To clarify the cause of the poor solubility of **PI-5**, wide-angle X-ray diffraction (WAXD) studies were performed because a crystalline domain can influence the solubility of the polymer [31]. The X-ray diffractograms of **PI-1** and **PI-3** showed broad diffraction curves without obvious peak features, indicating that the PIs have an amorphous morphology in principle (**Figure 10**). On the other hand, the X-ray diffraction curve of **PI-5** exhibited a relatively unambiguous peak around 17°, which indicated that **PI-5** has a more ordered phase compared to the other soluble PIs. This is likely related to the highly linear and rigid chain structure of **PI-5**, which induced high intermolecular interactions, resulting in a decrease of the solubility.

The soluble PIs of **PI-1**, **PI-2**, **PI-3**, and **PI-4** could be processed into flexible and tough films conveniently by casting from the DMAc polymer solutions. While **PI-2**, **PI-3**, and **PI-4** produced transparent films, **PI-1** generated a turbid film which appeared as the solvent was evaporated. **Table 3** shows the mechanical properties of the PI films. The PI films had tensile strength levels, elongation at break values, and a Young's modulus in the ranges of 92–145 MPa, 26–55%, and 2.1–3.2 GPa, respectively. These values were comparable to the mechanical strength of the PIs prepared from 2,2′-bis(trifluoromethyl)benzidine [22] and also indicated that the PI films are strong enough for use.

The thermal properties of the PIs were evaluated by TGA, DSC, and TMA, and these results are summarized in **Table 4**. The dynamic TGA result showed high thermal stability in which 5% weight loss occurred for the PIs in the range of

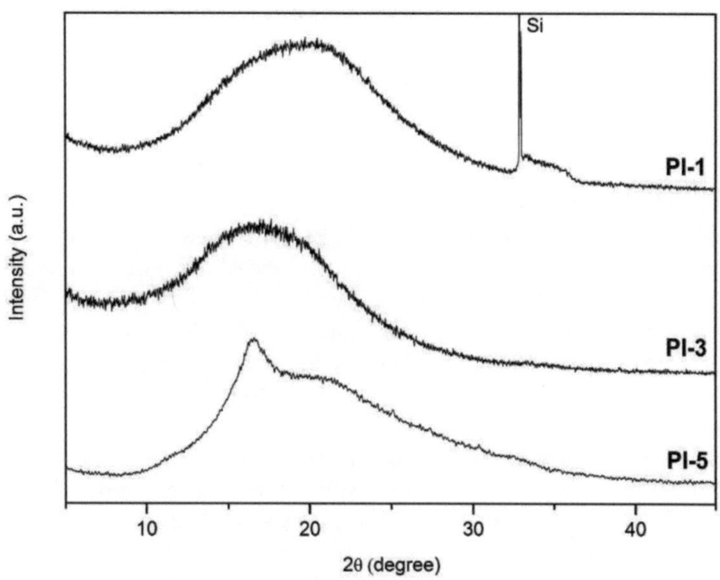

Figure 10.
*Wide-angle X-ray diffractograms of **PI-1** (film), **PI-3** (film), and **PI-5** (powder).*

Polymer code	Tensile strength (MPa)	Elongation at break (%)	Young's modulus (GPa)
PI-1	145	26	3.2
PI-2	125	55	2.8
PI-3	92	35	2.1
PI-4	95	39	2.6

Table 3.
Mechanical properties of the PI films.

Polymer code	T_{d5} (°C)[a]		T_g (°C)[b]	CTE (ppm/°C)[c]		Cutoff wavelength (nm)	Transmittance at 550 nm (%)
	In N_2	In air		2nd run	3rd run		
PI-1	604	594	366[d]	6.8	6.8	398	34
PI-2	581	553	348	42.0	42.5	371	87
PI-3	589	566	345	52.6	51.9	368	89
PI-4	535	523	362	63.1	63.0	354	90
PI-5	605	569	—[e]	—[f]	—[f]	—[f]	—[f]

[a]*5% weight loss temperature measured by TGA at a heating rate of 10°C/min.*
[b]*Measured by DSC (the second scan) in N_2 at a heating rate 10°C/min.*
[c]*A TMA analysis was conducted three times for each sample at a heating rate of 5°C/min in a heating range up to 300° C. Each CTE value was calculated from the mean coefficient of the linear thermal expansion over a specific temperature range between 50 and 250°C in the second and third runs, respectively.*
[d]*Measured by TMA at a heating rate of 5°C/min.*
[e]*Not detected.*
[f]*Not measured.*

Table 4.
Thermal and optical properties of the PIs.

535–605°C in nitrogen and 523–594°C in air (**Figure 11a** and **b**, respectively). DSC experiments were conducted at a heating rate of 10°C/min in nitrogen (**Figure 11c**). A survey of all of the PIs by DSC revealed that no endothermic peaks associated with melting were observed up to the temperature region investigated here. Moreover, while the glass-transition temperatures (T_g) of **PI-2**, **PI-3**, and **PI-4** were clearly detected, **PI-1** and **PI-5** did not show any discernible glass transition on the DSC thermograms. Therefore, the T_g of **PI-1** was measured by the TMA method after preparation of the polymer film (**Figure 11d**). All PIs exhibited high T_g values above 340°C which depended on the chemical structure of the aromatic dianhydride component. **PI-1** obtained from BPDA showed the highest T_g value (366°C) among the soluble PIs owing to the absence of a flexible linkage between the phthalimide units. Compared with PIs derived from 4-(4′-aminophenoxy)-3,5-bis(trifluoromethyl)aniline [40], the PIs based on 2,6-bis(trifluoromethyl)benzidine possess higher T_g values due to their greater chain rigidity. Even when compared to the PIs derived from 2,2′-bis(trifluoromethyl)benzidine [18–22, 25], the PIs synthesized here showed similar or higher T_g values. For example, the PIs based on BPDA and 6-FDA with 2,2′-bis(trifluoromethyl)benzidine exhibited T_g values of 287–373°C and 335°C, respectively. This result could be attributed to the unsymmetrical introduction of the two CF_3 groups, which further increases the rotational barrier of the polymer chains compared to the symmetrically introduced case.

The coefficients of thermal expansion (CTEs) of the PI films were found in the range of 6.8–63.1 ppm/°C. In general, polymers consisting of rod-like backbone structures together with a high chain alignment toward the direction parallel to the film plane have shown relatively low CTE values [24–28]. The relationship between the chain rigidity/degree of in-plane orientation and the CTE value can be applied to this study. The CTE value of the PIs decreased from 63.1 to 6.8 ppm/°C with an increase in the chain rigidity and the degree of in-plane orientation, as identified through the birefringence value (**Table 5**). Although the birefringence value of **PI-1** could not be measured due to the low transparency in this case, it can be speculated that **PI-1** has the highest degree of in-plane orientation because **PI-1** possesses the most rigid chain structure among the soluble PIs [18, 19, 24, 25, 28–30]. Therefore, **PI-1** gave the lowest CTE value (6.8 ppm/°C), displaying good dimensional stability.

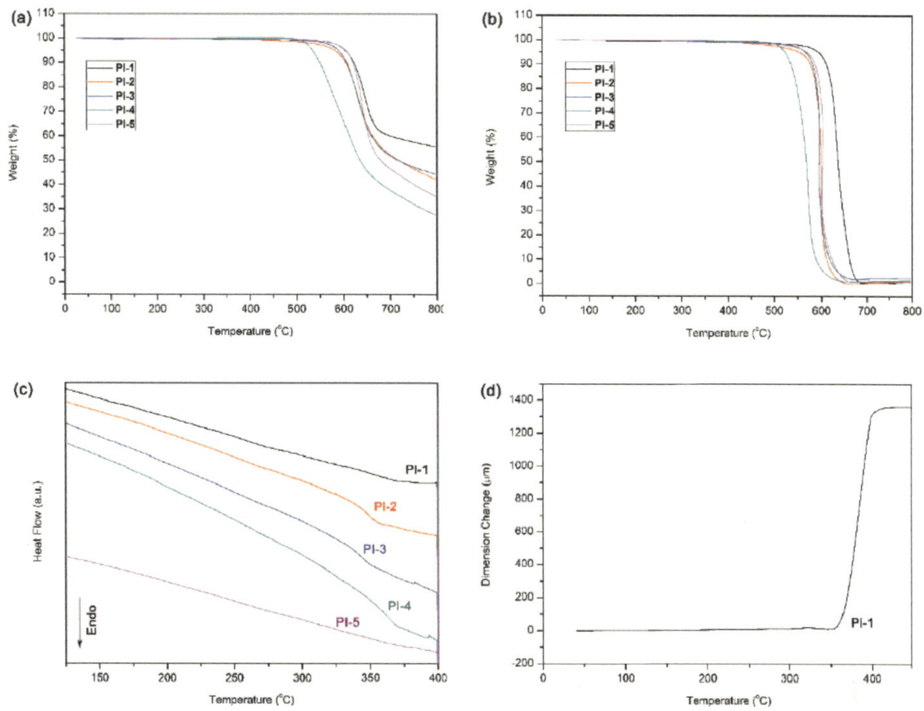

Figure 11.
TGA curves of the PIs in (a) nitrogen and (b) air at a heating rate of 10°C/min, (c) DSC curves of the PIs (the second heating run ranging from 0 to 400°C at a heating rate of 10°C/min in N_2) and (d) TMA CURVE Of ***PI-1*** *(measured at a heating rate of 5°C/min).*

Polymer code[a]	n_{TE}[b]	n_{TM}[c]	n_{av}[d]	Δn[e]	ε[f]	d (μm)[g]
PI-2	1.638	1.566	1.614	0.072	2.87	4.7
PI-3	1.619	1.561	1.600	0.058	2.82	3.1
PI-4	1.548	1.506	1.534	0.042	2.59	3.8

[a]*Measured at a wavelength of 633 nm at room temperature.*
[b]n_{TE}*: the in-plane refractive index.*
[c]n_{TM}*: the out-of plane refractive index.*
[d]n_{av}*: the average refractive index* ($n_{av} = (2n_{TE} + n_{TM})/3$).
[e]Δn*: birefringence* ($n_{TE} - n_{TM}$).
[f]*Dielectric constant estimated from the refractive index:* ε $1.10n_{av}^2$.
[g]*Film thickness for the refractive index measured.*

Table 5.
Refractive indices of the PIs.

The corresponding UV-vis spectra of the PI films with a thickness of about 60–80 μm are shown in **Figure 12a**. While the light transparencies of **PI-2**, **PI-3**, and **PI-4** were as high as 87–90% at a wavelength of 550 nm, **PI-1** film exhibited extremely low transparency of 34%. This outcome is attributable to the high degree of in-plane orientation caused by the most rigid chain structure compared to other PIs. The cutoff wavelengths (λ_o) of the PI films listed in **Table 4** range from 354 to 398 nm, indicating that they are light-colored films. The color of the PI films decreased in the following order: **PI-1** > **PI-2** > **PI-3** > **PI-4**. As shown in **Figure 12b**, **PI-3** and **PI-4** produced fairly transparent and almost colorless PI films compared to the other PIs. The good optical properties of **PI-3** and **PI-4** were attributed to the poor CTC formation ability caused by the ether chain of ODPA and the bulky hexafluoroisopropylidene group of

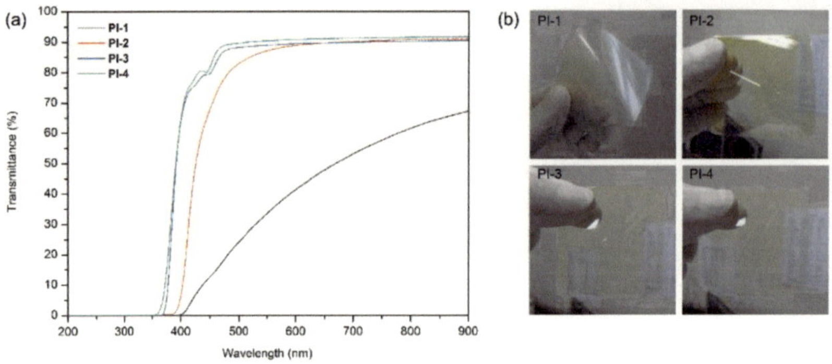

Figure 12.
(a) Transmittance UV-vis spectra and (b) photographs of the PI films.

6-FDA, respectively. The increased yellowness of **PI-2** resulted from the presence of the carbonyl group, attracting electrons in the dianhydride units.

The refractive indices and birefringence values of the **PI-2**, **PI-3**, and **PI-4** were measured using a prism-coupling method with a laser beam having a wavelength of 633 nm. The measurement of the **PI-1** film could not be performed due to the low transparency of the film. As shown in **Table 5**, the PIs showed low refractive indices (n_{av}) in the range of 1.534–1.614 due to the incorporation effect of the two CF_3 groups, which led to low molecular polarizability and density levels. The birefringence (Δn) values of the PIs were determined as the difference between n_{TE} and n_{TM} and were in the range of 0.042–0.072. The birefringence values among the PIs tested increased with an increase of the chain rigidity because the high chain rigidity led to high chain alignment in the direction parallel to the film plane [28–30]. **PI-4** showed the lowest birefringence value, indicating that the linear polarizability and segmental orientation of **PI-4** are the most isotropic among the PIs. The dielectric constant (ε) can be estimated from the refractive index n according to Maxwell's equation, $\varepsilon \approx n^2$. The ε value at 1 MHz was determined to be $\varepsilon \approx 1.10\ n_{av}^2$, including an additional contribution of approximately 10% due to infrared absorption [49]. The ε values of the PI films estimated from the average refractive indices ranged from 2.59 to 2.87. The low dielectric constants are also attributed to the existence of two CF_3 groups in the main chain.

In our previous work, the thermo-responsive behavior of PIs containing CF_3 groups was observed in acetyl-containing solvents [41]. Likewise, **PI-2** and **PI-3** also showed LCST behavior in the solvents used here. The transmittance of the ethyl acetate solutions of **PI-2** and **PI-3** was followed as a function of the temperature at a heating/cooling rate of 0.5°C/min. With an increase of the solution temperature, the solutions clearly turned turbid and opaque, indicating a phase transition. The clouding point temperatures (T_{cp}) of the **PI-2** and **PI-3** solutions were 44 and 48°C, respectively, at a concentration of 0.1 wt%, and the T_{cp} values decreased with an increase of the polymer concentrations. Compared to the polymers synthesized from 4-(4′-aminophenoxy)-3,5-bis(trifluoromethyl) aniline, **PI-2** and **PI-3** have lower T_{cp} values, presumably due to lower solubility originating from the increased chain rigidity. The solutions became clear and transparent again when they were cooled.

4. Conclusion

New fluorinated PIs were prepared from the benzidine monomer containing two trifluoromethyl groups on one aromatic ring, 2,6-bis(trifluoromethyl) benzidine.

Due to the rigid and twisted structure of the diamine monomer, the resulting PIs showed good solubility together with high thermal stability and excellent mechanical properties. The PIs also possessed low refractive indices and low dielectric constants due to the high fluorine contents. These PIs can be considered as promising processable high-temperature materials that can find applications in flexible electronics including substrates of flexible and rollable AMOLED displays and low-k dielectrics for microelectronics.

Acknowledgements

This work was supported by Technology Innovation Program (20007228, High transparent and heat-resistant fluorine polyimide technology for OLED substrate) funded by the Ministry of Trade, Industry & Energy (MOTIE, Korea).

Author details

Sun Dal Kim[1,2], Taejoon Byun[1], Jin Kim[2], Im Sik Chung[3] and Sang Youl Kim[1]*

1 Department of Chemistry, Korea Advanced Institute of Science and Technology (KAIST), Daejeon, Korea

2 Agency for Defense Development (ADD), Daejeon, Korea

3 Polymer and Nanomaterial Research Part, Korea Research Institute of Bioscience and Biotechnology (KRIBB), Daejeon, Korea

*Address all correspondence to: kimsy@kaist.ac.kr

References

[1] Ghosh MK, Mittal KL. Polyimides Fundamentals and Applications. New York: Marcel Decker; 1996

[2] Wilson D, Stenzenberger HD, Hergenrother PM. Polyimides. Glasgow and London: Blackie; 1990

[3] Wong WS, Salleo A. Flexible Electronics Materials and Applications. New York: Springer; 2009

[4] Liaw DJ, Wang KL, Huang YC, Lee KR, Lai JY, Ha CS. Advanced polyimide materials: Syntheses, physical properties and applications. Progress in Polymer Science. 2012;**37**:907-974

[5] Hasegawa M, Horie K. Photophysics, photochemistry, and optical properties of polyimides. Progress in Polymer Science. 2001;**26**:259-335

[6] Dhara MG, Banerjee S. Fluorinated high-performance polymers: Poly (arylene ethers) and aromatic polyimides containing trifluoromethyl groups. Progress in Polymer Science. 2010;**35**:1022-1077

[7] Ding M. Isomeric polyimides. Progress in Polymer Science. 2007;**32**: 623-668

[8] Kim SD, Lee S, Lee H, Kim SY, Chung IS. New soluble polyamides and polyimides containing polar functional groups: Pendent Pyrazole rings with amino and Cyano groups. Designed Monomers and Polymers. 2016;**19**: 227-235

[9] Hougham G, Tesoro G, Shaw J. Synthesis and properties of highly fluorinated polyimides. Macromolecules. 1994;**27**:3642-3649

[10] Hedrick JL, Carter KR, Cha HJ, Hawker CJ, Dipietro RA, Labadie JW, et al. High-temperature polyimide nanofoams for microelectronic applications. Reactive and Functional Polymers. 1996;**30**:43-53

[11] Sen SK, Banerjee S, Tg H. Processable fluorinated polyimides containing benzoisoindoledione unit and evaluation of their gas transport properties. RSC Advances. 2012;**2**: 6274-6289

[12] Chung IS, Park CE, Ree M, Kim SY. Soluble polyimides containing benzimidazole rings for interlevel dielectrics. Chemistry of Materials. 2001;**13**:2801-2806

[13] Lin SH, Li F, Cheng SZD, Harris FW. Organo-soluble polyimides: Synthesis and polymerization of 2,2-bis (trifluoromethyl)-4,4,5,5-biphenyltetracarboxylic dianhydride. Macromolecules. 1998;**31**:2080-2086

[14] Yang CP, Hsiao SH, Chen KH. Organosoluble and optically transparent fluorine-containing polyimides based on 4,4-bis amino-2-trifluoromethylphenoxy-3,3,5,5-tetramethylbiphenyl. Polymer. 2002;**43**:5995-6104

[15] Satoh A, Morikawa A. Synthesis and characterization of aromatic polyimides containing trifluoromethyl group from bis(4-amino-2-trifluoromethylphenyl) ether and aromatic tetracarboxylic dianhydrides. High Performance Polymers. 2010;**22**:412-427

[16] Cheng SZD, Wu Z, Eashoo M, Hsu SLC, Harris FWA. High-performance aromatic polyimide fibre: 1. Structure, properties and mechanical-history dependence. Polymer. 1991;**32**: 1803-1810

[17] Matsuura T, Yamada N, Nishi S, Hasuda Y. Polyimides derived from 2,2-bis (trifluoromethyl)-4,4′-diaminobiphenyl. 3. Property control for polymer blends and copolymerization of fluorinated

polyimides. Macromolecules. 1993;**26**: 419-423

[18] Matsuura T, Hasuda Y, Nishi S, Yamada N. Polyimide derived from 2,2-bis(trifluoromethyl)-4,4′-diaminobiphenyl. 1. Synthesis and characterization of polyimides prepared with 2,2-bis(3,4-dicarboxyphenyl) hexafluoropropane dianhydride or pyromellitic dianhydride. Macromolecules. 1991;**24**:5001-5005

[19] Feiring A, Auman BC, Wonchoba ER. Synthesis and properties of fluorinated polyimides from novel 2,2-bis(fluoroalkoxy)benzidines. Macromolecules. 1993;**26**:2779-2784

[20] Li F, Ge JJ, Honigfort PS, Fang S, Chen JC, Harris FW, et al. Dianhydride architectural effects on the relaxation behaviors and thermal and optical properties of Organo-soluble aromatic polyimide films. Polymer. 1999;**40**: 4987-5002

[21] Rozhanskii I, Okuyama K, Goto K. Synthesis and properties of polyimides derived from isomeric biphenyltetracarboxylic dianhydrides. Polymer. 2000;**41**:7057-7065

[22] Hergenrother PM, Watson KA, Smith JG Jr, Connell JW, Yokota R. Polyimides from 2,3,3,4-biphenyltetracarboxylic dianhydride and aromatic diamines. Polymer. 2002; **43**:5077-5093

[23] Matsuura T, Ishizawa M, Hasuda Y, Nishi S. Polyimides derived from 2,2-bis (trifluoromethyl)-4,4-diaminobiphenyl. 2. Synthesis and characterization of polyimides prepared from fluorinated benzenetetracarboxylic dianhydrides. Macromolecules. 1992;**25**:3540-3545

[24] Ishii J, Takata A, Oami Y, Yokota R, Vladimirov L, Hasegawa M. Spontaneous molecular orientation of polyimides induced by thermal imidization (6). Mechanism of negative In-plane CTE generation in non-stretched polyimide films. European Polymer Journal. 2010;**46**:681-693

[25] Arnold FE Jr, Cheng SZD, Hsu SLC, Lee CJ, Harris FW. Organo-soluble, segmented rigid-rod polyimide films: 2. Properties for microelectronic applications. Polymer. 1992;**33**: 5179-5185

[26] Eashoo M, Shen D, Wu Z, Lee CJ, Harris FW, Cheng SZD. High-performance aromatic polyimide fibres: 2 thermal mechanical and dynamic properties. Polymer. 1993;**34**:3209-3215

[27] Chuang KC, Kinder JD, Hull DL, McConville DB, Youngs WJ. Rigid-rod polyimides based on noncoplanar 4,4-biphenyldiamines: A review of polymer properties vs configuration of diamines. Macromolecules. 1997;**30**:7183-7190

[28] Terui Y, Matsuda S, Ando S. Molecular structure and thickness dependence of chain orientation in aromatic polyimide films. Journal of Polymer Science Part B: Polymer Physics. 2005;**43**:2109-2120

[29] Wakita J, Jin S, Shin TJ, Ree M, Ando S. Analysis of molecular aggregation structures of fully aromatic and Semialiphatic polyimide films with synchrotron grazing incidence wide-angle X-ray scattering. Macromolecules. 2010;**43**:1930-1941

[30] Matsuura T, Ando S, Sasaki S, Yamamoto F. Polyimides derived from 2,2-bis(trifluoromethyl)-4,4-diaminobiphenyl. 4. Optical properties of fluorinated polyimides for optoelectronic components. Macromolecules. 1994;**27**:6665-6670

[31] Kim SD, Ka D, Chung IS, Kim SY. Poly(arylene ethers) with low refractive indices: Poly(biphenylene oxide)s with trifluoromethyl pendant groups via a meta-activated nitro displacement

reaction. Macromolecules. 2012;**45**: 3023-3031

[32] Chung IS, Kim SY. Meta-activated Nucleophilic aromatic substitution reaction: Poly(biphenylene oxides) with trifluoromethyl pendent groups via nitro displacement. Journal of the American Chemical Society. 2001;**123**: 11071-11072

[33] Chung IS, Kim SY. Poly(arylene ether)s via nitro displacement reaction: Synthesis of poly(biphenylene oxides) containing trifluoromethyl groups from AB type monomers. Macromolecules. 2000;**33**:9474-9476

[34] Liu Y, Xing Y, Zhang Y, Guan S, Zhang H, Wang Y, et al. Novel soluble fluorinated poly(ether imide)s with different pendant groups: Synthesis, thermal, dielectric, and optical properties. Journal of Polymer Science Part A: Polymer Chemistry. 2010;**48**: 3281-3289

[35] Liu B, Hu W, Matsumoto T, Jiang Z, Ando S. Synthesis and characterization of organosoluble ditrifluoromethylated aromatic polyimides. Journal of Polymer Science Part A: Polymer Chemistry. 2005;**43**:3018-3029

[36] Long TM, Swager TM. Molecular design of free volume as a route to low-κ dielectric materials. Journal of the American Chemical Society. 2003;**125**: 14113-14119

[37] Lee B, Byun T, Kim SD, Kang HA, Kim SY, Chung IS. Soluble para-linked aromatic polyamides with pendent groups. Macromolecular Research. 2015; **23**:838-843

[38] Kim SD, Byun T, Lee B, Kim SY, Chung IS. Rigid-rod polyamides from 3,3-bis-(trifluoromethyl)-4,4-diamino-1,1-biphenyl. Macromolecular Chemistry and Physics. 2015;**216**: 1341-1347

[39] Chung IS, Kim KH, Lee YS, Kim SY. Poly(arylene ether)s with trifluoromethyl groups via meta-activated nitro displacement reaction. Polymer. 2010;**51**:4477-4483

[40] Kim SD, Kim SY, Chung IS. Soluble and transparent polyimides from unsymmetrical diamine containing two Trifluoromethyl groups. Journal of Polymer Science Part A: Polymer Chemistry. 2013;**51**:4413-4422

[41] Kim SD, Kim SY, Chung IS. Unprecedented lower critical solution temperature behavior of polyimides in organic media. Macromolecules. 2014; **47**:8846-8849

[42] Chern YT, Tsai JY. Low dielectric constant and high organosolubility of novel polyimide derived from unsymmetric 1,4-bis(4-aminophenoxy)-2,6-di-tert-butylbenzene. Macromolecules. 2008; **41**:9556-9564

[43] Choi H, Chung IS, Hong K, Park CE, Kim SY. Soluble polyimides from unsymmetrical diamine containing benzimidazole ring and trifluoromethyl pendent group. Polymer. 2008;**49**: 2644-2649

[44] Kim M, Kim SY. Curable aromatic polyimides containing enaminonitrile groups. Macromolecules. 2002;**35**: 4553-4555

[45] Chung IS, Kim SY. Soluble polyimides from unsymmetrical diamine with trifluoromethyl pendent group. Macromolecules. 2000;**33**: 3190-3193

[46] Hsiao SH, Guo W, Chung CL, Chen WT. Synthesis and characterization of novel fluorinated polyimides derived from 1,3-bis(4-amino-2-trifluoromethylphenoxy) naphthalene and aromatic dianhydrides. European Polymer Journal. 2010;**46**: 1878-1890

[47] Kim SD, Lee S, Heo J, Kim SY, Chung IS. Soluble polyimides with trifluoromethyl pendent groups. Polymer. 2013;**54**:5648-5654

[48] Kim SD, Lee B, Byun T, Chung IS, Park J, Shin I, et al. Poly(amide-imide) materials for transparent and flexible displays. Science Advances. 2018;**4**: eaau1956

[49] Wooten F. Optical Properties of Solids. New York: Academic Press; 1972

Chapter 3

Design, Fabrication, and Application of Colorless Polyimide Film for Transparent and Flexible Electronics

Wenlin Chen, Hui Ding, Jianshu Yu, Ying Zhang, Xuejiao Sun, Bin Chen, Yanya Jin, Rao Fu and Zhongfu Zhou

Abstract

Driven by the emerging development of transparent and flexible electronics, colorless polyimide (CPI) has been attracting much attention in recent years. As a key component for next generation electronics, CPI film will be well focused both on research and commercialization. In this chapter, we would like to provide a review and outlook to the field for the reference of scientists, engineers, and entrepreneurs. Topics being addressed are formulation/design, synthesis of the resin, fabrication, and characterization of the CPI films, as well as trends of the film application for the next generation of electronics. Attention will also be given to the current stage of manufacturing of CPI monomers and resin, industrial production of CPI films, etc.

Keywords: colorless polyimide, design, fabrication, electronics

1. Introduction

Polyimides (PIs) are one of the most important classes of polymers with high mechanical properties, thermal stability, high temperature resistance, good chemical resistance, and dielectric features [1, 2]. Since their mass production in 1955 [3], they enjoy a superior role in the application of electronics, aerospace, automobile, and military fields for their excellent combined properties under harsh environment [4, 5]. With the rapid development of electronic industry, PI film has even become indispensable in microelectronic and optoelectronic engineering due to their high demand of lightweight, durable, and reliable materials [6]. PI films are used in image display devices, optical films, organic photovoltaics, flexible printing circuit boards, and other optoelectronic devices [7]. However, conventional colored PI films which have lower optical transmittance can't fulfill the need of optical transparency in electronic devices, and a reliable colorless polyimide (CPI) is highly desired.

The lower optical transmittance of conventional PI is caused by the formation of intra- and intermolecular charge transfer complex (CTC) originated from their conjugated structures [8, 9]. Therefore, the basic principle to increase the transparency of CPI is to avoid or reduce the conjugated units and suppress the CTC. To achieve this, two manners of molecular design are adopted in most of the CPI studies. One is to choose diamines with lower electron-donating capability or dianhydrides with lower electron-accepting capability to decrease the CTC induced by them [9]. For example, 2,2′-bis(trifluoromethyl)-4,4′-diaminobiphenyl (TFMB) [10], which is used widely as the monomer of CPI material, has weaker electron-donating property than many diamines like p-phenylenediamine (PPD), 4,4′-oxydianiline (ODA) and will show lighter color if copolymerizing with the same dianhydride [9]. The similar phenomenon will happen to 3,3′,4,4′-biphenyl tetracarboxylic diandhydride (BPDA) with weaker electron-accepting property compared with pyromellitic dianhydride (PMDA) [9]. The other manner is to break the chemical regularity of the copolymer, like inducing the alicyclic structural unit [11, 12]. But in most cases, these two manners are combined together to improve the transparency of CPI. To be concrete, many CPIs are synthesized by introducing fluorine atoms, cyclic side groups, bulky substituents, flexible linkages, and so on [13–18].

Besides the monomers, the film preparation method can also influence the color of PI films by heat or solvents [15]. There are two conventional ways to prepare CPI films: PAA route and organo-soluble PI route. As heat treatment in air will cause coloration, it is suggested that the film preparation should be under inert gas atmosphere or in vacuum by PAA route, however, the high temperature about 300–350°C will still affect the transparency. If the CPIs are soluble, the solvent method will be a better way to prepare CPI compared to PAA route. But, it needs to be concerned that organo-soluble PIs may have relatively low heat and solvent resistance.

Since the discovery of CPI, quantities of studies have been done. In terms of patent layout, patent applications in the CPI field began in the 1970s and reached a peak in 2011. Among them, the number of patent applications in Japan ranks first, accounting for nearly half of the total number of applications.

After more than 30 years of development, the industrial production of CPI has been realized in recent years. While PI production has high technical barriers, there are only several main suppliers, including DuPont, Kaneka, Ube Industries, South Korea SKC, and Taimide, which occupy more than 90% of the global market. The number of CPI suppliers is even less, such as Kolon Industry, SKC, and Sumitomo Chemical. Sumitomo Chemical took up more than 95% of the global CPI films market in 2018. The main suppliers of CPI and their products are listed in **Table 1** [19].

Although CPI has been used in applications, like OLED and photovoltaics, there are still many aspects to be improved. For designing molecular structure, high-temperature stability, optical transparency, and other properties need to be balanced, as in many cases these properties contradict with each other. For PI film production, the major challenge is the high level of technical capabilities is needed to improve the yield. Other properties like water and gas properties need to be enhanced too.

Company	Product name	Transmission	T_g (°C)
MitsubishiGas Chemical	Neopulim®	89–90	300–489
DuPont-Toray	Colorless Kapton®	87	>300
Kolon Industry	—	>89	330–350
SKC	—	89	—
Sumitomo Chemical	—	>90	—
HiPolyking	—	>90	>330

Table 1.
The main suppliers of CPI and their products.

2. The formulation of CPI

Conventional aromatic PIs showing deep color by absorbing visible light intensely is not applicable in microelectronic and optoelectronic engineering. This is well known due to the formation of charge transfer complexes (CTCs) from the interactions between electron-donating diamine and electron-accepting dianhydride. Recent development shows that the suppression of CTC can be achieved by incorporating structures such as fluorinated group, alicyclic monomers, noncoplanar structure, meta-substitute structure, or sulpone into the main chain of PIs thus increasing the transparency of PI films [20]. According to the composition of main chain, colorless polyimides (CPIs) will be classified into fluorinated CPIs, alicyclic CPIs, noncoplanar CPIs, and other CPIs. The key monomers are listed below in tables to show the studies clearly (**Tables 2–6**).

2.1 Fluorinated CPIs

Examples of aromatic CPIs are largely dominated by fluorinated ones, as incorporating the charge negative fluorine atoms into the electron-donating diamine will suppress the formation of CTC and therefore increase the transparency of PI films. Furthermore, due to the larger volume of fluorine atoms, the free volume between the molecules are also increased which improves the dielectric properties of PIs. Despite the improvement of both optical and dielectric properties, the incorporation of fluorine atoms might lead to the lowering of mechanical strength and glass transition temperature (T_g).

2.2 Alicyclic CPIs

A more effective approach to synthesize CPIs is to use nonaromatic monomers either in diamines or dianhydrides. An alicyclic compound is an organic compound that contains one or more all-carbon rings, which may be either saturated or unsaturated but do not have aromatic character. By adopting the alicyclic moieties in the main chain of PI, the probability of undergoing inter- or intramolecular charge transfer (CT) is lower, which leads to an improvement in the optical properties of PIs. However, the incorporation of alicyclic unit generally reduces thermal stability and mechanical strength and both are crucial factors for PI. Therefore, it's important to consider the proportion of alicyclic structure without too much compromise on the thermal and mechanical properties.

2.3 Noncoplanar colorless PIs

Incorporation of noncoplanar structures into polymer chains is considered as one of the effective ways to improve the optical transparency of PIs without too much compromise of their thermal stability and mechanical strength. The nonplanar structure reduces the strong interaction and tight stacking of the molecular chains preventing the forming of CTC.

2.4 Other colorless PIs

Besides the monomers listed, there are also other structures that can be introduced into the PIs to improve the optical transparency of PI films. For instance, sulfonyl-substituted PIs exhibit good combined quality, including good thermal stability and high optical transparency within the visible light region. This is due to

Monomer	Chemical structure	Ref.
1,3-Bis(4-amino-2-trifluoromethyl-phenoxy)benzene	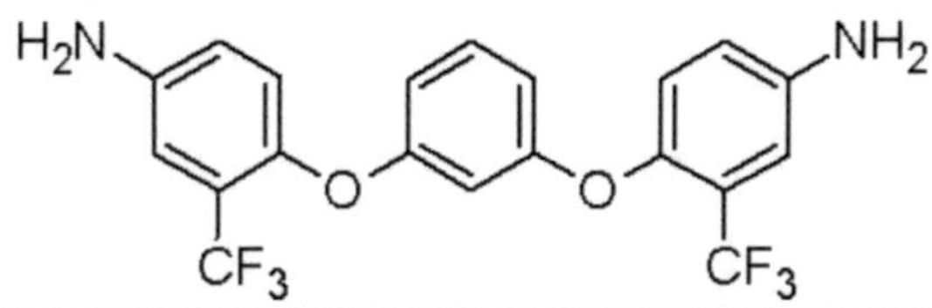	[21]
2,2-Bis[4-(2-trifluoro-methyl-4-aminophenoxy) phenyl] hexafluoropropane	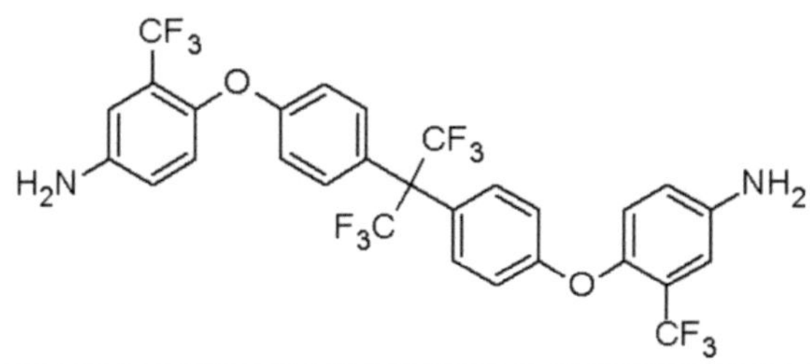	[22]
2,2′-Bis (trifluoromethyl)-4,4′-diaminobiphenyl (TFMB)	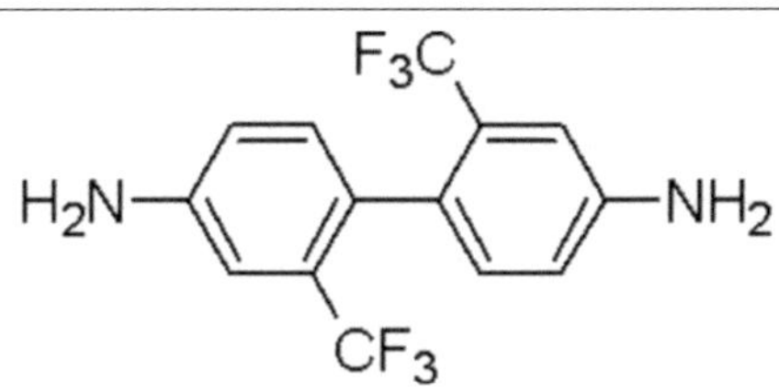	[23, 24]
4,4′-Bis(4-amino-2-trifluoromethylphenoxy) biphenyl	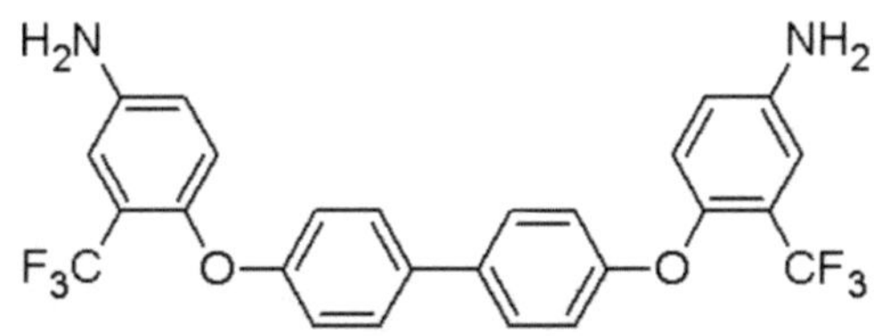	[25]
4,4′-((Perfluoro-[1,1′-biphenyl]-4,4′-diyl)bis (oxy))dianiline	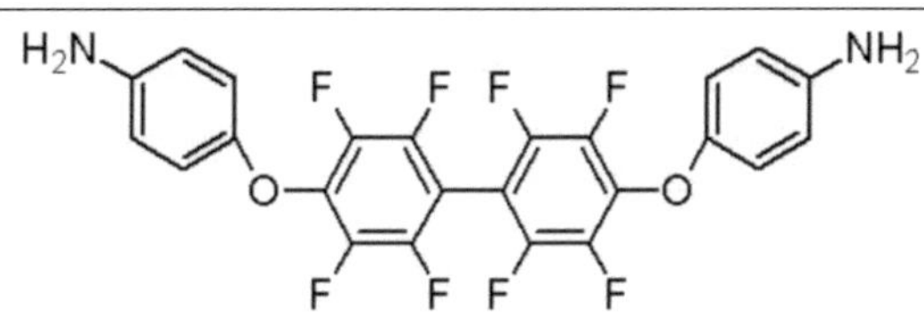	[26]
4,4′-((Perfluoro-[1,1′-biphenyl]-4,4′-diyl)bis (oxy))bis(2,6-dimethylaniline)	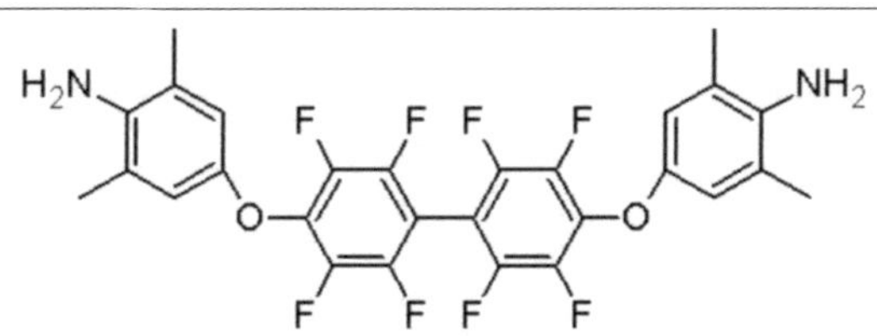	[26]
1,4-Bis(4-amino-2-trifluoromethylphenoxy) benzene	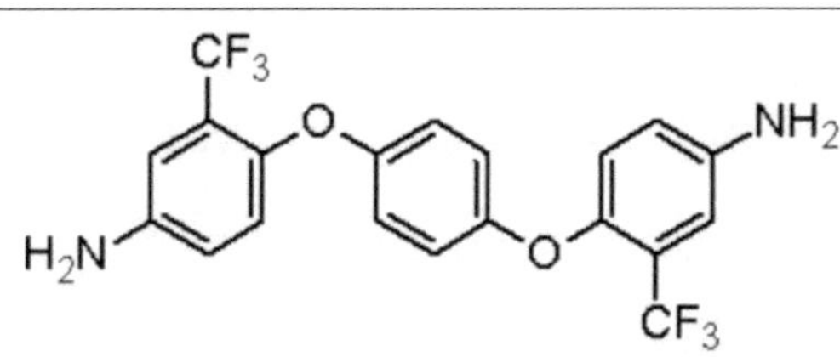	[25]
4,4′-Bis(4-amino-2-trifluoromethylphenoxy) diphenyl ether	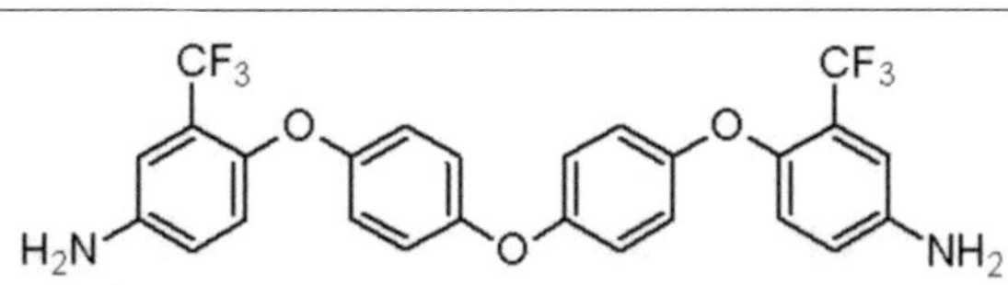	[27]

Monomer	Chemical structure	Ref.
2,4-Diamino-1-(1H,1H,2H,2H-perfluorodecathio) benzene	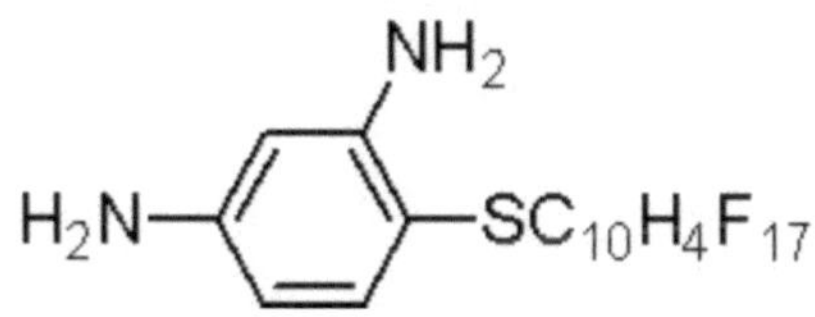	[28]
2,2′-Bis((1H,1H,2H,2H-perfluorodecyl)thio) [1,1’-biphenyl]4,4′-diamine	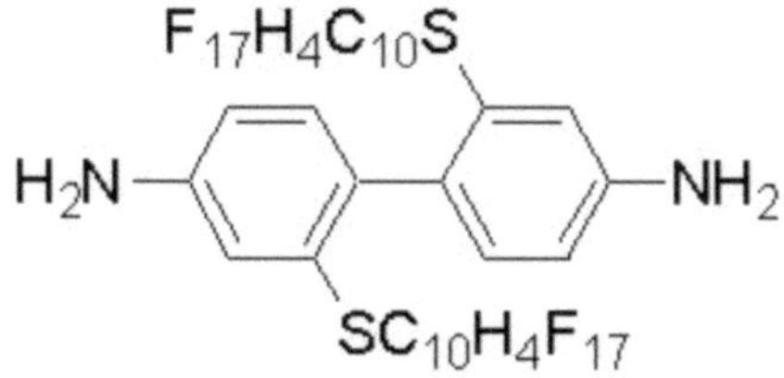	[28]
3,5-Di(2-trifluoromethyl-4-amino)-1-trifluoromethyl-tuluene	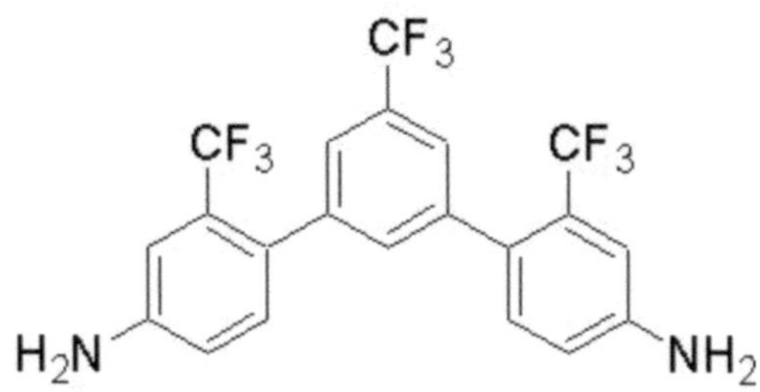	[29]
2,2-Bis(3,4-dicarboxyphenyl) hexafluoropropane dianhydride (6FDA)	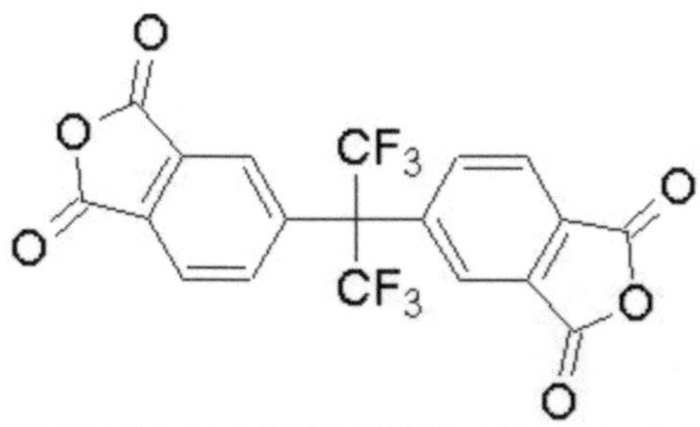	[23, 27, 28, 30–32]

Table 2.
Key monomers for fluorinated CPIs.

the existence of charge-accepting sulfur atoms in sulfonyl suppressing the formation of CTC. Another example is incorporating meta-substituted diamine to increase the free volume which lowers the interaction between molecules and effectively reduces the CTC. Moreover, hyperbranched polymer is considered to have low probability of forming CTC due to their asymmetrical spherical loose structures with defect (**Table 6**).

3. CPI film production

In general, polymer film manufacturing techniques include several types, such as melting extrusion, casting, and blowing. There are many factors that affect the choice of polymer film production process, including physical and chemical properties, color and appearance requirements of polymer resins, and the existing capabilities of film production equipment. For crystalline polymer resins with clear melting points, such as PET and PEN, non-solvent melt extrusion technology is mainly used. For amorphous polymers with low to medium melting points, such as

Monomer	Chemical structure	Ref.
4,4′-Methylenebis (cyclohexylamine) (MBCHA)	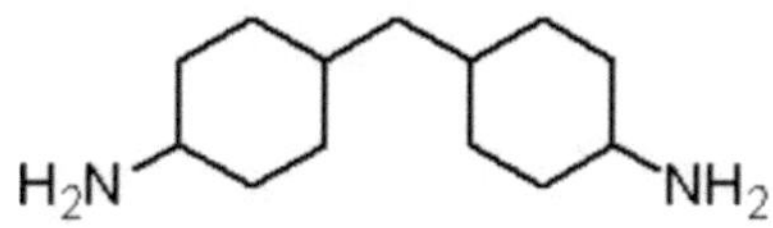	[11]
4,4′-Methylenebis(2-methylcyclohexylamine)	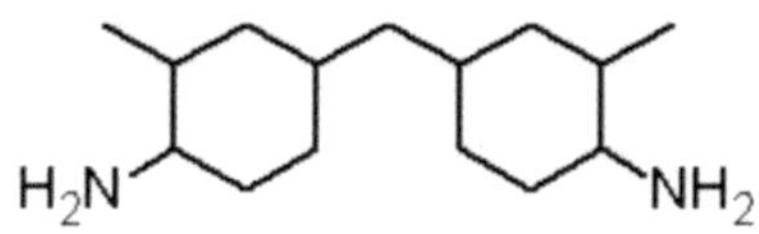	[33]
1,4-Cyclohexanediamine (CHDA)	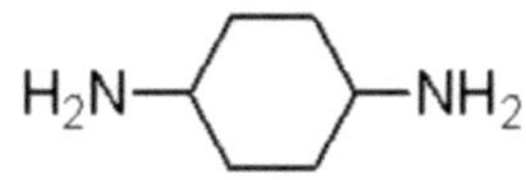	[11]
Cyclobutane tetracarboxylic dianhydride (CBDA)	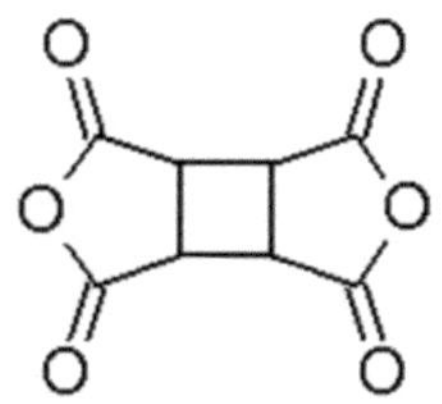	[12, 25, 34]
Cyclopentane tetracarboxylic dianhydride (CPTA)	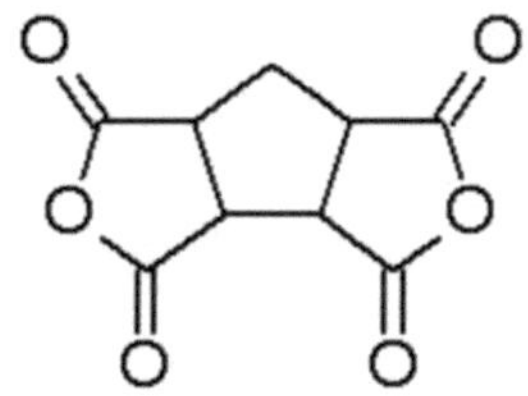	[25]
1,2,4,5-Cyclohexanetetracarboxylic dianhydride	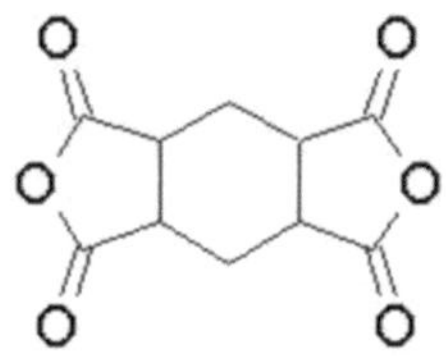	[25, 35]
Cyclohexanetricarboxylic–based dianhydrides	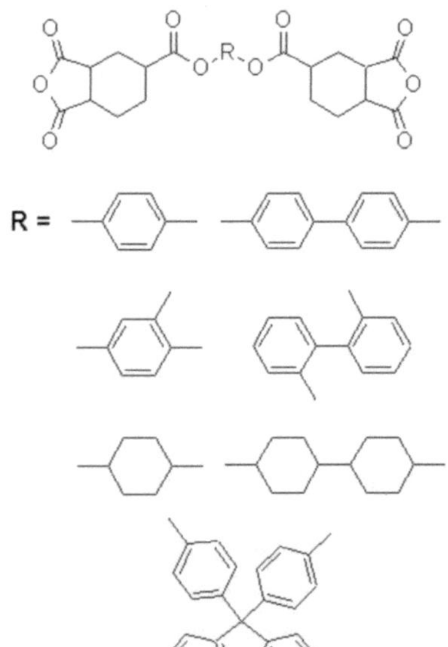	[11]

Monomer	Chemical structure	Ref.
1,4:3,6-Dianhydro-2,5-di-O-(4-aminophenyl)-D-mannitol	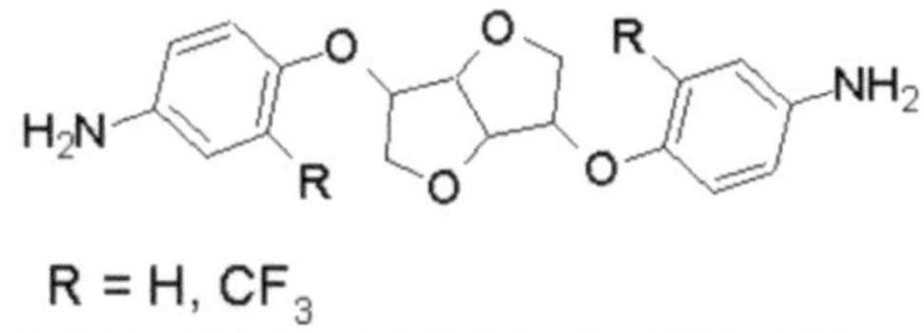	[36]
1,4:3,6-Dianhydro-2,5-di-O-(3,4-dicarboxyphenyl)-D-mannitol dianhydride	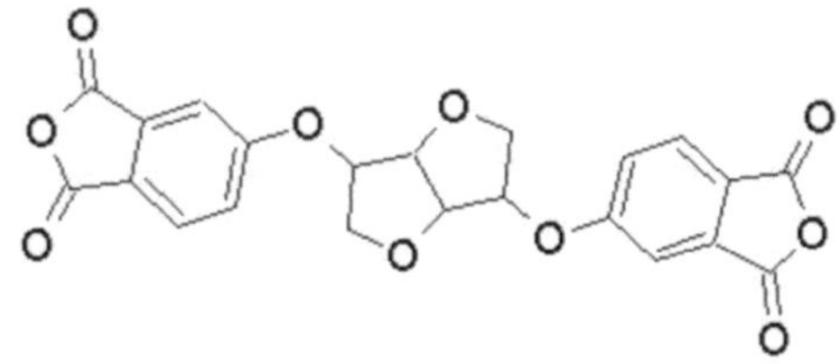	[36]
5,5′-(1,1,3,3-Tetramethyl-1,3-disiloxanediyl) bisnorbornane-2,3-dicarboxylic anhydride	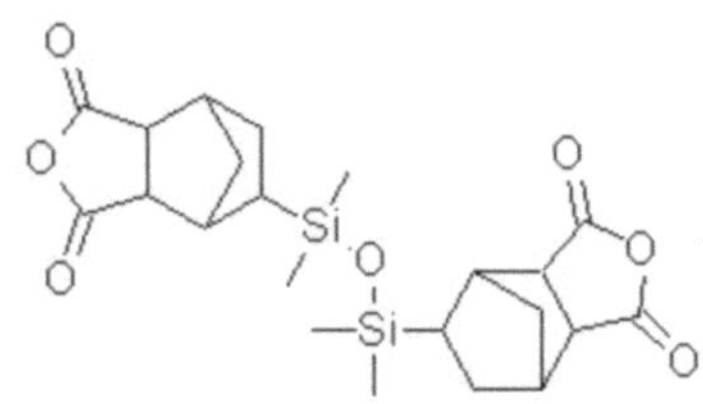	[27]
3,4-Dicarboxy-1,2,3,4-tetrahydro-1-naphthalene succinic dianhydride (TDA)	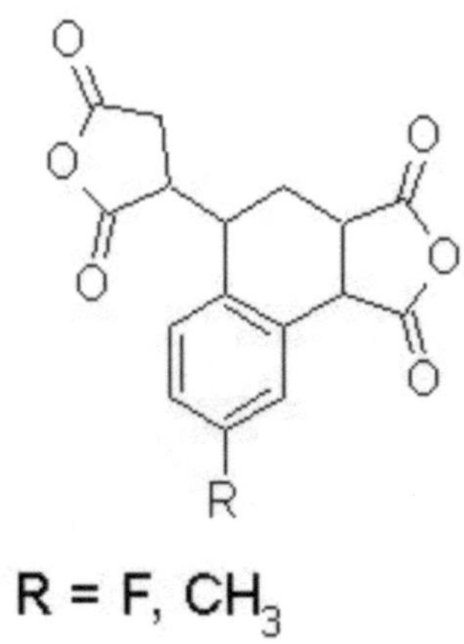	[37, 38]

Table 3.
Key monomers for Alicyclic CPIs.

PC and PES, melt extrusion and solution casting are used. As PIs are amorphous polymers with high melting points due to their hard molecular skeleton, the solvent casting process [47] (as shown in **Figure 1**) is usually the best choice, especially in the laboratory.

The solvent casting process to prepare colorless polyimide film in the laboratory can be realized by two conventional film-forming routes, PAA and organo-soluble PI route [48–54]. Similar to the preparation of ordinary PI film, the PAA method to produce CPI just needs to add salivation, drying, stretching, and other steps in the middle of the two steps of PI polymer production, synthesis of polyamide acid, and imidization. In detail, the monomers of dianhydride and diamine are first polymerized in DMAc or NMP to form a PAA solution. A PAA film is then formed by casting the solution on a clean substrate. Finally, an imidization treatment is performed to produce a colorless PI film. Because PAA membranes are prone to

Monomer	Chemical structure	Ref.
3,3′,5,5′-Tetramethyl-4,4′-diaminodiphenyl-butyltoluene (BADP)	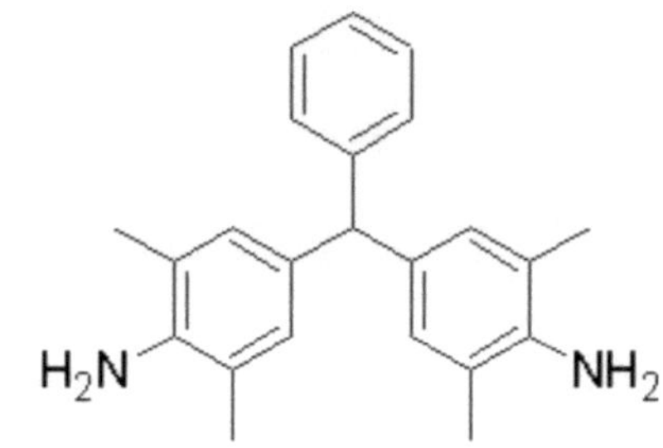	[39]
3,3′-Ditertbutyl-4,4′-diaminodiphenyl-4″-tertbutylphenylmethane	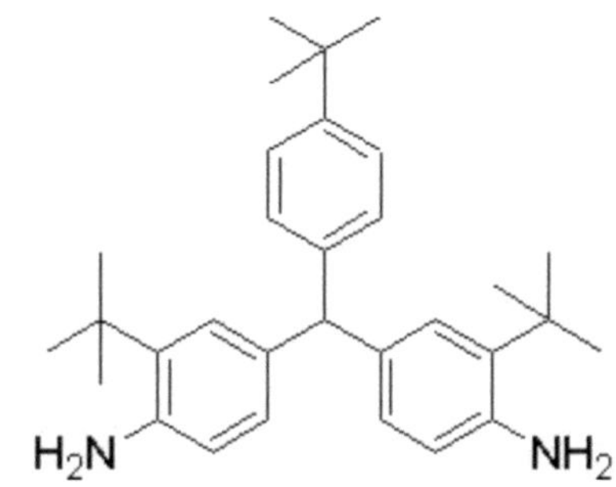	[40]
4,4′-(Naphthalen-1-ylmethylene)dianiline	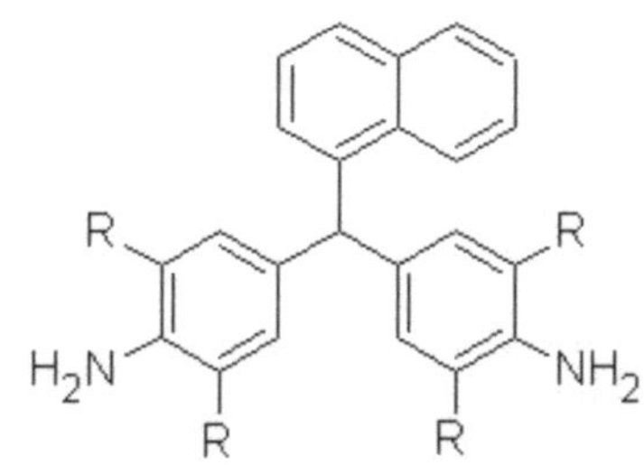 R = H, CH_3, $CH(CH_3)_2$	[30]
4-[1-(4-Aminophenyl) cyclohexyl]aniline	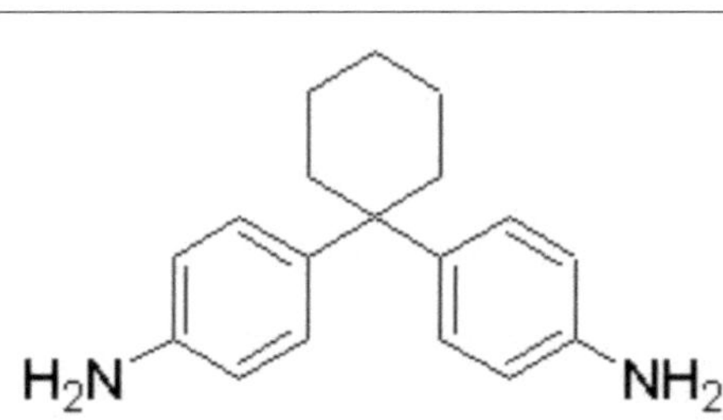	[24]
2,2′-Bis(4-chlorothiophenyl) benzidine	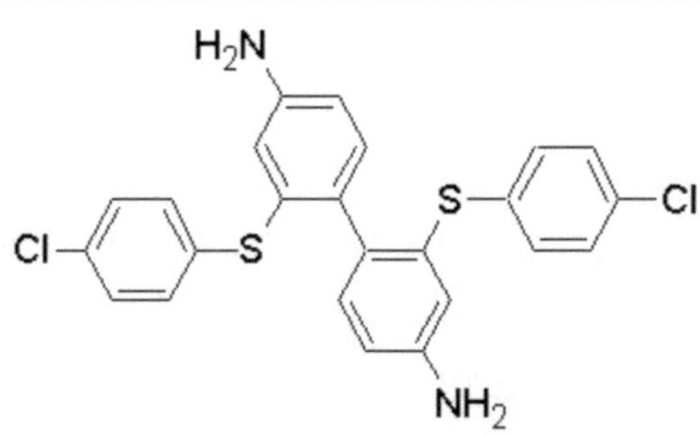	[41]

Table 4.
Key monomers with large substituent group or side group.

Monomer	Chemical structure	Ref.
2,3,3′,4’-Biphenyl tetracarboxylic dianhydride (a-BPDA)	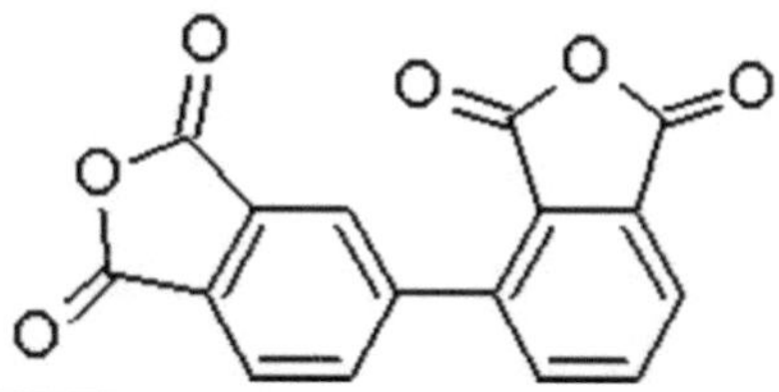	[25]
4,4′-(4,4′-Isopropylidenediphenoxy)bis (phthalicanhydride) (BPADA)	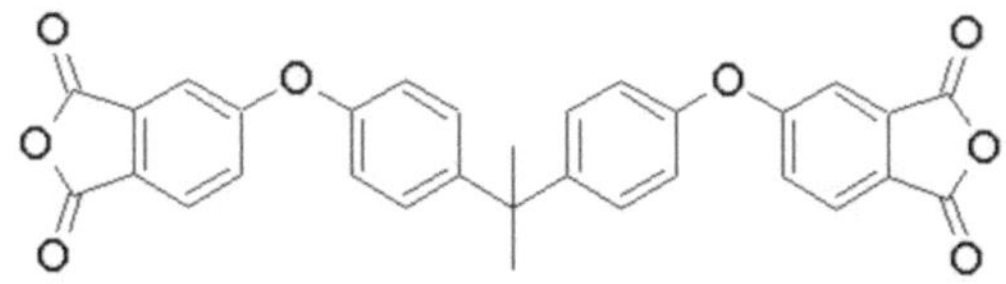	[17, 42]
2,2′-Bis[4-(5-amino-2-pyridinoxy)phenyl]-propane	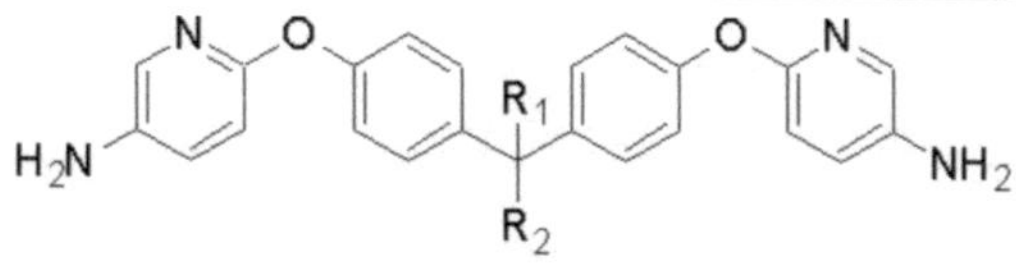 $R_1 = R_2 = CH_3$ $R_1 = R_2 =$ phenyl $R_1 = CH_3$, $R_2 =$ phenyl	[43]
1,3-Bis(3-aminophenoxy) benzene	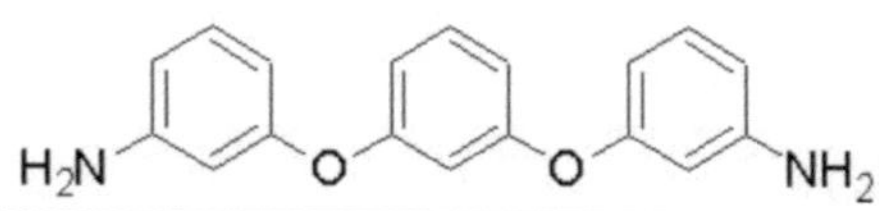	[42]
2,2′-Bis(5-amino-2-pyridinoxy)biphenyl	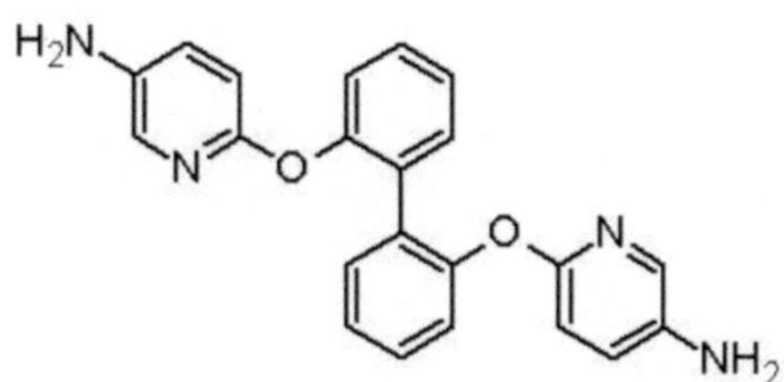	[31]
4,4′-Bis(5-amino-2-pyridinoxy)biphenyl	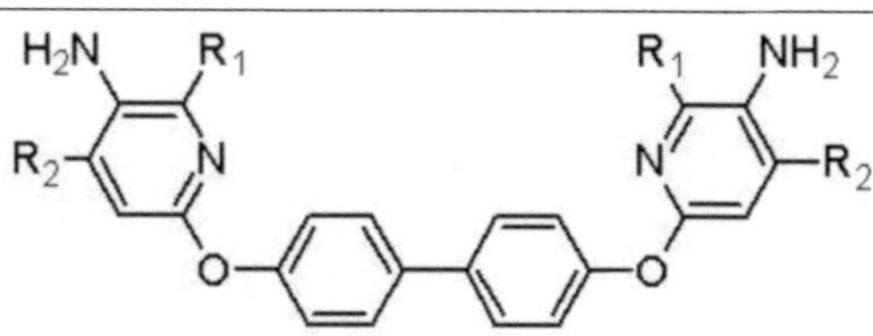 $R_1 = H$, $R_2 = CH_3$ $R_1 = CH_3$, $R_2 = H$	[31]

Table 5.
Key monomers in bended main chain of CPIs.

thermal decomposition or hydrolysis, freshly synthesized PAA is usually used to prepare PI films. High-temperature curing can cause obvious coloring, so the preparation of CPI film should be carried out under vacuum or inert atmosphere. The curing process includes both solvent evaporation and dehydration imidization. In

Monomer	Chemical structure	Ref.
4,4-Bis(4-amino-phenylenesulfany l)diphenylsulfone	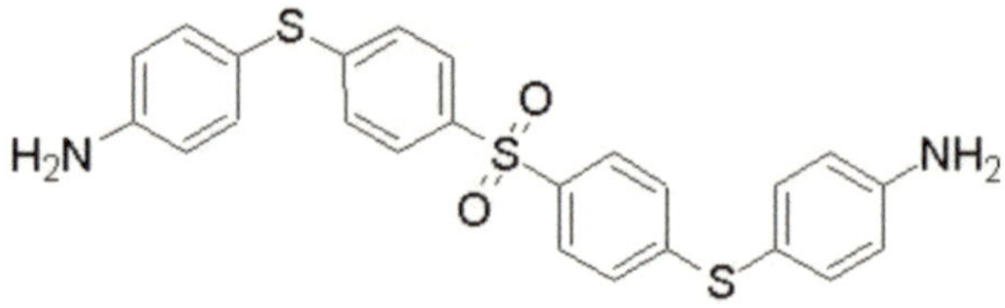	[44]
m-Phenylenediamine	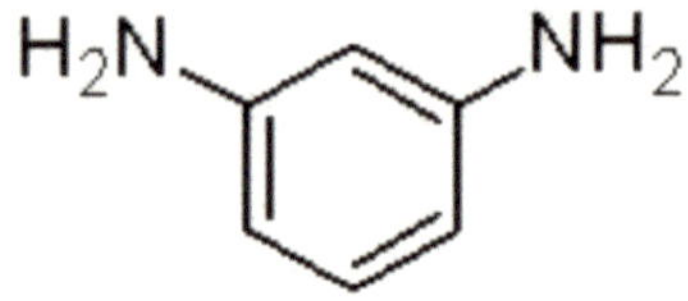	[45]
Bisphenol A bis (3,5-diaminobenzoate)	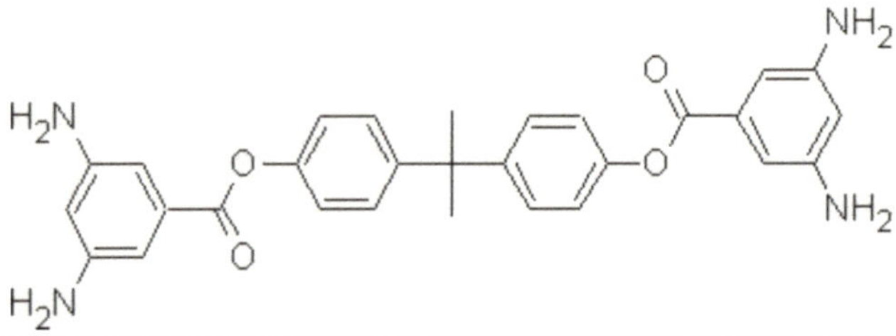	[46]

Table 6.
Sulfonyl-substituted, meta-substituted, and hyperbranched monomers.

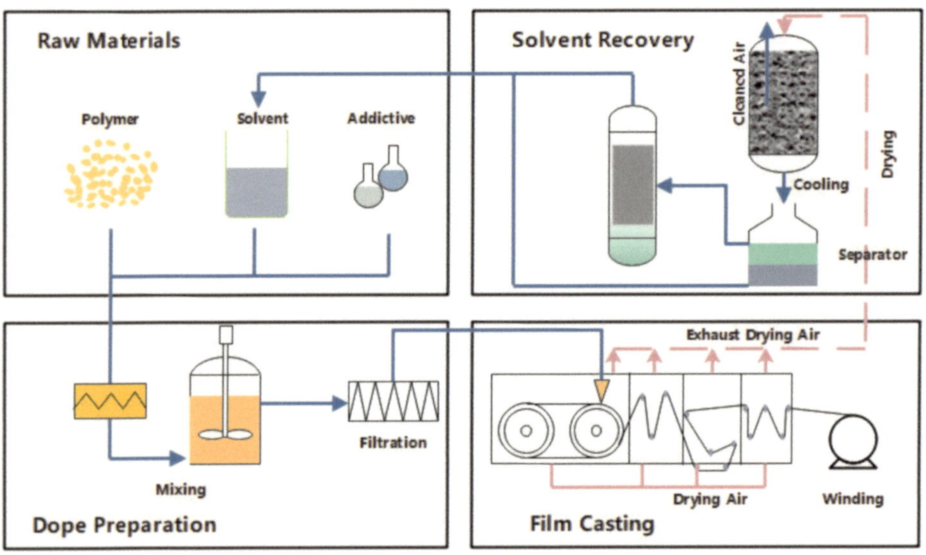

Figure 1.
Production of solvent cast films, adopted from [47].

most cases, the temperature of the imidization process needs to reach 300–350°C [55], under which, even in a vacuum or inert atmosphere, the PI system is prone to discoloration. At the same time, microdefects such as pinhole cracks are also prone to occur during the dehydration process. Therefore, in the preparation process of the colorless PI film, the PAA route is rarely used. The organo-soluble PI solution is only applicable to polyimides that can be dissolved in organic solvents. This method uses a stable high-solid polyimide solution to form a film on a clean substrate.

Because the curing process of the pre-iminated PI solution is almost a pure physical process of solvent evaporation, the curing process can be realized under a lower temperature (≤250°C) which will not cause coloring. CPI films prepared this way will have a good surface smoothness and better transparency.

The excellent performance of PI film is related to its heteroaryl molecular structure as well as their unique production technology. Compared with laboratory preparation, industrial production also use the solvent casting process, however, the biggest difference between them is the stretching process. There are two stretching methods, uniaxial stretching (machine direction, MD) and biaxial stretching (transverse direction, TD, and mechanical direction, MD) technologies, as shown in **Figure 2** [56]. Biaxial stretching is always used to ensure the evenness of the film. The gelation of the PAA membrane used for stretching can be achieved by partial evaporation of the solvent, or by chemical treatment of the dehydrating agent (acetic anhydride, dicyclohexylcarbodiimide, etc.) and its catalyst (pyridine). During the stretching process, whether it is uniaxial stretching or biaxial stretching, the gel-like PAA film will lead to a complete orientation and stretching of the PI molecular chain. From the perspective of polymer physics, stretching will greatly improve the mechanical properties of the resulting PI film. CPI films produced by biaxial stretching have good optical transparency, heat resistance, and reduced dimensional change.

In production, the monomer of polyimide is firstly sent to a polymerization reactor for polycondensation to obtain PAA solution. The solution is degassed and casted to a continuous film on a heated, rotating steel drum forming a self-supporting PAA film. The gel-like PAA film will be peeled off from the metal drum and stretched in the machine direction (MD) and lateral direction, which is performed at a temperature of about 350°C to promote the imidization of PAA. This method above has been widely used in the preparation of polyimide films industrially. Whereas, due to the high temperature (350°C) of full imidization of PAA, the above-mentioned manufacturing process may be difficult for producing colorless polyimide film. Therefore, a novel manufacturing technology has been developed in recent years. It uses soluble PI resin as the starting material instead of PAA. The key elements of this procedure include the following: (1) the PI resin must be soluble in volatile solvents; (2) the formed PI solution should be stable with a reasonable solids

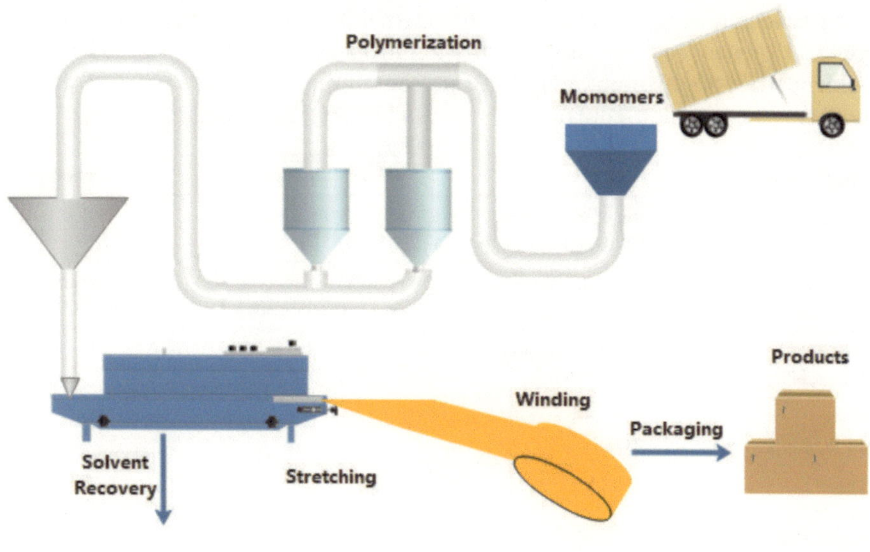

Figure 2.
Production process of polyimide films, adopted from [56].

content and viscosity; and (3) it must be able to form a uniform film and be releasable from the casting support. Mitsubishi Gas Chemical Company introduced a method [57, 58] for producing colorless and transparent PI films using this technology. The starting soluble PI resin was prepared from 1,2,4,5-cyclohexanetetracarboxylic dianhydride and aromatic diamine by one-step high-temperature polycondensation. The PI film was then stretched 1.01 times in the machine direction and 1.03 times in the transverse direction under the condition of nitrogen at a temperature of 250°C and dried. The obtained PI film showed good properties, with a thickness of is 200 mm, a light transmittance 89.8%, and a yellow index 1.9.

The major challenge of polyimide film market is the high level of technical capabilities involved in the film processing. Besides, the strong heat and chemical resistance of polyimide lead to processing difficulties, such as lack of solubility. This makes it difficult to integrate other materials into the polymer matrix. At the same time, there are some other key issues in processing of CPI film, high unevenness, the thickness decline caused by high stress (50–60% of the deposited value), and poor adhesion.

4. Challenges and solutions

4.1 Challenges of CPI for TaFEs

Targeting the applications of transparent and flexible electronics, it is of priority that CPI film owns satisfied optical, mechanical, thermal, chemical, optical, and electrical properties. For the key applications, the specifications of such properties are listed in **Table 1**. Like other materials, usually, the physical and chemical properties of CPI film are not independent; it is always the case that we need to adjust the composition and structure of the CPI film to balance the needs of the properties. However, there are major challenges we have to take, with the consideration of balancing the overall performance of the CPI film (**Table** 7).

4.2 Balance of the desired properties of CPI film for TaFE applications

Extensive studies have been made to improve the transmittance of CPI film. The major strategies to improve the transmission properties of light with the balance of other properties are by employing fluorine containing monomers [63–67], aliphatic monomers [68, 69], non-coplanar monomers [70, 71], adjusting conjugated

Specification	Parameters	Desired figures	Reference
Transmittance	Transmission of visible light around 400–700 nm (T)	~90%	[59]
Thermal expansion	Coefficient of linear thermal expansion (CTE)	<30 ppm/°C (as low as the inorganic components)	SiNx gas barrier (5–15 ppm/°C) [60]
Oxygen permeability	Water vapor transmission rate (WVTR)	$<10^{-5}$ cm^3/m^2 day at 23°C (OLED)	[61]
Water permeability	Oxygen transmission rate (OTR)	$<10^{-6}$ g/m^2 day at 23°C (OLED)	[61]
Thermal stability	T_g	~300°C (process temperature of α-Si TFTs is below 300°C)	[62]

Table 7.
The specifications of CPI film for transparent and flexible electronics.

electrochromophores which limit the formation of charge transfer complex [72]. The influence of these monomer molecular changes on the transmittance and thermal stability can be observed more clearly in PI with a similar structural unit [11] (**Figure 3**). As shown in **Figure 3**, introducing fluorine atoms, alicyclic moieties, and flexible main chain is a very effective way to improve the transparency of CPI. The transparency will grow with more of these groups added into the molecule. However, the thermal stability decreases fast with the increase of the transparency of CPI, which is what the researchers have tried for years to overcome. With the deepening of the research of CPI, a unique kind of monomers-isomerized monomers have drawn people's attention. **Figure 4** has shown that changing a CPI monomer to an isomerized one will enhance not only the transparency of the CPI but also the thermal property, which is different from other CPI monomers [73]. This means this kind of noncoplanar structure have no or less influence in the rigidity, and therefore will not sacrifice the mechanical and thermal properties. So it is suggested that using the noncoplanar in the main chain as well as bringing in fluorine atoms, alicyclic groups in the molecular design may be a very useful way to achieve a better comprehensive performance under the synergistic action of all the factors.

As discussed in the previous section, it is very promising to obtain CPI with good mechanical and thermal properties by using noncoplanar structure cooperating with fluorine and alicyclic groups. But these factors are associated with negative effects to the applications as gas barriers and transparent packaging materials. They will lead to worse water/oxygen permeability, even though the water/oxygen permeability is always a pain point of CPI. The usual methods to solve this problem is coating barriers like SiN_xO_y [74] and Al_2O_3 [75] or using composite fillers like graphene [75], montmorillonite [76], and other nanosheet fillers. Nanocomposite filler method is more practical in industrial production and can reduce the thermal expansion of CPI while keeping its transparency and thermal properties, so it is a more promising way in the future.

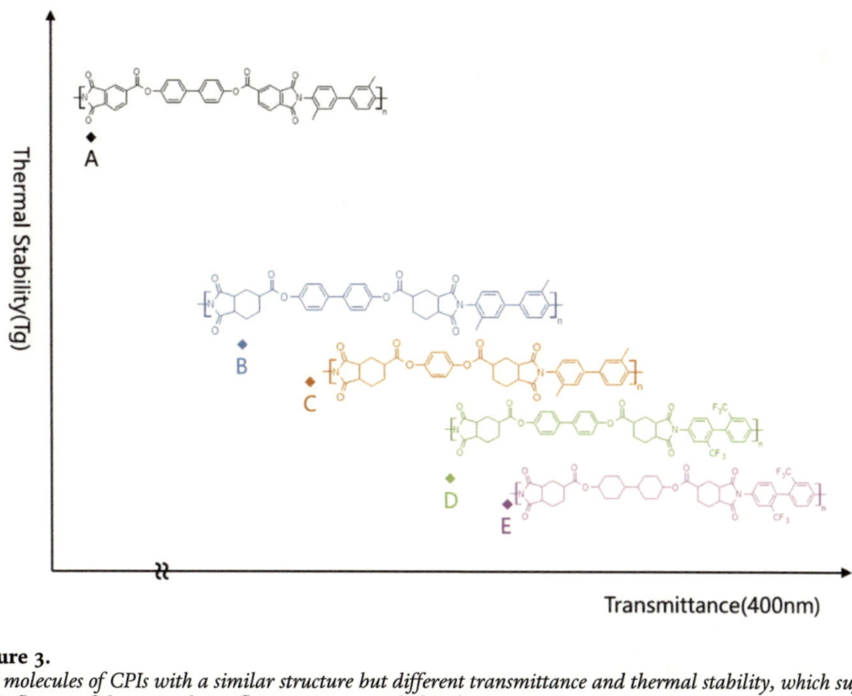

Figure 3.
The molecules of CPIs with a similar structure but different transmittance and thermal stability, which suggest the influence of the main chain, fluorine atoms, and alicyclic groups.

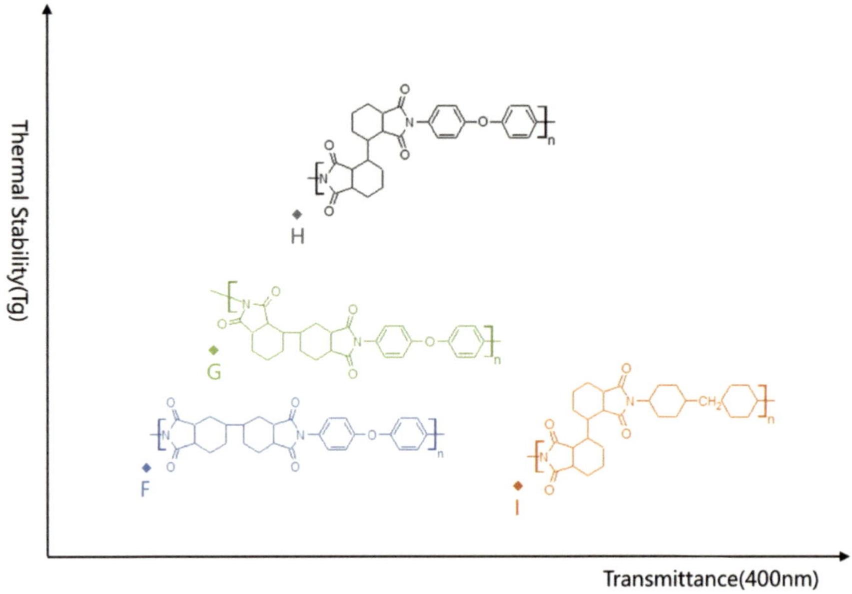

Figure 4.
The molecules of CPIs with isomeric structures have different transmittance and thermal stability.

In general, it is thought prospective to produce CPI with high comprehensive performance by using nanosheets as the filler to modify CPI with good thermal and mechanical properties. This kind of PI can be prepared by using the noncoplanar in the main chain as well as bringing in fluorine atoms, alicyclic groups.

Author details

Wenlin Chen[1], Hui Ding[1], Jianshu Yu[1], Ying Zhang[2], Xuejiao Sun[3], Bin Chen[2], Yanya Jin[3], Rao Fu[1] and Zhongfu Zhou[1,4,5,6]*

1 School of Materials Science and Engineering, Shanghai University, Shanghai, China

2 Shanghai Kington Technology Ltd., Shanghai, China

3 Techcelerate (Shanghai) Ltd., Shanghai, China

4 Department of Mathematics and Physics, Aberystwyth University, Aberystwyth, Ceredigion, UK

5 Inner Mongolia Enterprise Key Laboratory of New Carbon Materials, Baotou, China

6 Inner Mongolia Engineering Research Center for Graphene Industrialization, Baotou, China

*Address all correspondence to: zzz@aber.ac.uk

References

[1] Pantani R. Processing-Structure-Properties Relationships in Polymers. MDPI AG; 2019

[2] Ghosh M. Polyimides: Fundamentals and Applications. Taylor & Francis; 1996

[3] Edwards WM, Maxwell RI. Polyimides of pyromellitic acid. 1955

[4] Polyimides and Other High Temperature Polymers: Synthesis, Characterization and Applications

[5] Kirby AJ. Polyimides: Materials, Processing and Applications. iSmithers Rapra Publishing; 1992

[6] Ohya H, Kudryavsev VV, Semenova SI. Polyimide Membranes: Applications, Fabrications and Properties. CRC Press; 1997

[7] Matsuura T, Ando S, Sasaki S. Polyimides for Optical Applications. p. 46

[8] Liaw D-J et al. Advanced polyimide materials: Syntheses, physical properties and applications. Progress in Polymer Science. 2012;**37**:907-974

[9] Shinji A, Tohru M, Shigekuni S. Coloration of aromatic polyimides and electronic properties of their source materials. Polymer Journal. 1997;**29**:69-76

[10] Kanosue K et al. A colorless semi-aromatic polyimide derived from a sterically hindered bromine-substituted dianhydride exhibiting dual fluorescence and phosphorescence emission. Materials Chemistry Frontiers. 2019;**3**:39-49

[11] Hasegawa M, Kasamatsu K, Koseki K. Colorless poly(ester imide)s derived from hydrogenated trimellitic anhydride. European Polymer Journal. 2012;**48**:483-498

[12] Chen Y et al. Preparation and properties of novel colorless and transparent polyimide films. Insulation Materials. 2018;**51**:1-5

[13] Ni H, Liu J, Wang Z, Yang S. A review on colorless and optically transparent polyimide films: Chemistry, process and engineering applications. Journal of Industrial and Engineering Chemistry. 2015;**28**:16-27

[14] Yu X-H, Liu J-N, Wu D-Y. Colorless PI structure design and evaluation for achieving low CTE target. Materials Today Communications. 2019;**21**:100562

[15] Tapaswi PK, Ha C-S. Recent trends on transparent colorless polyimides with balanced thermal and optical properties: Design and synthesis. Macromolecular Chemistry and Physics. 2019;**220**: 1800313

[16] Shin HI, Chang J-H. Transparent polyimide/organoclay nanocomposite films containing different diamine monomers. Polymers. 2020;**12**:135

[17] Chang J-H. Equibiaxially stretchable colorless and transparent polyimides for flexible display substrates. Reviews on Advanced Materials Science. 2020;**59**: 1-9

[18] Yi C et al. High-temperature-resistant and colorless polyimide: Preparations, properties, and applications. Solar Energy. 2020;**195**: 340-354

[19] Colorless and Transparent High-Temperature-Resistant Polymer Optical Films—Current Status and Potential Applications in Optoelectronic Fabrications. IntechOpen. Available from: https://www.intechopen.com/books/optoelectronics-materials-and-devices/colorless-and-transparent-high-temperature-resistant-polymer-optical-films-current-status-and-potent

[20] Li Z, Yu X. Research progress of colorless and transparent polyimide films. Synth. Technol. Appl. 2019;**34**: 26-29

[21] Zhang L, Yu X, Xu Y, Fu J. Preparation and performance study of 1,3-bis (4-amino-2-trifluoromethyl-phenoxy) benzene and its transparent polyimide film. Insulation Materials. 2010;**43**:4-8

[22] Han Q et al. Synthesis of fluorinated diamine and properties study of transparent polyimide film. Insulation Materials. 2013;**46**:51-55

[23] Zhang M, Gao S, Wu Z, Cui H, Gao Y. Preparation and characterization of low expansion transparent copolyimide films. Materials Science and Technology. 2019;**27**:81-86

[24] Zhang L et al. Synthesis and properties of soluble and transparent polyimide films. Insulation Materials. 2019;**49**:14-18

[25] Liu J, Li Z, Gao Z, Yang H, Yang S. Synthesis and properties of fluorinated semi-alicyclic transparent polyimide films for optocommunication applications. Chinese Journal of Materials Research. 2008;**22**:615-618

[26] Yeo H, Goh M, Ku B-C, You N-H. Synthesis and characterization of highly-fluorinated colorless polyimides derived from 4,4′-((perfluoro-[1,1′-biphenyl]-4,4′-diyl)bis(oxy))bis(2,6-dimethylaniline) and aromatic dianhydrides. Polymer. 2015;**76**:280-286

[27] Gu Z et al. Synthesis and characterization of novel fluorosilicone polyimides with high optical transparency. High Performance Polymers. 2014;**26**:335-346

[28] Tapaswi PK, Choi M-C, Nagappan S, Ha C-S. Synthesis and characterization of highly transparent and hydrophobic fluorinated polyimides derived from perfluorodecylthio substituted diamine monomers. Journal of Polymer Science Part A: Polymer Chemistry. 2015;**53**:479-488

[29] Jiang J, Jiang P, Lu F, Tu G. Synthesis of novel rigid terphenyl diamine and application in transparent polyimide. Electro-Optical and Infrared Systems: Technology and Applications. 2020;**35**:49-55

[30] Li T, Huang H, Wang L, Chen Y. High performance polyimides with good solubility and optical transparency formed by the introduction of alkyl and naphthalene groups into diamine monomers. RSC Advances. 2017;**7**: 40996-41003

[31] Guan Y et al. High transparent polyimides containing pyridine and biphenyl units: Synthesis, thermal, mechanical, crystal and optical properties. Polymer. 2015;**62**:1-10

[32] Li T-L, Hsu SL-C. Preparation and properties of a high temperature, flexible and colorless ITO coated polyimide substrate. European Polymer Journal. 2007;**43**:3368-3373

[33] Xu Y, Wang Y, Pu Y, Wang K, Yu X. Preparation and properties of polyimide film with 3,3′-dimethyl-4,4′-diamino dicyclohexyl methane. Insulation Materials. 2017;**50**:40-45

[34] Suzuki H, Abe T, Takaishi K, Narita M, Hamada F. The synthesis and X-ray structure of 1,2,3,4-cyclobutane tetracarboxylic dianhydride and the preparation of a new type of polyimide showing excellent transparency and heat resistance. Journal of Polymer Science Part A: Polymer Chemistry. 2000;**38**(1):108-116

[35] Hasegawa M et al. Solution-processable colorless polyimides derived from hydrogenated pyromellitic dianhydride with controlled steric structure. Journal of Polymer Science

Part A: Polymer Chemistry. 2013;**51**: 575-592

[36] Yang G et al. Synthesis of novel biobased polyimides derived from isomannide with good optical transparency, solubility and thermal stability. RSC Advances. 2015;**5**: 67574-67582

[37] Zhang X, Song Y, Liu J, Yang S. Synthesis and properties of cost-effective light-color and highly transparent polyimide films from fluorine-containing tetralin dianhydride and aromatic diamines. Journal of Photopolymer Science and Technology. 2016;**29**:31-38

[38] Guo Y, Song H, Zhai L, Liu J, Yang S. Synthesis and characterization of novel semi-alicyclic polyimides from methyl-substituted tetralin dianhydride and aromatic diamines. Polymer Journal. 2012;**44**:718-723

[39] Mo X, Li G, Jiang J. Synthesis and characterization of polyimide with high solubility and optical transparency. Materials Reviews. 2012;**26**:67-71

[40] Liu C, Pei X, Huang X, Wei C, Sun X. Novel non-coplanar and tertbutyl-substituted polyimides: Solubility, optical, thermal and dielectric properties. Chinese Journal of Chemistry. 2015;**33**:277-284

[41] Tapaswi PK, Choi M-C, Jeong K-M, Ando S, Ha C-S. Transparent aromatic polyimides derived from thiophenyl-substituted benzidines with high refractive index and small birefringence. Macromolecules. 2015; **48**:3462-3474

[42] Xiao X et al. Optically transparent high temperature shape memory polymers. Soft Matter. 2016;**12**: 2894-2900

[43] Yao J et al. Highly optical transparency and thermally stable polyimides containing pyridine and phenyl pendant. Designed Monomers and Polymers. 2017;**20**:449-457

[44] Liu J, Zhang X, Li Z, Yang H. Synthesis and properties of sulfonyl-substituted polyimides with high refractive indices and high transparency. Journal of Functional Materials. 2008;**39**:460-464

[45] Chen X, Yu X, Xu Y. Preparation and properties study of mPDA/ODPA-PI film. Insulation Materials. 2012:19-21

[46] Zhou Z, Sun Z, Zhang Q. Highly transparent hyperbranched polyimide liquid crystal alignment layers. Journal of Huangshan University. 2013;**15**:36-39

[47] Stribeck N, Smarsly B. Scattering Methods and the Properties of Polymer Materials. Berlin Heidelberg: Springer; 2005

[48] Kim GY, Choi M-C, Lee D, Ha C-S. 2D-aligned graphene sheets in transparent polyimide/graphene nanocomposite films based on noncovalent interactions between poly (amic acid) and graphene carboxylic acid. Macromolecular Materials and Engineering. 2012;**297**:303-311

[49] Mathews AS, Kim I, Ha C-S. Fully aliphatic polyimides—Influence of adamantane and siloxane moieties. Macromolecular Symposia. 2007;**249–250**:344-349

[50] Mathews AS, Kim I, Ha C-S. Fully aliphatic polyimides from adamantane-based diamines for enhanced thermal stability, solubility, transparency, and low dielectric constant. Journal of Applied Polymer Science. 2006;**102**:3316-3326

[51] Choi M-C, Wakita J, Ha C-S, Ando S. Highly transparent and refractive polyimides with controlled molecular structure by chlorine side groups. Macromolecules. 2009;**42**: 5112-5120

[52] Kim J et al. Itraconazole, a commonly used antifungal that inhibits Hedgehog pathway activity and cancer growth. Cancer Cell. 2010;**17**:388-399

[53] Tapaswi PK et al. Synthesis and characterization of fully aliphatic polyimides from an aliphatic dianhydride with piperazine spacer for enhanced solubility, transparency, and low dielectric constant. Journal of Polymer Science Part A: Polymer Chemistry. 2014;**52**:2316-2328

[54] Mathews AS, Kim D, Kim Y, Kim I, Ha C-S. Synthesis and characterization of soluble polyimides functionalized with carbazole moieties. Journal of Polymer Science Part A: Polymer Chemistry. 2008;**46**:8117-8130

[55] Ghosh A, Kumar Sen S, Banerjee S, Voit B. Solubility improvements in aromatic polyimides by macromolecular engineering. RSC Advances. 2012;**2**: 5900-5926

[56] Yang S-Y, Yuan L-L. Advanced Polyimide Films. Advanced Polyimide Materials. Elsevier; 2018. pp. 1-66

[57] Jitsuo O, Sotaro H, Shuta K. Process and apparatus for production of colorless transparent resin film. U.S. Patent 8,357,322. 2013

[58] Jitsuo O, Takashi M, Ko K, Shuta K. Process for producing polyimide film. U.S. Patent 7,871,554. 2011

[59] Yan M et al. A transparent, high barrier, and high heat substrate for organic electronics. Proceedings of the IEEE. 2005;**93**:1468-1477

[60] Dumont P, Tornare G, Leterrier Y, Månson J-AE. Intrinsic, thermal and hygroscopic residual stresses in thin gas-barrier films on polymer substrates. Thin Solid Films. 2007;**515**:7437-7441

[61] Amberg-Schwab S, Langowski H-C. Flexible Barrier Materials for Technical Applications. ITG FACHBERICHT. 2004. pp. 173-178

[62] Pang H et al. Recent progress of flexible AMOLED displays. In: Advances in Display Technologies; and E-papers and Flexible Displays. Vol. 7956 79560J. International Society for Optics and Photonics; 2011

[63] Behniafar H, Sefid-girandehi N. Optical and thermal behavior of novel fluorinated polyimides capable of preparing colorless, transparent and flexible films. Journal of Fluorine Chemistry. 2011;**132**:878-884

[64] Damaceanu M-D et al. Highly transparent and hydrophobic fluorinated polyimide films with ortho-kink structure. European Polymer Journal. 2014;**50**:200-213

[65] Hsiao S-H, Guo W, Chung C-L, Chen W-T. Synthesis and characterization of novel fluorinated polyimides derived from 1,3-bis(4-amino-2-trifluoromethylphenoxy) naphthalene and aromatic dianhydrides. European Polymer Journal. 2010;**46**: 1878-1890

[66] Takizawa K et al. Enhancing photoconductivity of aromatic polyimide films by incorporating fluorinated dianhydrides and main chain triphenylamine structure. Polymer. 2018;**157**:122-130

[67] Yang Y et al. Effect of fluorination on haze reduction in transparent polyimide films for flexible substrates. Journal of Applied Polymer Science. 2017;**134**

[68] Takemasa C, Chino T, Ishige R, Ando S. Anisotropic photoconductivity of aromatic and semi-aliphatic polyimide films: Effects of charge transfer, molecular orientation, and polymer chain packing. Polymer. 2019; **180**:121713

[69] Yu H-C, Jung J-W, Choi J-Y, Oh SY, Chung C-M. Structure-property relationship study of partially aliphatic copolyimides for preparation of flexible and transparent polyimide films. Journal of Macromolecular Science, Part A. 2017;**54**:97-104

[70] Itamura S, Yamada M, Tamura S, Matsumoto T, Kurosaki T. Soluble polyimides with polyalicyclic structure. 1. Polyimides from bicyclo[2.2.2]oct-7-ene-2-exo,3-exo,5-exo,6-exo-tetracarboxylic 2,3:5,6-dianhydrides. Macromolecules. 1993;**26**:3490-3493

[71] Matsumoto T, Kurosaki T. Soluble and colorless polyimides from bicyclo [2.2.2]octane-2,3,5,6-tetracarboxylic 2,3: 5,6-dianhydrides. Macromolecules. 1997;**30**:993-1000

[72] Zhang Q, Tsai C-Y, Li L-J, Liaw D-J. Colorless-to-colorful switching electrochromic polyimides with very high contrast ratio. Nature Communications. 2019;**10**:1-8

[73] Hu X, Mu H, Wang Y, Wang Z, Yan J. Colorless polyimides derived from isomeric dicyclohexyl-tetracarboxylic dianhydrides for optoelectronic applications. Polymer. 2018;**134**:8-19

[74] Ueno S, Konishi Y, Azuma K. Highly water-impermeable SiNx film prepared using surface-wave-plasma chemical vapor deposition and improvement of the barrier performance for organic light-emitting displays. Thin Solid Films. 2015;**580**:116-119

[75] Tseng I-H, Tsai M-H, Chung C-W. Flexible and transparent polyimide films containing two-dimensional alumina nanosheets templated by graphene oxide for improved barrier property. ACS Applied Materials & Interfaces. 2014;**6**:13098-13105

[76] Liu Y et al. Enhanced gas barrier and thermal properties of polyimide/montmorillonite nanocomposites as a result of "dual-plane" structure effect. Polymer Composites. 2017;**39**:E1725-E1732

Section 3

Polyimide in Space Applications

Chapter 4

Charging and Discharging Mechanism of Polyimide under Electron Irradiation and High Voltage

Xiaoping Wang, Daomin Min and Shengtao Li

Abstract

Polyimide has been widely used as insulating and structural materials in spacecraft due to its excellent electrical, thermal and mechanical properties. However, its charging and discharging problem in harsh space environment has been a major limit to the development of high-voltage and high-power spacecraft. In this chapter, charging and discharging phenomena of dielectric materials under electron irradiation environment were presented. First, the electrical properties of polyimide consisting of dielectric properties, trap properties, conductivity and electrical breakdown properties were investigated, which have great influences on charging and discharging characteristics. Then, a surface charging model under relatively low-energy electron irradiation was proposed for polyimide, based on the synergistic effects of electron movement above surface and charge transport in surface layer. The DC surface flashover of polyimide under electron irradiation with different energies, fluxes and incident angles was investigated. Furthermore, a deep charging model under high-energy electron irradiation with the Fluence Model for Internal Charging (FLUMIC) spectrum was established. The effects of electron flux enhancement and operating voltage on charging characteristics were discussed in different grounding modes. It indicates that the processes of discharging under electron irradiation have a close link with the charge transport characteristics of polyimide.

Keywords: charging and discharging, charge transport, electron irradiation, high voltage, polyimide

1. Introduction

The charging and discharging of dielectric materials under space radiation environment are the main factors that cause anomalies in a spacecraft. Koons et al. counted the abnormal failures of the spacecraft, suggesting that 54.2% from the total 299 cases were caused by the charging and discharging of dielectric materials [1]. A spacecraft is inevitably exposed to space plasma, energetic particles radiation, extreme temperature, cosmic rays, etc. [2]. A situation has to be taken into consideration that partial accumulation of space charges and high electric field occur when energetic electrons penetrate through the aluminum shield and deposit in the surface or deep layer of insulating materials. When the maximum electric field of insulating material exceeds a certain threshold, electrostatic discharge (ESD) will occur. Consequently, it will lead to the deterioration of insulating materials and even the failure of the whole electronic equipment. With the rapid increasing interest on space exploration, several countries are making efforts to build a Space Solar Power Station (SSPS) with megawatts or even gigawatts [3]. The reliability of the spacecraft becomes a very important problem. Polyimide is widely used in spacecraft system because of its good insulating, mechanical and antiaging properties [4]. Therefore, the charging and discharging mechanism of polyimide under electron irradiation and high voltage is a research focus in the field of spacecraft reliability.

Surface dielectric charging and deep dielectric charging are two kinds of dielectric charging, which are divided by the incident electron energy range and discharge position [2]. Surface dielectric charging refers to the deposition of low-energy electrons (e.g., 1–50 keV) on the dielectric surface and the induction of surface potential, while the deep dielectric charging refers to the penetration of high-energy electrons (e.g., 0.1–10 MeV) from the dielectric surface, deposition within the insulating materials and establishment of internal electric field [5]. Modeling the dielectric charging based on secondary electron yield, surface potential decay processes and characteristic parameters is the research focus in surface dielectric charging [2, 6–10]. While for deep dielectric charging, the charge transport properties of insulating materials irradiated by energetic electrons are key issues, and several models have been proposed to investigate it [11–14]. There are two types of typical models: the radiation-induced conductivity (RIC) model and the charge generation-recombination (GR) model. RIC model describes the transport processes of electrons in insulating materials under the irradiation of electron beam. It is a macroscopic model in which the parameters are given by the measurement of radiation-induced conductivity [14]. GR model describes the generation and recombination processes of electron-hole pairs in insulating materials. It is a microscopic model in which some specific parameters are difficult to be determined.

Charge behavior on the dielectric surface layer or the deep layer under electron irradiation has an important influence on discharging properties. As to DC surface flashover, it implies that the essence of surface flashover is the charge transport behavior across gas-solid interface under high electric field, which involves charge trapping and de-trapping properties in dielectric surface layer, secondary electron emission properties, impact ionization of gas molecules and electron multiplication properties in gaseous phase (or desorbed gas). The development process and formation of surface flashover is a coupling effect of the above factors. The vacuum surface flashover voltage of dielectric material irradiated by electrons is much lower than that in vacuum or gaseous atmosphere. At present, several theories have been postulated to explain the surface flashover phenomenon in vacuum, among which the theory of secondary electron emission avalanche (SEEA) is dominant [15]. The flashover of insulating material in vacuum under electron beam irradiation is also closely related to the field-emission electrons emitted from the cathode-dielectric-vacuum triple junction (CTJ) and secondary electrons (SE) [16]. A large number of experimental studies emphasize the effects of deposited charges in the dielectric surface layer, while few data can be obtained about the effect of kinetic electron from the electron beam on surface flashover [17]. On the aspect of dc electrical breakdown mechanism of polyimide, it has been proven that under the action of a high electric field, charges are injected into the insulating materials, and space charges are accumulated [18–20]. The electric field distortion appears inside the insulating materials caused by the accumulated space charges. When the maximum local electric field exceeds a threshold value, the electrical breakdown will occur [19, 21].

In this chapter, the charging and discharging phenomena of dielectric materials under electron irradiation environment were introduced. The electrical properties of polyimide were investigated. The surface and deep charging process and model of polyimide radiated by electrons were analyzed. Then, the experimental results of DC surface flashover during electron irradiation with different energies, fluxes and incident angles were investigated.

2. Charging and discharging phenomena of dielectric materials under electron irradiation environment

2.1 Charging and discharging phenomena and hazards

In the field of spacecraft charging and discharging, the potential of the spacecraft is relative to the zero potential of the space plasma. Although the density of space plasma is fluctuating, it is much faster than the change of the spacecraft potential on the time scale. The spacecraft potential is floating. The spacecraft operates in a harsh space environment, such as plasma, high-energy electrons, atomic oxygen, etc., and charging and discharging phenomena will occur in the surface or deep layer of the spacecraft [2]. **Figure 1** depicts a schematic diagram of the spacecraft floating potential.

When the energy of the incident particles is low, the charge exchange process will appear between spacecraft and the surrounding environment. Environmental electrons or ions interacting with target atoms on the surface of the dielectric material will generate the secondary electrons and backscattered electrons. In addition, when the spacecraft is operating on the sunny side, photoelectrons are generated on the surface of the dielectric material. The combined effects of the above processes will cause charging and discharging phenomena on the surface of the dielectric materials. For different dielectric materials, due to their different secondary electron emission coefficients, backscattering coefficients and photoelectron coefficients, the surface charge exchange processes are different. Consequently, different surface potentials appear on the dielectric materials, which will cause unequal charge between the dielectric materials [2].

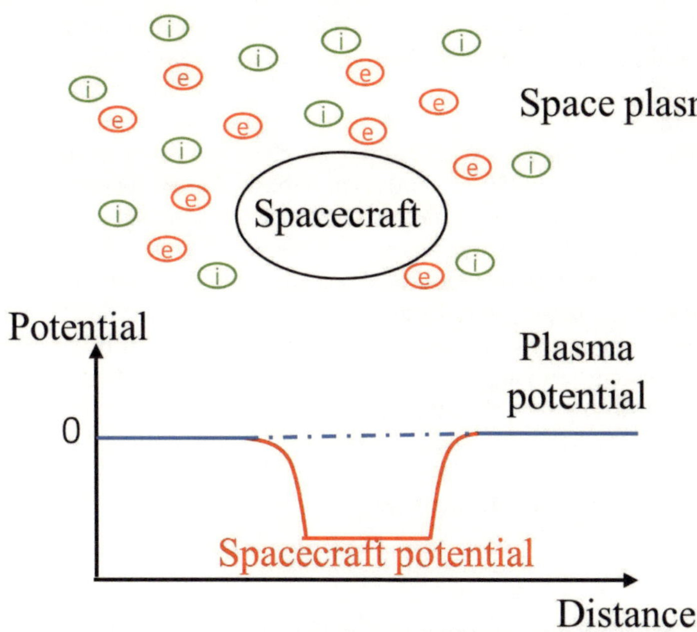

Figure 1.
The schematic diagram of spacecraft floating potential in space plasma environment [2].

The deep dielectric charging refers to the process that high-energy electrons (MeV) penetrate through the dielectric surface and deposit within the insulating materials [22]. Incident electrons penetrate into insulating materials, and their energy will gradually transfer into target atoms, owing to the physical mechanism of elastic scattering or inelastic scattering. For high-resistivity polymer, the intrinsic conductivity is very low. High-energy electrons penetrate the surface and deposit inside the material. These charges are called deposited electrons. Under the radiation of the space electron spectrum, electrons of different energies have different penetration distances inside the material, resulting in the formation of deposited charge layers of different depths. The charge accumulation will cause distortion of the electric field, which is likely to cause internal electrical breakdown of the dielectric materials [23].

From 1980 to 2005, the statistics of 156 anomalies of orbiting spacecraft showed that 45% of spacecraft anomalies were caused by the failure of the power system of the spacecraft [24], among which the insulating materials and structure of solar array and its drive assembly are most likely to discharge. The spacecraft power system fails once the solar array or its drive assembly fails. Even worse, the spacecraft will be out of control. A Nigerian satellite launched by China in November 2008 completely failed due to the failure of solar array drive assembly [25]. Especially with the increase of spacecraft operating voltage and power requirements, the coupling effect of high operating voltage and space radiation environment will pose a greater threat to the insulation system of spacecraft.

2.2 Research process of dielectric charging and discharging

As early as the 1920s, Mott-Smith and Langmuir began the initial theoretical exploration of the electrostatic charging of isolated bodies in space [2]. With the launch of the first artificial satellite in 1957, humankind entered the era of space, and the related issues of space dielectric charging have gradually attracted researchers' attention. Before 1980, it was believed that the charging and discharging of the dielectric surface was the main cause of spacecraft anomalies, and related research focused on the surface charging phenomenon [26]. With the occurrence of abnormal spacecraft failures and the launch of CRRES satellite (Combined Release and Radiation Effects Satellite) in the 1990s, deep dielectric charging of the spacecraft came into focus and research on spacecraft charging entered a new era [27]. H.B. Garrett published two review papers in 1981 and 2000 [26, 27], which summarized the research progress of spacecraft surface charging before 1980 and research development of surface charging and deep charging between 1980 and 2000. Lai published a review paper in 2003 [28], which summarized the suppression methods of dielectric charging.

Since the twenty-first century, great achievements have been made in space environment exploration, basic theoretical research and ground simulation experiments. However, the charging and discharging of dielectric materials is still the main factor threatening the safe operation of spacecraft. Especially with the development of high-voltage and high-power spacecraft, the field of dielectric charging and discharging is facing new challenges.

3. Electrical properties of polyimide

3.1 Dielectric properties

The complex permittivity of polyimide with thickness of 100 μm was measured at room temperature using a broadband dielectric spectrometer (Concept 80, Novocontrol Technologies, Germany). The applied voltage was 1 V_{rms} and the frequency was from 10^{-2}–10^{5} Hz. **Figure 2** depicts the real and imaginary parts of the relative complex permittivity, obtained from polyimide sample at room temperature, which is a function of frequency in semi-logarithmic coordinates [29]. **Figure 2** shows that the real part of relative complex permittivity increases slightly as frequency decreases. In the frequency range of 10^{-2}–10^{5} Hz, the imaginary part is lower than 3.6×10^{-3}. The small dielectric relaxation strength of the relaxation peak around 30 Hz reveals that the dipolar moment is very low. The dielectric loss, $\varepsilon''/\varepsilon'$, is very low, which indicates that in the dc electrical breakdown experiments at room temperature, the Joule heating generated by the dipole orientation is negligible [29].

3.2 Bulk and surface trap properties

Thermally stimulated depolarization current (TSDC, Concept 90, Novocontrol technologies, Germany) was carried out on a polyimide sample with a thickness of 100 μm to investigate its trap distribution characteristics. **Figure 3** shows the results of TSDC experiments for polyimide [29]. Thermally stimulated relaxation processes can be observed in the temperature range of 10–170°C. One obvious relaxation peak is around 69°C, while another relaxation peak may be located near 135°C. The experimental results were analyzed using the classical TSDC theory to reveal the thermally stimulated processes and their activation energies [30].

The TSDC experimental results were fitted and four relaxation peak components could be obtained. As shown in **Figure 3**, it can be seen that the fitting results are in good agreement with the experiments. We can determine the peak temperature, activation energy and relaxation time for the four relaxation processes listed in **Table 1**. The activation energies of four peaks at 69, 87, 109 and 135.5° are 0.60, 0.65, 0.70 and 0.83 eV, respectively. As the temperature at the relaxation peak increases, the corresponding activation energy increases. The three peaks at 69, 87 and 109°C may correspond to shallow traps that assist carriers hopping process in polyimide, while the peak at 135.5°C may correspond to deep traps that can capture mobile carriers and accumulate space charges. The energy of deep traps is

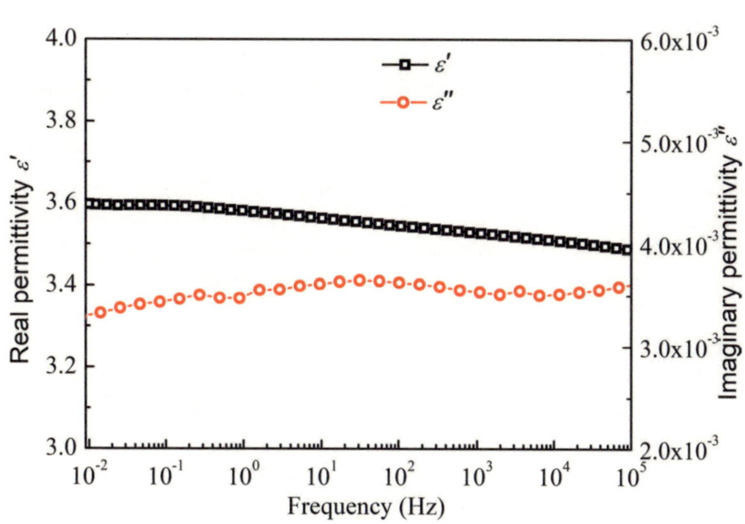

Figure 2.
The real and imaginary parts of relative complex permittivity, ε' and ε", of polyimide as a function of frequency at room temperature [29].

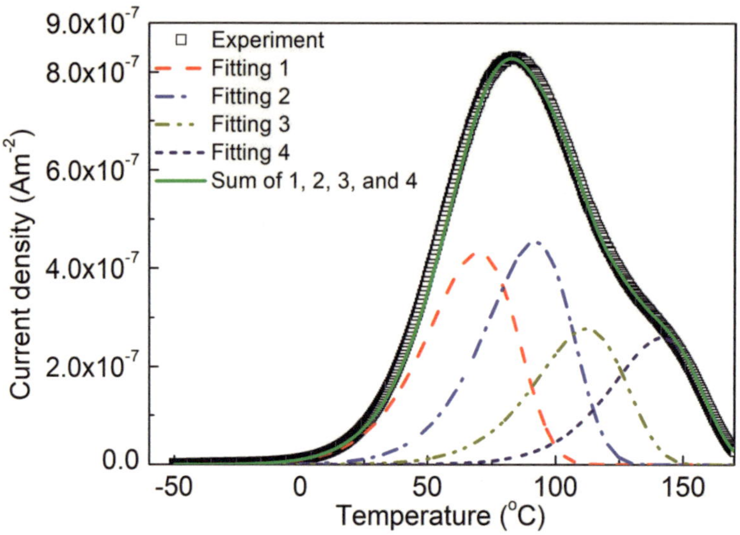

Figure 3.
TSDC experimental results of polyimide after being polarized at an applied voltage of 250 V at 180°C for 30 min. The classical TSDC theory was used to fit the experimental results. Symbols and solid curves represent experimental and fitting results, respectively [29].

Peak temperature (°C)	E_A (eV)	B (Am^{-2})	τ_0 (s)
69	0.60	2.63×10^{-4}	7.50×10^{-7}
87	0.65	2.01×10^{-4}	4.23×10^{-7}
109	0.70	1.80×10^{-4}	3.23×10^{-7}
135.5	0.83	1.60×10^{-4}	3.09×10^{-8}

Table 1.
Parameters for relaxation processes extracted from TSDC experimental results [29].

consistent with the results obtained from the Arrhenius relation between conductivity and temperature [4].

Surface potential decay was carried out on a polyimide sample under electron radiation to investigate its surface trap distribution characteristics. In this experiment, charging process takes a very short time, about 25 s, as the electron flux density was so high. We set that with a filament emission current of 10 μA and a radiation distance of 300 mm, and the charging process was completed within 30 s. After radiating for 30 s, we turned off the electron gun and then moved the probe over the sample to measure the surface potential. **Figure 4(a)** gives the surface potential decay curves of polyimide under electron radiation of different energy levels (3–11 keV) [31].

It can be seen that the initial surface potential gradually increases with the increase of electron energy. This indicates that the charging process and properties are different under electron radiation of different energy levels. Hence, the dielectric properties during the charging process can be investigated by analyzing the initial surface potential of the dielectric after the charging process.

The surface trap distribution of polyimide can be obtained from surface potential decay model, as shown in **Figure 4(b)** [31]. There are two types of traps, defined as shallow and deep traps, respectively. It can be seen that the trap charge density related to shallow traps is more than that of deep traps under the

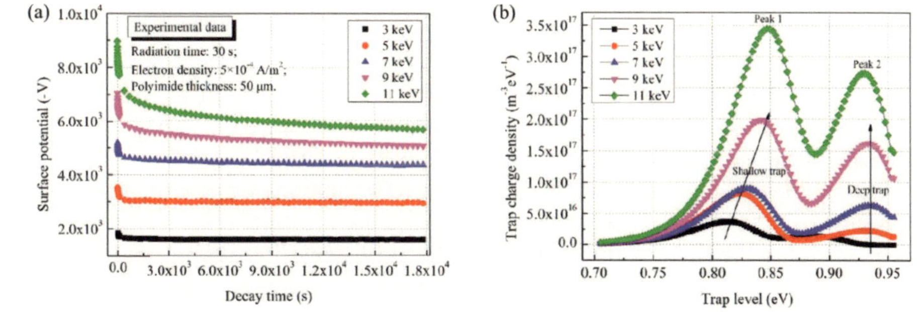

Figure 4.
Surface potential decay curves (a) and surface trap distributions (b) of polyimide after irradiation by electron beam with different energies [31].

same electron energy radiation. The charges captured in relatively shallow traps can escape the trap center in a short interval, which is demonstrated by the rapid decay of surface potential. With the time increases, these de-trapped electrons will migrate to the grounded electrode under the effect of the internal electric field. By contrast, deeply trapped charges remain in the trap center for a longer period. The density of deep traps determines the steady surface potential, and the stabilization time depends on the energy level of the deep traps.

The surface trap distribution of polyimide presents different behavior under radiation from electrons of different energy levels. The shallow trap level increases slightly with the increase of electron energy, while the deep trap level remains unchanged about 0.94 eV. Under the radiation of different electron energy, the depth of the electron deposition layer and the range of electrons are different. The higher the electron energy, the deeper the deposition layer. These trapped charges need to overcome a much higher potential barrier to escape the trap center. Therefore, the shallow trap energy level increases with the gradual increase of the electron energy. In addition, the total trap charge density gradually increases with increasing electron energy. Due to the increased electron energy, the distance from the electron deposition layer to the dielectric surface is longer, and much more charges will be captured by the trap centers [31].

3.3 Shallow trap-controlled carrier mobility and conductivity

The surface potential experimental results of samples charged by negative corona discharging and positive corona discharging as a function of time are shown in **Figure 5** [29]. Negative and positive charges are deposited on the surface of polyimide, and electric field is established inside the polyimide during the charging process. After charging, surface charges are injected into polyimide, and the migration of charges toward the grounded electrode in the bulk leads to the decay of surface potential. The decay rate of surface potential varies before and after the injected charge carriers flow out of the dielectric material, as shown in **Figure 5** [29]. The time when the front charge carriers arrive at the grounded electrode is defined as transit time t_T. The transit point existing at the beginning of the potential decay curve can represent the mobility of carriers controlled by shallow traps [29]. In order to obtain the transit time t_T, the potential decay results were fitted by an exponential function, and then we obtained the relation between $td\phi_s/dt$ and t. The time corresponding to the peaks can be regarded as transit time t_T, and it is used to calculate carrier mobility controlled by shallow traps according to the following Eq. (1) [29, 32]:

$$\mu_{0(e,h)} = d^2/\phi_{s0} t_T \quad (1)$$

Here, μ_0 is the carrier mobility controlled by shallow traps in $m^2V^{-1}s^{-1}$, d is the thickness of sample in m, and ϕ_{s0} represents the initial surface potential in V. The subscripts have the following meaning: (e) for electrons and (h) for holes. By calculation, the hole and electron mobilities controlled by shallow traps are 1.80×10^{-14} and 3.67×10^{-14} $m^2V^{-1}s^{-1}$, respectively.

For studying the surface and volume charge transportation properties, the isothermal surface potential decay (ISPD) experiment on space-grade polyimide was carried out at various temperatures from 298 to 338 K. In high vacuum, the charge was accumulated on the surface of polyimide under low-energy electron irradiation. After irradiation, the charge was transferred to the grounding electrode on the surface through the volume. Through the three parameters of surface resistivity, volume ohmic resistivity and charge carrier mobility, the leakage rate of electron was determined. And the three parameters were revealed by a two-dimensional ISPD model established by using genetic algorithm (GA), as shown in **Figure 6** [33].

As shown in **Figure 6**, the carrier mobility increased with temperature, while the surface resistivity and volume ohmic resistivity of polyimide decreased with temperature. The surface resistivity, volume ohmic resistivity and carrier mobility were obtained. For example, at 298 K, they were 1.02×10^{19} Ω, 2.87×10^{17} Ω m and 1.49×10^{-19} m^2/V s, respectively. The calculated errors were all not more than 0.9%, which showed that there was a good consistency between the experimental and simulated 2D ISPD results [33].

3.4 Electrical breakdown properties

The influencing mechanism of sample thickness on electrical breakdown of polyimide is not very clear until now. The dc electrical breakdown fields of polyimide films with different thicknesses from 25 to 250 μm were measured using a computer-controlled voltage breakdown test device. The dc electrical breakdown experiments were carried out under 30°C using spherical copper electrodes with a diameter of 25 mm in transformer oil. The rate of dc voltage increase is 1 kVs^{-1}. For

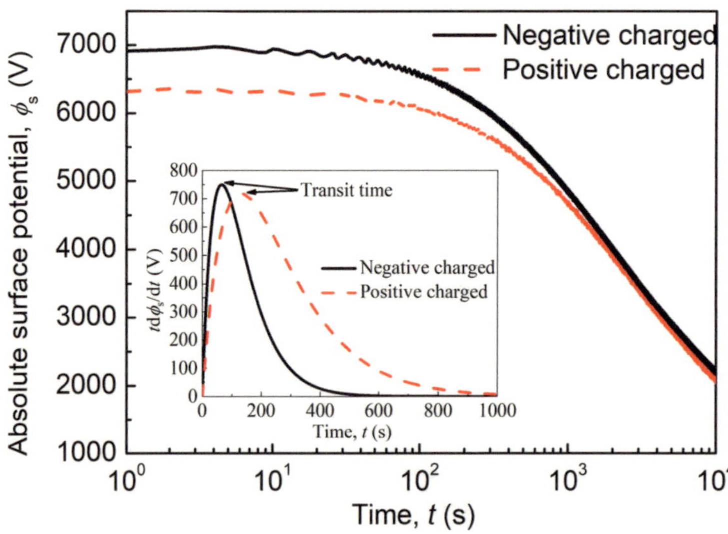

Figure 5.
Surface potentials of polyimide charged by negative corona discharging and positive corona discharging as a function of time at room temperature [29].

each thickness of the sample, at least 15 times breakdown tests are performed. The average value of all data is taken as the breakdown electric field of the sample.

Figure 7 shows the experimental results of the dc electrical breakdown field of the polyimide film, F_b, as a function of thickness, d, at room temperature [29]. As shown in **Figure 7(a)**, the dc electrical breakdown field of polyimide films decreases with an increase in sample thickness. In addition, the derivative of dc electrical breakdown field with respect to sample thickness $\mathrm{d}F_b/\mathrm{d}d$ decreases with the increase in sample thickness. The relation between the dc electrical breakdown field and sample thickness looks like an inverse power function. Accordingly, we change the linear coordinates in **Figure 7(a)** to double logarithmic coordinates in **Figure 7(b)**. It can be seen from **Figure 7(b)** that the dc electrical breakdown field of polyimide is linear with sample thickness under double logarithmic coordinates [29].

The influence of sample thickness on polymer breakdown can be explained by electron avalanche breakdown, electromechanical breakdown, free volume breakdown and space charge modulated electrical breakdown [29–34]. In electron avalanche breakdown, the energy gain of electron can be obtained by free electron movement in the conduction band of dielectric material under the action of electric field [34]. When the energy exceeds the band gap energy, the electrons in the valence band may be excited to the conduction band, resulting in the chemical bond breaking. The avalanche effect is caused by further collision and ionization of the released electrons with other matrix atoms, which results in the doubling of local current and finally triggers the breakdown. The electric breakdown field decreases

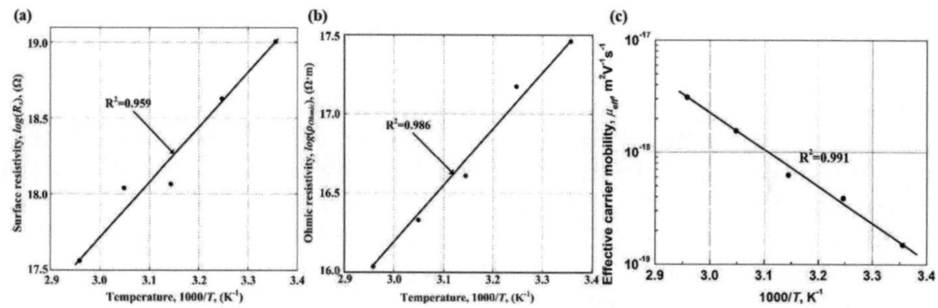

Figure 6.
Arrhenius plot of surface resistivity (a), volume ohmic resistivity (b), and charge carrier mobility (c) of polyimide. The linear fitting errors, R^2, were respectively 0.959, 0.986 and 0.991 from (a) to (c) [33].

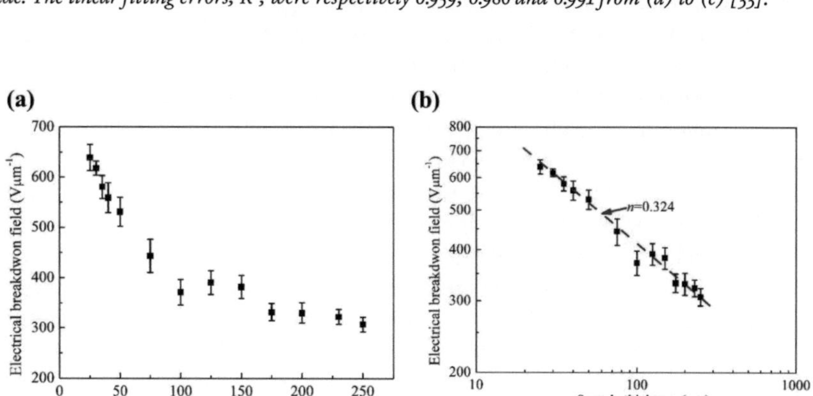

Figure 7.
Experimental results of dc electrical breakdown field of polyimide at various thicknesses in linear coordinates (a) and in double logarithmic coordinates (b) [29].

with the increase of sample thickness, which is due to the critical number of electrons produced in the whole sample thickness by collision ionization [29, 34]. The Stark-Garton model of mechanical and electrical breakdown has been widely used to predict the breakdown strength of thermoplastics, while the Young's modulus and dielectric constant of temperature-sensitive polymers determine the mechanical and electrical breakdown strength [34]. Because the thickness of the sample determines the electrostatic compressive stress and the opposite elastic stress produced by the electrostatic attraction of the two electrodes, the electric breakdown field is a decreasing function of the sample thickness. In the theory of free volume breakdown, it is assumed that the electric breakdown field of polymer depends on the longest mean free path of electron. Electrons are accelerated in the free volume, and their average free path depends on the maximum length of the free volume. When enough energy is obtained by electrons in the free volume to overcome the potential barrier, the local current will be multiplied, so that the material is heated to a very high temperature and finally causes the phenomenon of electrical breakdown. From a statistical point of view, the longest free path is a function of sample size, so the electric breakdown strength is related to sample thickness.

4. Surface charging and discharging mechanism of polyimide

4.1 Surface charging process and model of polyimide radiated by low-energy electrons

The synergistic effect of surface electron movement and charge transport in dielectric surface layer should be taken into account when studying the charging process under low-energy electron radiation (1–50 keV). A schematic diagram of charge transport on polyimide surface and in its surface layer under low-energy electron radiation is shown in **Figure 8**. 'Surface layer' refers to the area inside the material that is about a few micrometers from the surface of the dielectric material.

The intrinsic conductivity of polyimide with high resistivity is very low, but its total conductivity will increase due to the radiation-induced conductivity (RIC). The incident electrons are mainly deposited in a dielectric surface layer of about a few microns [16], and they will migrate to the interior of polyimide. However, the charge in the surface layer will continue to be accumulated, because the charge conduction velocity is far lower than that of deposition [5].

The surface potential is very low in the initial stage of electron radiation, whose reverse effect on the incident electron energy is very weak. Rather than being released by the secondary electrons, the incident electrons will be deposited on the surface. On the one hand, the change of the distribution of the deposited electrons in the surface layer and the change of the charge transfer characteristics occur due to the change of the incident electron energy and density on the dielectric surface, and it will further affect the negative potential and the induced reverse electric field on the surface in turn. On the other hand, these deposited electrons will generate an internal electric field, whose intensity will gradually increase with the radiation duration. A reverse-acting force will be produced by this field on the moving electrons reaching the dielectric surface [16]. As a result, the incident trajectory and the kinetic energy of the incident electron can be changed by the reverse electric field, by which the secondary electron yield characteristics of dielectric surface will be greatly affected.

The charging process will be stable, if the incident electron current is equal to the sum of the conduction current in the surface layer and the secondary electron generation current on the surface. Therefore, the key to the study of the charging

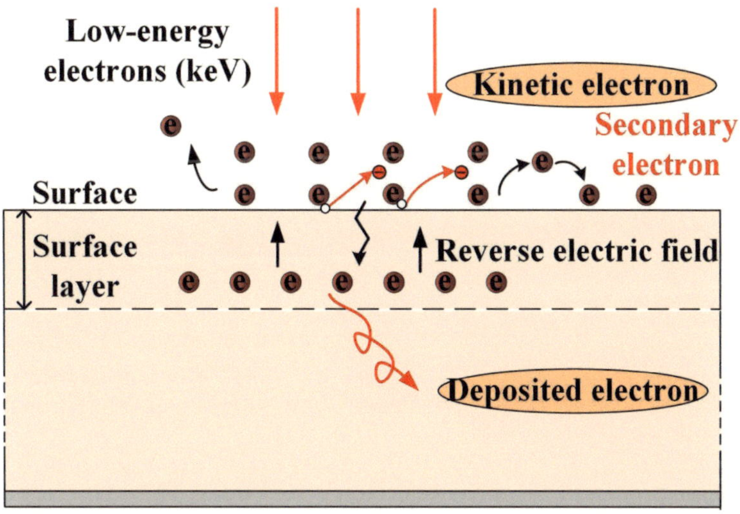

Figure 8.
Schematic diagram of charge transport on polyimide surface and in the surface layer under low-energy electron radiation [31].

process is a thorough understanding of the charge transfer properties in the dielectric surface layers and kinetic electrons in the surface [31].

4.1.1 Surface kinetic electron properties

A reverse electric field will be formed in the process of electron radiation by the electrons accumulated in the polyimide surface layer, by which the trajectory of the incident electrons will be changed, and thus there will be a dynamic impact on the density and energy of the electrons reaching the polyimide surface. The characteristics of the subsequent incident electrons are different from those of the initial electrons. They will change with time, thus affecting the yield attributes of the surface secondary electrons. **Figure 9(a)** gives the energy and density of electrons reaching the polyimide surface over radiation time. **Figure 9(b)** shows the current density of secondary electrons emitted from polyimide surface and the surface conduction current against time [31].

Due to the repelling effect from the electric field forming in the surface layer, with the radiation time increasing, the energy and density of electrons reaching the polyimide surface gradually decrease, as shown in **Figure 9(a)** [31]. It can also be observed that, with radiation time increasing, the energy and density of electrons reaching the polyimide surface become a whole range of values from single values, resulting in a great impact on the dynamic processes of secondary electron movement and electron deposition, transport and accumulation behavior in the dielectric surface layer. Secondary electron yield coefficient gradually increases with the drop of the energy of kinetic electrons reaching the polyimide surface, and correspondingly the secondary electron yield current gradually increases, as shown by the red curve in **Figure 9(b)** [31]. In addition, the phenomenon that some of the incident electrons deposit in the surface layer after penetrating the dielectric surface will occur, especially at the initial stage. With different radiation time and material position, the distributions of deposited electrons are different. The change of charge conduction current density on the polyimide surface is shown by the blue curve in **Figure 9(b)** [31]. In the initial radiation stage, the charge conduction process can be overcome by most of the incident electrons under the radiation-induced conductivity, after they penetrate the dielectric surface. On the contrary,

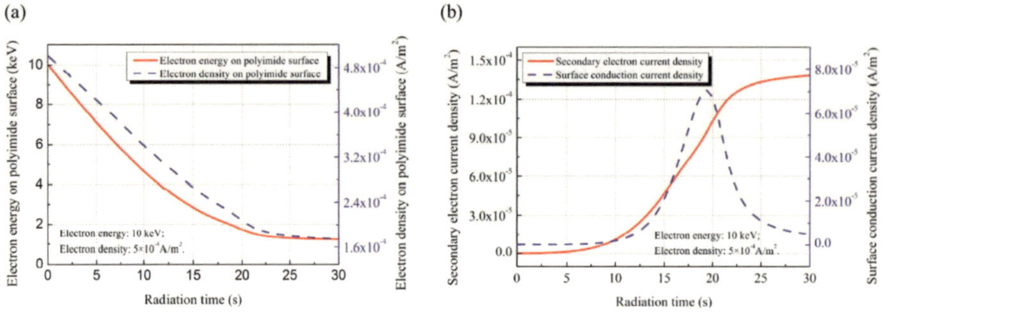

Figure 9.
Surface kinetic electron properties. (a) The energy and density of electrons reaching the polyimide surface over the radiation time and (b) secondary electron emission and charge conduction on polyimide surface over the radiation time [31].

the production process of secondary electron is very weak. The secondary electron yield current increases with the energy of the kinetic electron to the surface of polyimide decreasing over radiation time. With the radiation time increasing, the conduction current density on the polyimide surface will gradually decrease, resulting in most of the incident electrons on the polyimide surface being released by the secondary electrons, and only a few electrons penetrating the surface. In the case of low-energy electron radiation, the influence of secondary electron generation process is more obvious than that of deposition electron transport process. The dynamic process of charge transport in the dielectric surface layer plays a leading role in the initial stage of radiation, so it cannot be ignored [31].

4.1.2 Charge transport properties in the surface layer of polyimide

A non-uniform distribution of potential and electric field is caused by the different spatial distribution of charge in polyimide surface layer under low-energy electron radiation. By solving charge balance equation, current conduction equation and Poisson equation, the distribution of electric potential and electric field can be obtained. **Figure 10(a)** and **(b)** depicts the spatial and temporal distributions of the internal potential and electric field of polyimide under electron radiation. The electron energy is 10 keV and the flux density is 5×10^{-4} A/m^2.

Figure 10(a) shows that with the radiation time increasing, the surface potential increases gradually, and the maximum potential appears at about 25–30 s. Meanwhile, with the material depth increasing, the potential decreases. It can be seen in **Figure 10(b)** that the electric field intensity increases with the radiation time increasing, which is due to the electrons accumulating in the polyimide surface layer. The electric field tends to be stable when the radiation time is more than 25 s. It can be obtained that the electric field decreases gradually from the maximum electron range to the dielectric surface, on which the electric field is equal to zero, according to Poisson's equation. The distribution of the maximum potential and the maximum electric field over the radiation time is depicted in **Figure 10(c)**. It can also be seen from **Figure 10(c)** that the maximum surface potential increases with the radiation time increasing and tends to stabilize at 25–30 s. When the radiation time is 30 s, the stable potential reaches −8778 V. The corresponding experimental result that was measured by the non-contact surface potentiometer was −8424 V, which is slightly lower than the simulated value. Correspondingly, the maximum electric field is 1.78×10^8 V/m, which is very high, but does not cause damage to the material. Once electron radiation stops, the electric field value will drop sharply.

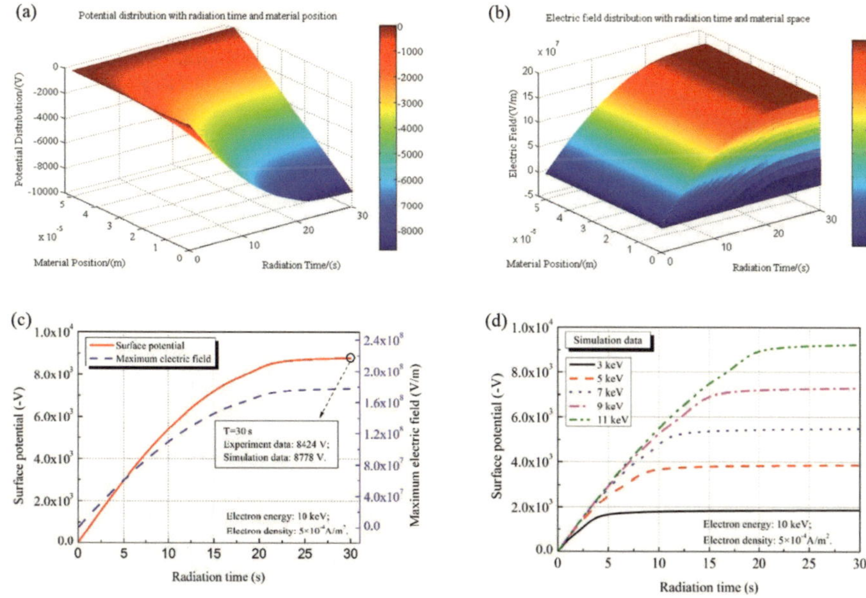

Figure 10.
Charge transport properties in polyimide surface layer. Distributions of internal potential (a) and internal electric field (b) at various material positions and radiation times. Maximum potential and maximum electric field (c) and surface potential (d) as a function of radiation time [31].

Figure 10(d) shows the distribution of surface potential with over the radiation time under different incident electron energy levels [31].

4.2 DC surface flashover mechanism of polyimide irradiated by electrons

Li et al. measured the DC surface flashover voltage of insulating material in vacuum under electrons irradiation by controlling the energy, emission flux and incident angle of the electron beam [16, 35]. Combining the common effects of deposited electrons and kinetic incident electrons, they proposed a physical model of surface flashover under electrons irradiation.

Figure 11(a)–(c) depicts the effect of electron energy, incident angle and electron flux on DC surface flashover voltage of polyimide during electron irradiation. The surface flashover voltage of polyimide irradiated by electron beam is determined not only by the deposited electrons in the surface layer of the dielectric but also by the kinetic incident electrons striking the dielectric surface [35].

During low-energy electron irradiation, for one thing, deposited electrons will reduce the electric field in the vicinity of CTJ; thus, the field-emission effect is suppressed, hindering the initiation of SEEA. For another, the surface potential established by deposited electrons is proportional to the electron energy. The secondary electrons will be repelled away from the polyimide surface, hindering the development process of SEEA. Both of these two effects will promote the surface flashover voltage.

However, during high-energy electron irradiation, the kinetic incident electrons will strike the polyimide surface to generate secondary electrons, which promotes the development of SEEA. If the impact points of kinetic incident electrons are close to the CTJ, they will be an alternative to field-emission electrons as the seed of SEEA. Thus, a high voltage to generate field-emission electrons and initiate the SEEA is no longer needed. A lower applied voltage can provide energy for secondary

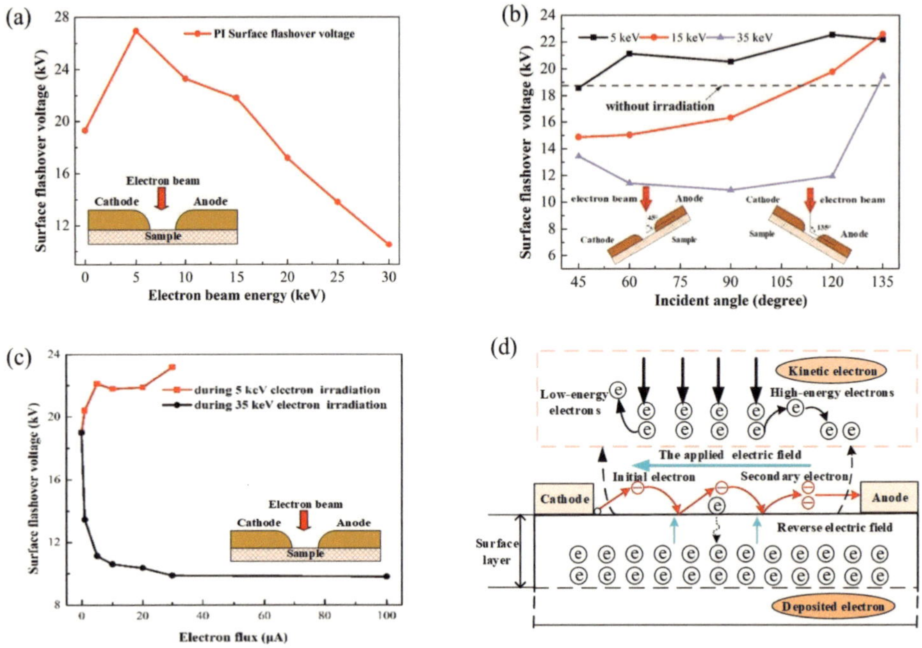

Figure 11.
DC surface flashover properties of polyimide under electron irradiation. Effects of electron energy (a), incident angle (b), and electron flux (c) on surface flashover voltage. (d) The surface flashover model for dielectric materials under electron irradiation [35].

electron multiplication. In other words, the applied voltage for electron multiplication is much lower than that for the field-emission–initiated SEEA. For another, the electron beam bombardment will release the adsorbed gases on the irradiated area of polyimide surface. Considering the shielding effect of the cathode, when the applied voltage is the same value, the irradiated area of the case during high-energy electron beam irradiation is larger than that of the case during low-energy electron beam irradiation. When enough adsorbed gases are released, ionization may be caused by electron beam bombardment as well as secondary electrons that gain enough energy from the applied electric field. If the electron beam can approach the polyimide surface, the effects of deposited electrons will be suppressed by those kinetic incident electrons. The model of surface flashover under electrons irradiation is shown in **Figure 11(d)** [35].

5. Deep charging and discharging mechanism of polyimide

5.1 Deep charging model of polyimide radiated by electrons

Energetic electrons are difficult to conduct when they are deposited inside polyimide due to its low conductivity, resulting in deep charging of insulation. Under the condition of typical electron radiation environment in geosynchronous orbit (GEO), deep charging of polyimide normally does not cause discharge risk. However, during the energetic electron storm, the electron flux will increase by 2–3 orders of magnitude within a few days and last for 10 days or so. At this point, the incident electron flux will exceed the threshold of 0.1 pA/cm^2, resulting in a great risk of ESD [5, 36].

FLUMIC model, proposed by Rodgers et al., based on spacecraft data of GOES/SEM and STRV-1b/REM was utilized in this paper to manifest the electron radiation environment in GEO [37, 38]. It is commonly agreed that FLUMIC model is suitable for charging risk assessment and spacecraft design due to its complete demonstration of seasonal and annual variations in energetic electron flux. **Figure 12(a)** depicts FLUMIC spectrum under typical and extreme space environment [36].

The penetration depth of energetic electrons in polyimide can be obtained from the Weber semi-empirical equation. The charge conduction process consists of inherent charge conductivity and radiation-induced conductivity. Charge transport process satisfies the current conduction equation, the charge continuity equation and Poisson's equation [36].

Assume that electrons irradiate a plate polyimide from the upper side. HV is applied to one side of the sample, and the other side is suspended or grounded. Four cases of the sample are considered altogether, that is, (A) suspended-HV; (B) HV-suspended; (C) grounded-HV; and (D) HV-grounded, as shown in **Figure 12(b)** [36]. The condition before the hyphen indicates condition on the upper surface, and the latter indicates condition on the lower surface.

Here, the first case will be discussed: HV is 0 V, that is, and the electrode is grounded. Case A becomes suspended-grounded, case B becomes grounded-suspended, and cases C and D are merged into grounded-grounded. We take the condition with enhancement of 100 and radiation time of 5 days for an example.

5.2 Simulation results and discussion

In case A, the maximum electric field strength reached 5.00×10^7 V/m, appearing near the lower electrode. Most of the charge deposited near the radiated surface, though part of the charge mitigated toward the lower electrode driven by the electric field, as shown in **Figure 13(a1)–(a3)** [36]. In case B, the maximum electric field strength reached 4.39×10^7 V/m, appearing near the upper electrode. Vast charges are accumulated at the region near the radiated surface. As **Figure 13(b1)–(b3)** shows, compared with case A, the electric field in case B tends to move downward, inhibiting the migration of electrons from the field to the bulk of sample, which leads to deposition of the charges near the surface and formulate a local high-space charge area [36]. When both electrodes are grounded, it is clear that, similar to the results in case B, the electric field near the upper electrode is at a lower position vertically, restricting the transformation of the electrons to the bulk of the sample and electrons accumulated at the region near upper electrode. In addition, as **Figure 13(c1)–(c3)** shows, the electric field close to the downward electrode tends to move up vertically, fostering the electron migration downward [36].

Furthermore, the impact of electron flux promotion on the charging of polyimide is addressed. Here, with four cases considered, we take the HV of 500 V and radiation time of 10 days for an example. It is shown in **Figure 14(a)** and **(b)** that

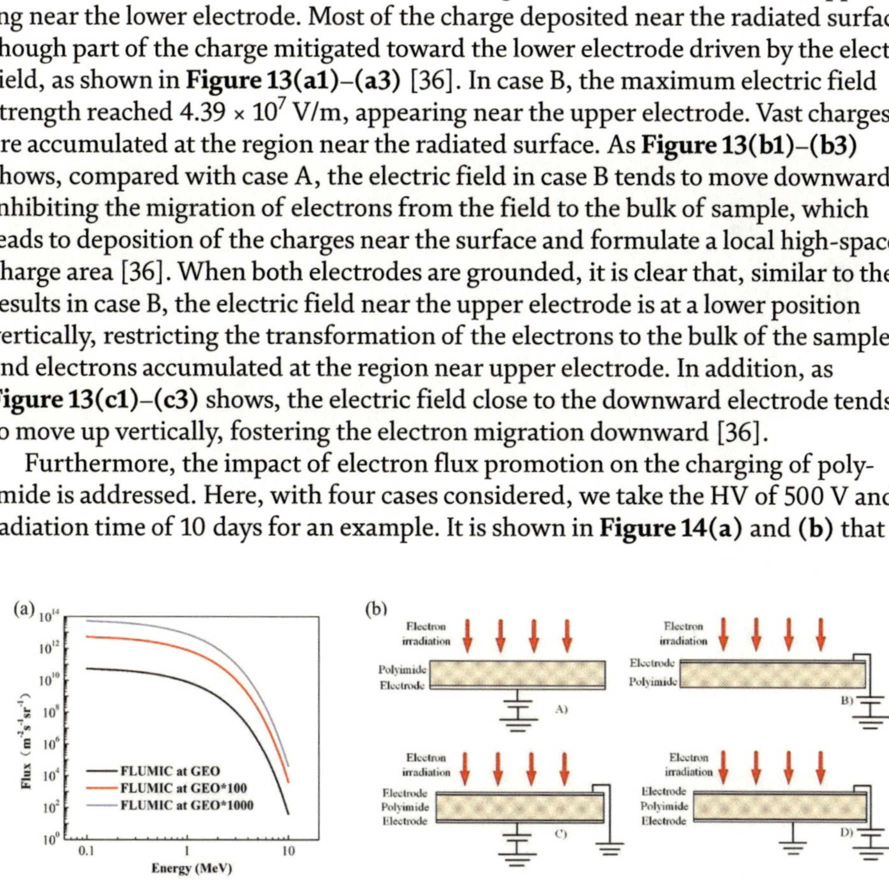

Figure 12.
(a) The FLUMIC model value at GEO environment. (b) Four cases of the sample [36].

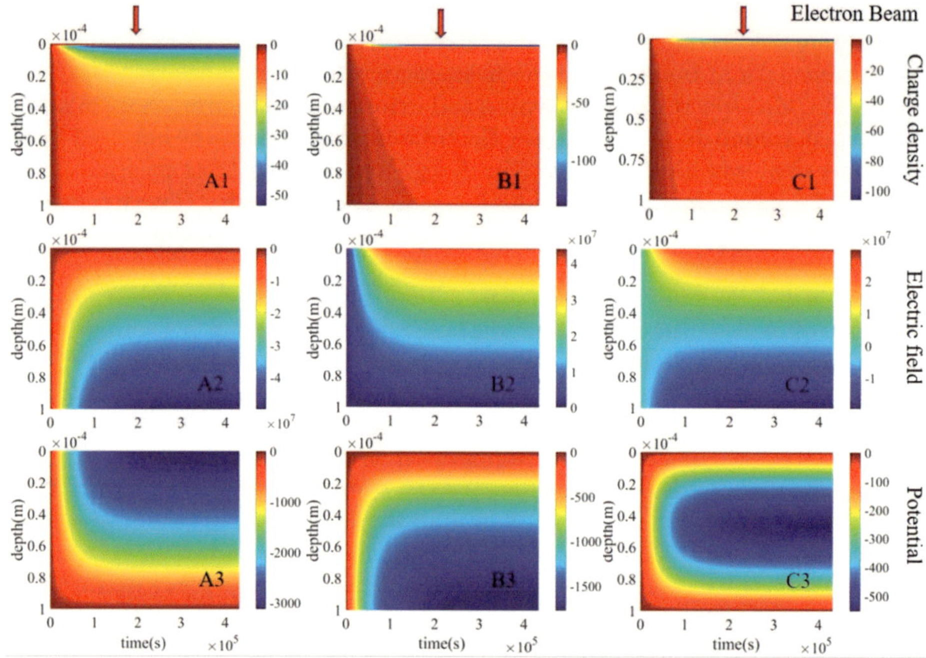

Figure 13.
Distribution of charge density, electric field and potential. (a) Suspended-grounded, (b) grounded-suspended, (c) grounded-grounded [36].

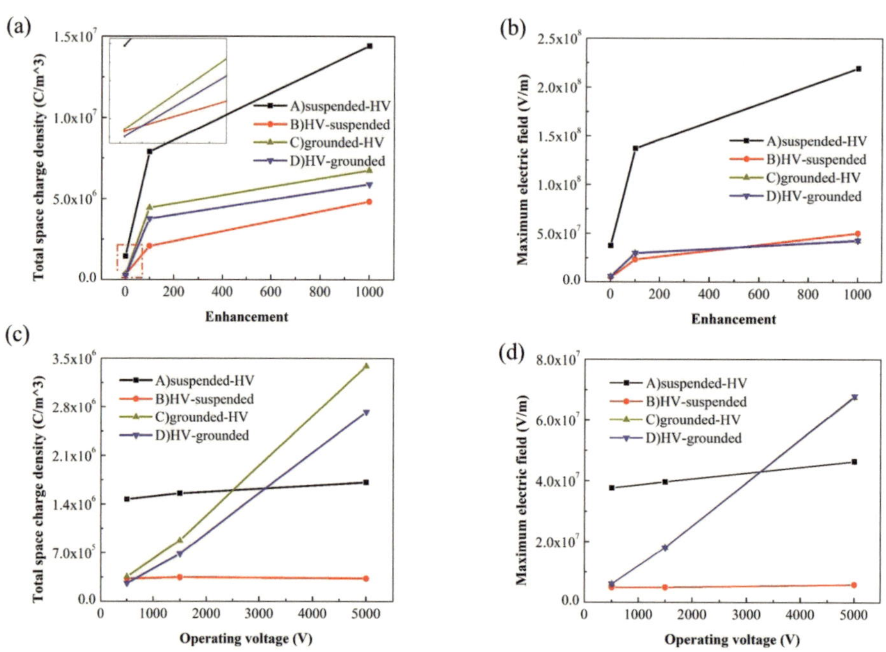

Figure 14.
(a) Influence of electron flux enhancement on total space charge density. (b) Influence of electron flux enhancement on maximum electric field. (c) Influence of operating voltage on total space charge density. (d) Influence of operating voltage on maximum electric field [36].

case A has the highest total space charge density and maximum electric field, which are significantly higher than those in other three cases at the same enhancement [36]. With flux enhancement increases, total space charge density reaches the valley value under case D when enhancement is 1, while when enhancement increases to 100 and 1000, lowest charge density is seen in case B. Additionally, though charge density in cases B and D is varied, lines representing maximum electric field almost overlap. In case A, considering the voltage is applied to the lower electrode and electric field moves upward, accumulated electrons are attracted to the lower electrode; therefore, more electrons may be injected into the sample. On the contrary, in case B, the voltage is applied to upper electrode and electric field moves down; hence, vast charges are accumulated at the region near upper electrode, inhibiting further electron injection. In cases C and D, the electric field moves down and up at the region near upper and downward electrode, respectively. Based on the previous analysis, it can be determined that with an increase in flux enhancement, its impact on case A is more obvious than that in other cases.

At last, the influence of operating voltage on the charging of polyimide is discussed. Take the enhancement of 1 and the radiation time of 10 days as an example; we discuss the influence of operating voltage on the charging of polyimide in the four cases. As can be seen from **Figure 14(c)** and **(d)**, the increase of operating voltage has a small influence on cases A and B, since the virtual electrode is at infinity in both cases A and B [36].

6. Conclusions

Charging and discharging problem of polyimide in harsh space environment has been a major limit to the development of high-voltage and high-power spacecraft. Electrical and charge transport properties have great influences on the surface and deep charging-discharging characteristics. The conclusions drawn are as follows:

1. The parameters obtained from the electrical experiments can be used in the simulation of charge transport process, such as permittivity, trap energy level, trap density, the activation energy and so on. The electrical breakdown field decreases with an increase in sample thickness in the form of an inverse power function. The elongation of free volume caused by the displacement of the molecular chain associated with the accumulation of space charges and the distortion of electric field may play important roles in the breakdown characteristics of polyimide.

2. In terms of the surface electron properties of polyimide under electron radiation, the electrons deposited in dielectric surface layer will form a reverse electric field, which has a great impact on the dynamic process of the secondary electron movement and the process of deposition, transport and accumulation of electrons in the dielectric surface layer. In terms of charge transport properties in polyimide surface layer, the electrons deposited in dielectric surface layer will migrate to the inside under the action of the RIC. The charge conduction velocity is much lower than that of deposition, so the charge will continuously accumulate in the surface layer, which will cause a reaction force on kinetic electrons flowing to the dielectric surface.

3. Negative surface charge accumulation can increase the flashover voltage, to some extent. Since kinetic incident electrons in the vicinity of the CTJ can initiate the surface flashover at a much lower voltage, the shield of the spacecraft is of great importance. If a trade-off must be made on the shielding layer, the region of CTJ should be ensured. Moreover, narrow and deep gap between the electrodes can shield the kinetic incident electrons with non-normal incidence and may promote the surface flashover voltage.

4. The use of the suspended-HV insulation should be limited to reduce the influence of electron flux enhancement when designing a spacecraft. To increase the operating voltage of a large spacecraft like SSPS in the future, the rapid increase of space charge density and maximum electric field in grounded-HV and HV-grounded cases should be further considered.

Acknowledgements

This work was supported by the National Natural Science Foundation of China (NSFC) under Project with No. 51337008, the National Basic Research Program of China (973 Program) under Project with No. 2015CB251003, and NSFC under Projects with Nos. 11575140, 11275146, 51323012 and 51221005.

Conflict of interest

The authors declare no conflict of interest.

Author details

Xiaoping Wang, Daomin Min* and Shengtao Li*
State Key Laboratory of Electrical Insulation and Power Equipment, School of Electrical Engineering, Xi'an Jiaotong University, Xi'an, Shaanxi, China

*Address all correspondence to: forrestmin@xjtu.edu.cn and sli@xjtu.edu.cn

References

[1] Koons H, Mazur J, Selesnick R. The impacts of the space environment on space systems. In: Proceedings of the 6th Spacecraft Charging Conference; 1 September 2000; Hanscom. 2000. pp. 7-11

[2] Lai S. Fundamentals of Spacecraft Charging: Spacecraft Interactions with Space Plasma. 1st ed. Princeton University Press: Princeton; 2011. p. 272 DOI: 10.1515/9781400839094

[3] Wang L, Hou X. Key technologies and some suggestions for the development of space solar power station (in Chinede). Spacecraft Environment Engineering. 2014;**31**:343-350. DOI: 10.3969/j.issn.1673-1379.2014.04.001

[4] Sessler G, Hahn B, Yoon D. Electrical conduction in polyimide films. Journal of Applied Physics. 1986;**60**:318-326. DOI: 10.1063/1.337646

[5] Li G, Li S, Min D. Deep dielectric charging characteristics of ring structure irradiated by energetic electrons. IEEE Transactions on Dielectrics and Electrical Insulation. 2015;**22**:2349-2357. DOI: 10.1109/TDEI.2015.004840

[6] Ohya K, Inai K, Kuwada H. Dynamic simulation of secondary electron emission and charging up of an insulating material. Surface and Coatings Technology. 2008;**202**:5310-5313. DOI: 10.1016/j.surfcoat.2008.06.008

[7] Davies R, Dennison J. Evolution of secondary electron emission characteristics of spacecraft surfaces. Journal of Applied Physics. 1986;**60**: 318-326. DOI: 10.1063/1.337646

[8] Hodges J, Dennison J, Dekany J. In situ surface voltage measurements of dielectrics under electron beam irradiation. IEEE Transactions on Plasma Science. 2014;**42**:255-265. DOI: 10.1109/TPS.2013.2291862

[9] Fujii H, Okumura T, Takahashi M. Low-energy electron beam induced charging and secondary electron emission properties of fep film used on satellite surfaces. Electrical Engineering in Japan. 2014;**188**:9-17. DOI: 10.1002/eej.22501

[10] Fitting H, Touzin M. Secondary electron emission and self-consistent charge transport in semi-insulating samples. Journal of Applied Physics. 2011;**110**:44111. DOI: 10.1063/1.3608151

[11] Le R, Baudoin F, Griseri V. Charge transport modelling in electron-beam irradiated dielectrics: A model for polyethylene. Journal of Physics D: Applied Physics. 2010;**43**:315402. DOI: 10.1088/0022-3727/43/31/315402

[12] Perrin C, Griseri V, Inguimbert C. Analysis of internal charge distribution in electron irradiated polyethylene and polyimide films using a new experimental method. Journal of Physics D: Applied Physics. 2008;**41**:205417. DOI: 10.1088/0022-3727/41/20/205417

[13] Baudoin F, Le R, Teyssedre G. Bipolar charge transport model with trapping and recombination: An analysis of the current versus applied electric field characteristic in steady state conditions. Journal of Physics D: Applied Physics. 2008;**41**:25306. DOI: 10.1088/0022-3727/41/2/025306

[14] Sessler G, Figueiredo M, Leal FG. Models of charge transport in electron-beam irradiated insulators. IEEE Transactions on Dielectrics and Electrical Insulation. 2004;**11**:192-202. DOI: 10.1109/TDEI.2004.1285887

[15] Miller HC. Flashover of insulators in vacuum: the last twenty years.

IEEE Transactions on Dielectrics and Electrical Insulation. 2015;**22**:3641-3657. DOI: 10.1109/TDEI.2015.004702

[16] Li G, Li S, Pan S. Effect of electron irradiation on dc surface flashover of polyimide in vacuum. IEEE Transactions on Dielectrics and Electrical Insulation. 2016;**23**:1846-1853. DOI: 10.1109/TDEI.2016.005429

[17] Mengu C, Daniel E. Dielectric charging processes and arcing rates of high voltage solar arrays. Journal of Spacecraft and Rockets. 1991;**28**: 698-706. DOI: 10.2514/3.26302

[18] Zha J, Dang Z, Song H. Dielectric properties and effect of electrical aging on space charge accumulation in polyimide/tio2 nanocomposite films. Journal of Applied Physics. 2010;**108**:94113. DOI: 10.1063/1.3506715

[19] Matsui K, Tanaka Y, Takada T. Space charge behavior in low density polyethylene at pre-breakdown. IEEE Transactions on Dielectrics and Electrical Insulation. 2005;**12**:406-415. DOI: 10.1109/TDEI.2005.1453444

[20] Laurent C, Teyssedre G, Le R. Charge dynamics and its energetic features in polymeric materials. IEEE Transactions on Dielectrics and Electrical Insulation. 2013;**20**:357-381. DOI: 10.1109/TDEI.2013.6508737

[21] Takada T, Hayase Y, Tanaka Y. Space charge trapping in electrical potential well caused by permanent and induced dipoles for ldpe/mgo nanocomposite. IEEE Transactions on Dielectrics and Electrical Insulation. 2008;**15**:152-160. DOI: 10.1109/T-DEI.2008.4446746

[22] Henry B, Albert C, Joseph H. Guide to Mitigating Spacecraft Charging Effects. Jet Propulsion Laboratory California Institute of Technology; Hoboken NJ: Wiley; 2012. p. 196. DOI: 10.1002/9781118241400

[23] Li G, Min D, Li S. Research of deep dielectric charging characteristics of polytetrafluoroethene irradiated by energetic electrons. Acta Physica Sinica. 2014;**63**:454-461. DOI: 10.7498/aps.63.209401

[24] Tafazoli M. A study of on-orbit spacecraft failures. Acta Astronautica. 2009;**64**:195-205. DOI: 10.1016/j.actaastro.2008.07.019

[25] Xie F. Several global satellites in orbit failed in 2008. Satellite TV & IP Multimedia. 2009;**3**:20-21. DOI: CNKI:SUN:WSDS.0.2009-03-006

[26] Henry B. The charging of spacecraft surfaces. Reviews of Geophysics and Space Physics. 1981;**19**:577-616. DOI: 10.1029/RG019i004p00577

[27] Henry B, Albert C. Spacecraft charging, an update. IEEE Transactions on Plasma Science. 2001;**28**:2017-2028. DOI: 10.1109/27.902229

[28] Lai ST. A critical overview on spacecraft charging mitigation methods. IEEE Transactions on Plasma Science. 2003;**31**:1118-1124. DOI: 10.1109/TPS.2003.820969

[29] Min D, Li Y, Yan C. Thickness-dependent dc electrical breakdown of polyimide modulated by charge transport and molecular displacement. Polymers. 2018;**10**:1012. DOI: 10.3390/polym10091012

[30] Turnhout J. Thermally stimulated discharge of electrets. In: Sessler G, editor. Electrets. 2nd ed. Springer-Verlag: Berlin; 1987. pp. 81-215. DOI: 10.1007/3540173358_11

[31] Li G, Li S, Pan S. Dynamic charge transport characteristics in polyimide surface and surface layer under low-energy electron radiation. IEEE Transactions on Dielectrics and Electrical Insulation. 2016;**23**:2393-2403. DOI: 10.1109/TDEI.2016.7556518

[32] Sonnonstine T, Perlman M. Surface-potential decay in insulators with field-dependent mobility and injection efficiency. Journal of Applied Physics. 1975;**46**:3975-3981. DOI: 10.1063/1.322148

[33] Min D, Cho M, Khan AR. Surface and volume charge transport properties of polyimide revealed by surface potential decay with genetic algorithm. IEEE Transactions on Dielectrics and Electrical Insulation. 2012;**19**:600-608. DOI: 10.1109/TDEI.2012.6180255

[34] Chadband W. Electrical degradation and breakdown in polymers. IEE Review. 1992;**38**:11-12. DOI: 10.1049@ir19920168

[35] Li S, Pan S, Li G. Influence of electron beam irradiation on dc surface flashover of polyimide in vacuum. IEEE Transactions on Dielectrics and Electrical Insulation. 2017;**24**:1288-1294. DOI: 10.1109/TDEI.2017.006103

[36] Pan S, Min D, Wang X. Effect of electron irradiation and operating voltage on the deep dielectric charging characteristics of polyimide. IEEE Transactions on Nuclear Science. 2019;**66**:549-556. DOI: 10.1109/TNS.2018.2889167

[37] Wrenn G, Rodgers D, Buehler P. Modeling the outer belt enhancements of penetrating electrons. Journal of Spacecraft and Rockets. 2000;**37**: 408-415. DOI: 10.2514/2.3575

[38] Rodgers D. The FLUMIC electron environment model. In: Proceedings of the 8th Spacecraft Charging Technology Conference; 20-24 October 2003; Huntsville, Alabama. 2003. pp. 846-858

Chapter 5

Polyimide Used in Space Applications

Virginie Griseri

Abstract

Polyimide (PI) is an interesting material for space applications as it offers excellent thermal properties. However, due to its dielectric properties, charge storage and release can be at the origin of electrostatic discharges that are hazardous for the surrounding electrical equipment. Depending on the spacecraft orbit, it is necessary to study the impact of specific surrounding environment. In any cases, the effect of vacuum and temperature variations can be combined with electrons and protons' irradiation, atomic oxygen erosion, and photons impact from UV exposure. On the market, there exist many types of PI, and since several years, composite are also developed. The main properties that are usually observed are the conductivity that is analyzed from surface potential decay, the photoemission and the ability to initiate and propagate surface flashover. Since several years, the space charge storage analysis by the pulse electro-acoustic method has been developed as an interesting complementary tool. It is important to remember that experimental characterization needs to be representative to the space environment especially because it has been observed that PI can recover its original properties in air in a couple of hours depending on the ageing degree.

Keywords: electrostatic discharges, electron and proton irradiation, conductivity, space charge, surface potential decay, secondary emission, photoemission

1. Introduction

Spacecraft evolves in very high-energy radiation environment that is directly dependent on the orbit and sun activity. This harsh environment is composed of high vacuum, energetic electrons and protons' radiations, atomic oxygen, UV exposure, and thermal cycling [1, 2]. A good understanding of the charge accumulation dynamics in dielectrics subjected to the charging space radiative environment is necessary to ensure spacecraft operation reliability [3, 4].

Among the dielectrics that are used in satellites conception, the polyimide (PI) has been selected since quite a long time because it offers excellent electrical and thermal properties [5]. However, the modification of their properties with time under such a specific environment needs to be studied carefully. To do so, large facilities have been developed in laboratories to reproduce the complex spectra that may be encountered in worst configurations [6–8]. Many tests arrangements have been developed and adapted to this specific environment to be able to analyze the properties evolution of materials submitted to various types of external aggression.

The aim of this chapter is to get an overview on experimental results obtained on polyimide (PI) used in space environment. The amount of studies is quite important and always undergoing so that is why we will focus on main techniques and tendency. Many PIs are available on the market. In order to improve some properties, the production of composite material [9] appears quite interesting as it might allow to reach in a near future a better control on dielectric properties for such specific applications. Qualifications are necessary before sending such a new material in the space environment. This characterization and phase test can take quite a long time.

2. Environment effect on dielectric properties

As already mentioned, when spacecrafts are evolving into space, they are submitted to a large number of charged particles and encounter several operational anomalies. It has been demonstrated that the harmful anomalies are related to spacecraft charging and the risk of electrostatic discharges (ESDs). For instance, Koons et al. [10] established a list of spacecraft accidents between 1973 and 1997, where he shows that 54.2% of the 326 cases analyzed were directly due to ESDs. Other studies confirm the link between ESDs and anomalies recorded on satellites and try to establish links with surface and/or internal charging [11, 12]. The final idea is always to recommend solutions so as to improve satellite's design that mitigate the occurrence of ESD on future spacecraft even if they are submitted to extreme environmental conditions. Many other studies tend to give access to a broad panel of information concerning the origin of those anomalies through the space weather analyses [11, 12].

Usually spacecraft charging is attributed to three main categories called absolute charging, surface charging, and deep-dielectric charging. The absolute charging is associated to the charging of the whole structure versus its environment; the surface charging refers to the external charging and can be associated to the differential charging between the various part of satellites; and at last the deep-dielectric charging concerns the charging in bulk (**Figure 1**). Deep-dielectric charging can occur in materials located on the satellite surface or inside the main structure. It is due to higher energy particles that can cross thick materials on the way without being stopped. In most of the cases, it is surface materials that are studied as they are directly exposed to various sources of ageing and at the origin of most of the discharges.

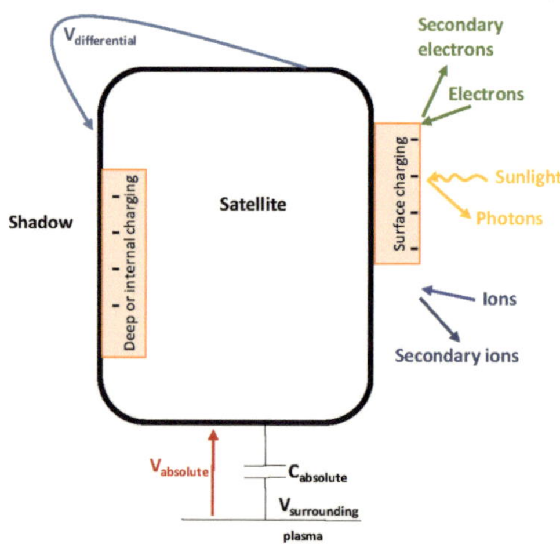

Figure 1.
Schematic representation of satellite charging.

As satellites are made of a large number of materials, it is important to determine surface-material properties and their role in spacecraft surface charging. The main properties that are usually analyzed are:

- the conductivity (surface and bulk)—the accumulation of charges into a dielectric can be responsible for the creation of critic potential difference between adjacent materials from which discharge phenomena can be initiated. In another words, if the conductivity of a dielectric is increased, the differential charging will be limited.

- the dielectric constant as the capability of charge storage will increase if thin samples with high dielectric constant are selected. For instance, 25 μm PI has a dielectric constant close to 3.4.

- the secondary electron emission (SEE) usually dielectric generates more secondary emissions than metals under identical electronic irradiation. Actually the material potential equilibrium will tend toward the second crossover point E_2 (**Figure 2**). This is in favor of a reduction in a dielectric surface potential versus a neighboring metal and might create a disequilibrium at the origin of a discharge.

- the photoemission as the photoelectric effect tends to drive sunlight surfaces more positive than shadowed surfaces and might create differential charging.

2.1 Conductivity and resistivity

The determination of the resistivity of insulating material used in spacecraft conception is very important as it determines how charges will accumulate and will be redistributed during flight. The estimation of an accurate decay time is necessary for the buildup of appropriate spacecraft charging models. As the ohmic resistivity of the PI is too high, the American Society for Testing and Materials (ASTM) method cannot be applied. This is due to the fact of the poor accuracy of pico ampere meter at low current (<1 pA). However, a charge decay resistivity test method was proposed in order to study the ohmic resistivity of high insulation materials [13, 14] and has been adopted by the community. Using this method and fitting surface potential decay (SPD) curves, the ohmic resistivity of PI was estimated to be in the order of 10^{17} Ω.m. It occurs to be several order higher than the value estimated by ASTM method.

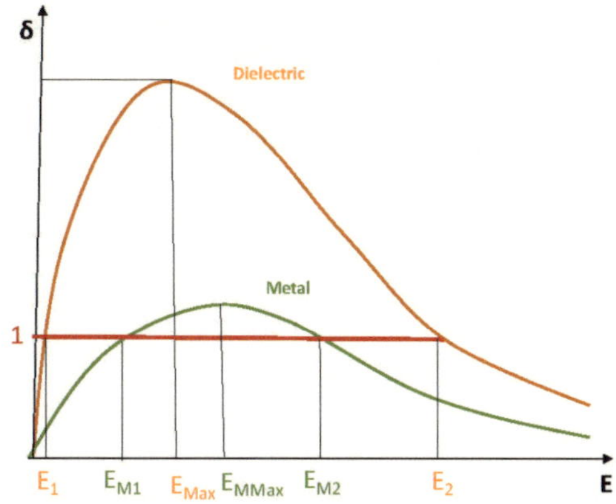

Figure 2.
Secondary emission efficiency for dielectric and metal.

2.1.1 Isothermal surface potential decay experiment

Experimental conditions always need to be carefully controlled. Isothermal SPD is in the range 298–338 k, while the PI sample is irradiated by low-energy electron beam that has been initially recorded. It has been observed that the conductivity is modified as the irradiation goes on and the trapping state gets filled with electrons. That is why depending on the way it is measured, data might be slightly modified.

Besides, we can consider, to some extent that for PI, the photoconduction is a form of radiation-induced conductivity (RIC) as its conductivity can be increased by several decades when exposed to solar lighting with respect to the conductivity in the dark. It has been reported that thin PI films will not charge in sunlight provided that its back surface is grounded. It is therefore recommended to keep PI sample into the dark while SPD is recorded.

The ISPD curve can be separated into three zones: in the part I (transient process), charge trapping and detrapping are competing, whereas in part III (steady state process), the conduction is predominant. In part II, all phenomena are competing [15]. This region limits are difficult to identify that is why most of the time only the initial transient and final steady-state regions are reported on graphs. As surface resistance of spacecraft dielectrics is higher than volume resistances [16], at least one order higher in the case of PI and because external aggression such as atomic oxygen seems to increase the surface resistance without modifying the bulk resistance [17], it is acceptable to neglect the surface charge transport to estimate the bulk conductivity [15]. It is also acceptable to consider that the relative permittivity of PI remains quite stable with the temperature and the electric field during ISPD. However, it was noticed that the surface potential decay is much faster if the initial surface potential is increased. The steady-state current density obtained for a time $t = 3 \times 10^5$ s plotted versus the applied voltage show clearly two regime: ohmic (at low voltage below −950 V) and space charge limited current (at high voltage above −950 V). From the slope, the ohmic resistivity was estimated to 1.2×10^{17} Ω.m and the effective charge carrier mobility to 1.9×1^{-19} $m^2/V.s$ and the trap density estimated to $1.3 \times 10^{21} m^{-3}$. Obviously this type of experiment realized at 298 k needs to be repeated at different temperatures.

Using the 2D ISPD model [18], it is possible to estimate the average surface resistivity, the volume resistivity, and the charge mobility of PI by a genetic algorithm from 298 up to 338 k. In the results reported in **Table 1**, we can observe that the values at 298 k differ a little from the previous estimation by the same authors. This means that these values remain relatively difficult to determine precisely even with the same equipment and the same material. However, the authors managed to calculate the PI surface and volume activation energy and the trap energy which are estimated to 0.3, 0.32, and 0.54 eV, respectively. To get these values, they fitted the surface and volume ohmic resistivity and charge carriers versus the temperature curves using the Arrhenius law.

T (k)	298	308	318	328	338
Surface resistivity ($\times 10^{17}$ Ω)	101.51	42.75	11.66	10.96	3.62
Volume resistivity ($\times 10^{16}$ Ω.m)	28.70	14.96	4.06	2.14	1.08
Charge mobility ($\times 10^{-19}$ $m^2/V.s$)	1.49	3.88	6.24	15.60	30.95

Table 1.
Surface and volume resistivity estimated from ISPD and simulation with an error of 0.9% [18].

2.1.2 Charge transport model based on deep and shallow traps

In order to model the charge transport in PI, complementary experiment such as thermally stimulated current (TSC) is often performed [19] to investigate the relaxation polarization properties that cannot be easily obtained by dielectric spectroscopy. The charge transport is controlled by high field mechanism as soon as the charging potentials exceed the transition voltage of the ohmic regime. It was reported that the time-dependent permittivity $\varepsilon_r(t)$ in PI obeys to Cole-Cole equation rather than Debye ones. Besides, the trap depth could be estimated to be near 1.35 eV [20]. Once released from the surface traps, the charges migrate through the shallow trap to the rear electrode. At 298 k, the carrier residence time was estimated between 7.92×10^{-12} and 1.34×10^{4} s in trap depth of 0.1 and 1 eV, respectively [21]. The carriers can easily hop in the shallow traps but will stay in deep traps much longer. Therefore, shallow traps assist conduction processes, whereas deep traps will control the space charge dynamics [22].

It is considered that the shallow traps will control the temperature-dependent hopping at low temperature; then at high temperature, deep traps can also assist the conduction processes as the residence time carriers drops (at 400 k, the residence time carriers in 1 eV trap depth is reduced to 0.648 s). At room temperature, it is reasonable to develop a unipolar charge transport model with a single deep trap level in a first stage (**Figure 3(a)**). It was notice that the surface electrons are easily released from traps surface center whereas they can remain into the bulk for a very long time when they are stored into deep traps. The steady state was reached after 94.8 h; at that time, 39.72% of electrons deposited close to the surface creating a surface potential of –2056 V [21] were released. The resistivity of PI was estimated in the range 8.08×10^{16}–9.40×10^{16} Ω.m. However, in order to improve the model, it would be better to refine the trap distribution characteristics (**Figure 3(b)**) and consider the density of localized state [23, 24].

2.1.3 Effect of radiation on radiation-induced conductivity

As already mentioned, the RIC is an important phenomenon occurring in PI materials [25]. The RIC can contribute the sample charge decay and prevent discharges. It has been shown that when a step function and uniform irradiation

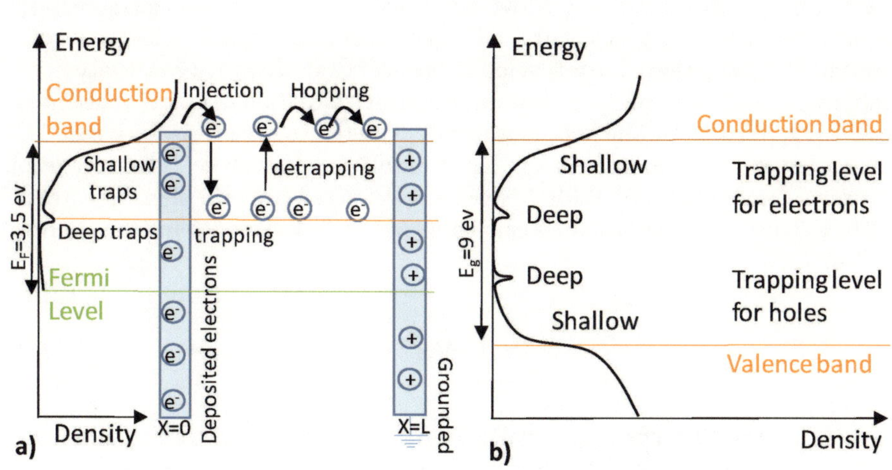

Figure 3.
Schematic representation of state density in dielectric materials. (a) Model with one level of deep and shallow traps. (b) Model of shallow and deep traps for electrons and holes that are related to physical and chemical disorder.

are applied, RIC initially rises to a maximum then decreases continuously [26]. In such a case, the accumulation of space charge is not prevented and ESD can be observed.

A different behavior has been observed in various PI (Kapton® 25 μm produced by Dupont© and PM-1-OA 15 μm or PM-1 13 μm produced by Russian services) submitted to large dose rate and long irradiation time [27]. In this study, 40 mm diameter films were Al coated (electrode characteristics: 32 mm diameter and 50 nm thick). They used fresh sample each time even if they mention that an annealing dose effects was observed in PI after 4 h in air at 393 K. For instance, measurements on PM-1 irradiated from 4 up to 10 MeV protons in vacuum (1–4 Pa) and room temperature shows a classic behavior up to a dose of 10^5 Gy (the dose rate could be in the range 100–5000 Gy/s). Above this dose, the signal increases by two or three orders of magnitude. Even after the end of irradiation, this dose-modified RIC remains much higher than the dark conductivity during a quite a long period of time. An interesting point is that this slow process is completely extinguished as soon as air is introduced into the chamber. The DM RIC effect was attributed to a metastable state in polymers with a strong donor-acceptor interaction that can be easily cancelled by the introduction of atmospheric oxygen. This metastable conjugated structure might be due to the presence of side or inner-chain molecular groups with local conjugation. This strong DM RIC produced by electron of the ambient plasma seems quite useful to reduce the bulk charging of PI used on the outside spacecraft but it is not well controlled. This DM RIC which is a PI intrinsic property prevents the RIC to decay drastically after long irradiation exposure. However, new materials providing, thanks to the introduction of nanoparticles, a better control of the RIC are expected as replacement in the future.

2.1.4 Effect of temperature on conductivity

The effect of temperature has also been studied as spacecraft charging in plasma, and radiative environment is highly sensitive to the temperature decrease which is accompanied by a reduction of the electrical conductivity in dielectric materials [28]. Spacecraft orbiting around the earth are submitted to a large temperature range between 120 and 400 K. It is therefore easy to understand that when dielectric materials are exposed to low temperature, the charge storage increases; whereas when a warming up occurs, a dissipation of this charge can be produced. Experimental setups have been developed in order to study the PI resistivity combining the effect of temperature and electron irradiation to take into account the RIC [29]. First of all, the temperature effect on the volume resistivity was investigated on a Kapton® 200H irradiated for 60 s under a 20 keV electron beam. The useful data reported in **Table 2** have been extracted from the signal recorded for 240 h. The volume resistivity is calculated in the dark region using the expression (Eq. (1)):

$$V(t) = V_0 e^{\left(-\frac{t}{\tau_d}\right)} \quad (1)$$

where V(t) is the surface potential, V_0 is the initial surface potential, and τ_d is the decay time constant in the dark current region. As the temperature increases, the volume resistivity decreases exponentially.

To determine the volume resistivity in short-time region where the polarization current is dominant, Eq. (2) has been used:

$$V(t) = V_0 \left[\varepsilon_r^\infty + \left(1 - \varepsilon_r^\infty\right) e^{\left(-\frac{t}{\tau_p}\right)} \right]^{-1} \quad (2)$$

where τ_p is the decay time constant for the decrease in potential due to polarization current; and ε_r^∞ corresponds to the relative permittivity when the complete polarization is achieved. The volume resistivity in short time seems to be independent from the temperature, while the volume resistivity in the dark region drops a lot with the increase of the temperature.

Another experiment consists in increasing the energy of the electron beam in addition to the temperature. It shows that when the electrons are injected deeper, the surface potential decay rate increases. This is due to an enhancement of the conductivity produced by trapped electrons in the irradiated area that are responsible for polymer chain scission reaction and the RIC effect. It is particularly true for electrons above 40 keV. It is always important to remember that several effects are combined in real situations that is why the analysis remains quite complex.

2.2 Secondary emission yield

As the electron emission related to irradiated electrons influences the satellite surface charge accumulation, the measurement of the secondary electron emission (SEE) from metal and insulating material used for satellites is quite important. Studies have been performed on material to determine the effect of surface degradation on SEE. The SEE yield is calculated as the ration of the primary incident electron current over the secondary electron current. The shape of the curve represented in **Figure 2** shows E_1 and E_2, the crossover energy values, where the yield is equal to 1 and the maximum of δ that corresponds to the primary electron energy E_{max}. The SEE Yield (SEEY) in solid depends therefore mainly on the primary electron energy Ep, the injection angle, the material density, and the surface status.

2.2.1 Measurement difficulties in polymers

Because of the difficulty in measurement, yield is often neglected as an important contributor into spacecraft charging and therefore the resistivity which is easier to be measured is taken into consideration. Indeed, a full study on the effect

Kapton® 200H 50 μm Temperature (K)	Volume resistivity (Ω.m)	
	Dark current (×10^17^)	Short time (×10^13^)
233	3.3	0.49
253	1.1	1
273	0.83	1.1
299	0.42	1.7
323	0.33	1.3
353	0.11	0.45

Table 2.
Volume resistivity data obtained on Kapton® 200H films irradiated with an 20 keV electron beam for 60 s at different temperatures [29].

of low-fluence electron yield [30] confirms that the electron provided by the measurement system cannot be easily extracted in insulators as in conductor and can affect the measurements. Furthermore, they come to the conclusion that in insulators with modest yield, the incident pulse does not produce enough SE to appreciably charge the specimen under studies. However, in the case of PI with a maximum yield $\sigma_{max} < 3$ due to the RIC and its persistent effect after the end of the irradiation, charge dissipation is possible and the incident pulse amplitude does not need to be reduced so much. Fortunately, since several years, research is made to improve the measurement system and make it more reliable for polymers and ceramics. The measurement method described in paper [31] shows that very accurate SEEY values can be recorded on Kapton-HN for instance.

2.2.2 Analysis based on frontier molecular orbital theory

Also it was reported that the SEEY of Upilex®-S was smaller than Kapton®-H [32]. This property was explained by a variation in the potential energy bound that can be estimated by quantum chemical calculation which is based on the density function theory. In this representation, the highest occupied molecular orbital (HOMO), corresponding to the conduction band, and the lowest unoccupied molecular orbital (LUMO), corresponding to the valence band, are calculated. For Upilex®-S, the gap between HOMO and LUMO was found to be smaller than for Kapton®-H. The ionization energy that corresponds to the energy difference between the vacuum level and the HOMO was found to be 5.36 eV for Upilex®-S and 5.91 eV for Kapton®-H that comfort the fact that it might be easier to get SEE in the case of Upilex®-S. Such approach needs further investigation.

2.2.3 Effect of atomic oxygen and UV radiations

SEEY is really dependent on the surface status of the material under studies. Among all external factors, the atomic oxygen (AO) plays the most important role in the erosion processes of organic materials on the low earth orbit (LEO). It is therefore important to determine the effect of such degradation on PI surface as it might affect quite a lot its SEE properties with time.

To simulate the collision in laboratory, many sources are available [33]. Unfortunately, many AO sources produce VUV radiation during their operation, a fact that should be taken into consideration when comparing results. It was reported that the maximum SEEY of PI film is 1.1 when primary electron energy is 600 eV. Specific work on PI films shows that when the fluence of the AO was increased, the SEE yield was decreased. However, the results were different if the AO was deliver by a laser detonation AO beam source or by the plasma asher method [34]. Usually it is considered that an exposure between 12 and 24 h with a fluence of 3.5×10^{19} and 6.9×10^{19} atom/cm^2 is respectively equivalent to 6 month and 1 year AO erosion in LEO. The SEE yield was increased in the second case. The difference in the result is due to the source of AO production. It is mentioned that the asher method compared to laser detonation generates AO more easily and avoids contamination which is more representative to what happens in space. The conclusion is that the charging effect of the space plasma should be less effective with time due to AO effect on LEO.

In many cases, the effect of UV and AO are studied simultaneously [35]. In some studies, the total electron emission yield (TEEY) is reported. It corresponds to the sum of the SEEY and the backscatter electron emission yield. The contribution of this phenomenon depends on the electron energy and the material properties [36]. In many cases, the SEEY remains the main source of electron in the TEEY. Similar results are obtained on virgin sample as reported in **Table 3**.

Reference	E_2 (eV)	E_{max} (eV)	σ_{max}
Virgin—Kapton HN [37]	500	150	1.7
[31]		180	1.95
Virgin—Kapton 100 H [35]	670	150	1.69
UV dose—Kapton 100 H D = 2000 ESH		150	2.1 Saturated D > 500 ESH
AO eroded—Kapton 100 H			
12 h		500	1.2
24 h		700	1.0

Table 3.
Secondary emission values of cross over energy E_2, maximum energy E_{max} and yield σ_{max}.

The effect of UV was to increase the TEEY maximum. A saturation was observed above 500 equivalent sun hours (ESH) UV exposure. In PI, UV creates bond scission and provides high concentration of free radicals that are remaining stable for several hours under vacuum as their lifetime is of about 20 h in air. The non-bonded electrons are active and can be excited more easily than bonded electrons that contribute to the increase of the TEEY. However, the UV can penetrate only about 100 nm in materials, whereas energetic electrons (> 3 keV) can go further. The effect of UV is mainly efficient close to the surface where low-energy electrons with a weak penetration depth are supposed to be located.

On addition, the AO erosion acts on both E_{max} and TEEY. The first one increases, whereas the second decreases with the exposure time. The AO acts on the surface roughness. Usually E_{max} and TEEY are expected to increase with the injection angle increase. However, in the case the roughness becomes too important, the secondary electrons might become the source of new primary electrons with lower energies, but the new secondary electrons might not have enough space to escape the surface. That is why as the AO exposure increases, TEEY decreases and E_{max} is significantly increased. The effect of surface roughness was clearly highlighted in the study tempting to demonstrate the effect of surface modification on spacecraft charging parameters [38]. An analysis on Kapton® HN shows that the presence of Dow corning DC 704 diffusion pump oil as surface contaminant oil or scratched produced at the surface during a polishing operation makes the reflectivity to reduce and the absorption coefficient to increase.

2.3 Photoemission yield

When a high frequency light illuminates a dielectric material, the photons interact with the orbital electrons of the atoms. The energy provided to the electrons might be large enough to make them overcome the material work function and become free in vacuum. These electrons are photoelectrons. The photo emission yield (PEY) is the number of photoelectrons to the incident photons. It depends mainly on the incident photon energy, wavelength, and incidence angle but also on the material properties such as absorbance and reflectivity.

It is also important to remember that the photoconduction plays an important role in PI as it is contributing to the RIC. It has been noticed that under constant solar lighting, the conductivity of PI can increase by several decades with respect to conductivity in the dark. Theoretically, PI should not pose any problems of charging under illumination and therefore no electrostatic should be expected. However, it is the surrounding that needs to be considered carefully.

As already mentioned even if spacecraft are operating in the sunlight, some parts will remain in the shadow. The photoemission phenomena maintain the spacecraft frame and polymers in sunlight to low potential as the photoemission current between the spacecraft and the ambient plasma dominates the current balance equation. On the contrary, polymers in the shadow are charging negatively because the photoemission does not occur and they are mainly impacted by the fast moving electrons of the surrounding plasma.

On the GEO orbit, mainly electron and proton encountered into the Van Allen Belt can lead to the material degradation with time. On LEO, the effects of atomic oxygen and ultraviolet rays are more that need to be considered as harmful as mentioned in the previous section. These UV radiations are responsible for photoemission that might be affected by other degradation with time.

2.3.1 Effect of proton and electron irradiation

Usually the photoemission induced a positive charging at the surface of dielectric that can influence the emission as photoelectron can be attracted back to the surface. That is why during measurements, it is important to choose new area each time by shifting the sample and using short-time exposure to UV pulse. It was shown on Kapton® 100H sample [39] that the proton ageing (in the range 1–15 years equivalent exposure on GEO with 50 keV proton in which penetration depth is estimated by Casino software to be of 600 nm and various flux) makes the PEY to raise. On the contrary, the same ageing with electrons (of 500 keV that can cross the 25 μm sample thickness) seems to make the PEY decay slightly. In both cases, the energy loss by the particles along the way is transmitted to the material, however in the case of proton, this energy is concentrated into the surface shallow layer that might induce high ionization and act in favor of photoemission. In the case of high-energy electron, the energy is deposited in the entire bulk and the surface ionization that might occur is not predominant. It is also important to remember that air exposure can help the sample recovery in a few hours [40]; it is recommended to perform all these measurements while the sample is maintained under vacuum.

2.3.2 Effect atomic oxygen and UV exposure

The UV ageing in the range 25–1000 ESH seems to make the PEY increase but with a saturation above 500 ESH. The same saturation was noticed on secondary electron emission experiments [39]. A study on Kapton®HN, Kapton®E, and Upilex®S showed increases in solar absorptance and the α/ε ratio under VUV radiation exposure, whereas emittance changes were not significant [41]. As more photons can be absorbed by damaged PI, the PEY is expected to increase. This effect of absorbance is probably coupled with the chemical degradation and the production of free radical during the UV irradiation [42]. Actually free radicals provide activated electrons that can enhance PEY. At last the AO effect seems to be limited even with the equivalent exposure of 1 year on LEO. In fact, AO increases the surface roughness of the PI film and the photoelectron that are emitted are probably recapture before they can really escape and be detected that is why the PEY tends to decrease with long AO beam exposure.

2.4 Surface flashover

Surface flashover has been identified at triple junction locations on spacecraft [43]. A triple junction is characterized by a metal electrode, a dielectric, and the surrounding environment (air or vacuum). The main theory is based on the secondary electron emission avalanche (SEEA) [44]. The electrons are emitted from the cathode due to field emission. The electrons hit the dielectric surface and produce secondary electrons. Those electrons continue to move into the direction of the anode under the effect of the electrostatic field. The production of other secondary electrons is made on the way and can be at the origin of an avalanche that produces a degassing and ionization phenomenon on the dielectric surface at the origin of the discharging channel (**Figure 4**).

To understand surface flashover phenomena in space environment, electron irradiation and dc voltage should be considered at the same time. It is also important to take into account the surrounding vacuum [45]. The number of studies combining both effects is growing and has confirmed that the experimental conditions are really sensitive.

The simultaneous action of dc voltage and electron irradiation (**Figure 4**) can be described as follows:

- During electronic irradiation, there is a charge injection and storage in the dielectric bulk. This SC leads to the buildup of an electric field in the bulk that can influence flashover performance.

- During the dc voltage application, a distortion of the electric field at the triple junction (metal/dielectric/vacuum) occurs and electron emission from the cathode is enhanced. The "classic" flashover propagation will occur.

Both phenomena are influencing each other; on one side, the internal electric field (due to the injected charges) creates a reverse-acting force on the kinetic electron flowing to the dielectric surface, and on the other side, the surface electric field might affect the electron injection.

2.4.1 Effect of electron irradiation

It was reported that after electron irradiation, the dc surface flashover of PI was increased due to the electron stored below the surface [46]. Due to the presence of electron in the bulk, the electric field at the triple junction is lowered and then the flashover initiation prevented unless the dc voltage is increased. For instance, the dc voltage flashover on raw polyimide is recorded at 19.2 and 23.9 kV after an irradiation under 20 keV. When the flashover is initiated, the secondary electrons are deviated from the surface and the propagation of the flashover is inhibited. It was noticed that electrons irradiation with an energy above 20 keV up to 30 keV seem to produce the same effect as the penetration depth increases the influence on the surface flashover initiation and propagation might have reached a limit. This tendency was also pointed out on deeply studied PI-type SKPI-MS30 provided by Changzhou

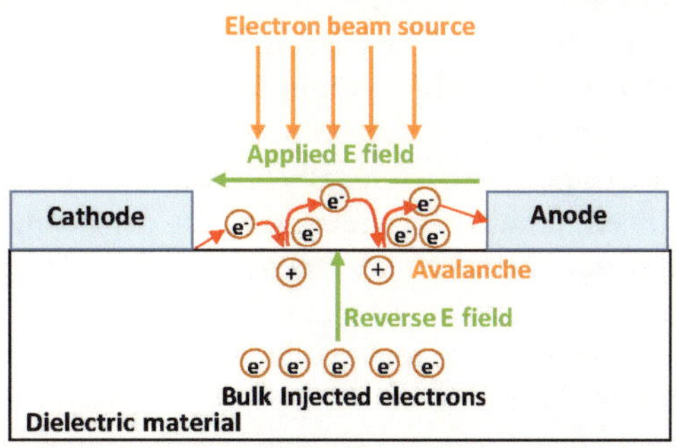

Figure 4.
Surface flashover propagation and combine effect due electron beam irradiation.

Sunchem High Performance Polymer company [47]. It has been observed that above a 20 keV electronic irradiation, the dc surface voltage flashover decreases slightly with the increase of the electron energy. It was also reported in this work that the previously irradiated PI presented higher flashover voltage than the initial sample. They suggest that irradiation could induce chemical-physical modifications such as degradation with the generation of low molecular compounds, gases, and crosslinking. The dielectric constant usually decreases in favor of the diminution of the electric field distortion and deep trap amount increases under low-energy irradiation. These deep traps prevent electrons from migration and favor a charge accumulation that is at the origin of the increase of the dc voltage flashover.

2.4.2 Effect of the electron beam flux

In addition, if the flux of the electron beam is increased (from 0.175 to 1.75 $\mu A/cm^2$), the dc surface flashover decreases exponentially with the electron energy. This is explained by the RIC phenomena. During the electron irradiation, electron-holes pairs created can exceed the number of intrinsic carriers and make the RIC to increase. This increase is generally following the energy increase. But when the dose rate increases, the RIC effect decreases and then the effect of accumulated charges increases and contributes to the reduction of the flashover voltage.

2.4.3 Behavior under continuous irradiation

Besides, when measurements are performed during irradiation on the previously irradiated samples, the dc voltage flashover varies. It decreases as the energy of radiation in increased. Starting at 26.9 kV under 5 keV electron irradiation, the dc flashover voltage is of about 10.5 kV under 30 keV electron irradiation. The dc flashover voltage was equal to the value recorded on raw PI under 17.5 keV electron irradiation. Below this energy, the electrons from the beam are repelled and the dc flashover voltage remains higher than on Virgin PI. On the other hand, when the energy is higher, the electrons can reach the surface and contribute to the flashover phenomenon and the initial dc flashover voltage is decaying. The effect of attraction of the electron of the beam toward the anode is not so important when the energy of the beam is high. The contribution of injected electron below the surface on the electric field distortion is in favor of the reduction of the dc voltage flashover decay.

2.4.4 Influence of vacuum

As mentioned earlier, the effect of vacuum variation in non-negligible. The surface flashover voltage rarely varies in the range of vacuum going from 1.10^{-6} to 6.10^{-1} Pa. However, the decay of the surface flashover voltage is extremely important when the vacuum levels drops. The Paschen's law is controlling the surface voltage flashover above 6.10^{-1} Pa. It could be thought that if the system is maintained in vacuum, as it is the case for satellite environment, these effects could be neglected. However, the pressure can be modified locally above some materials due to a degassing phenomenon. Indeed, it has been shown that the degassing effect can enhance flashover process and should be taken into account in the material selection [48]. Fluorination surface treatment was tested in order to prevent hydrogen degassing from PI samples. The fluorination is realized into a reactor containing a mixed gas composed of 12.5% of fluorine and nitrogen. It was observed that after this surface treatment, the surface flashover voltage was increased by 10.5%. The C–H and C–O bonds are replaced by C–F bonds at the surface which are more stable during the surface flashover and make degas process less effective. Besides, the electronegativity of the fluorine tends to weaken the avalanche phenomenon and facilitate the electron absorption. This fluorination treatment might be considered as an effective process to reduce flashover due to degassing, but the effect over time should be considered. It might fade with time due to the surrounding aggressions in space.

3. Complementary tools

As discussed in the previous section, there is a large number of conventional tests that have been selected and validated to study material used in a space environment. New tools have been developed to provide complementary information. For instance, the pulse electro acoustic (PEA) method that was commonly used since the 1980s to study the material encountered in electrical engineering application has been adapted to characterize the material under vacuum. This method has been developed to be also efficient to study materials under various conditions such as electron or proton irradiation and under various temperatures in the last 18 years.

3.1 PEA signal recorded after electron irradiation

At first, it was necessary to demonstrate that it is was possible to combine measurements of the surface potential and the distribution of charges in the volume by the PEA method. Initial PEA measurements were performed ex situ by using a classical system. Then, a specially adapted PEA cell to perform in situ measurements during the irradiation was developed [49, 50]. The analysis of surface potential and PEA data provides additional information.

Thanks to the PEA technique, it is possible to follow the buildup of charge into the bulk during an irradiation and then follow the relaxation. The direct observation of the SC distribution with time is providing quite a lot details on the dynamics of the charges migration with time that cannot be obtained by surface potential measurements alone. Depending on their energy, the electrons are expected to be stored at a specific penetration depth that can be predicted by various online programs such as ESTAR [51] that provides, for instance, the range and stopping power in PI. In **Figure 5**, an example of PEA signal recorder on a Kapton®-H film irradiated under a 100 keV electron beam shows a negative peak of injected charges near the back surface in agreement with the theoretical calculation. Thanks to PEA results, it has been shown that the relaxation of these negative charges is quite

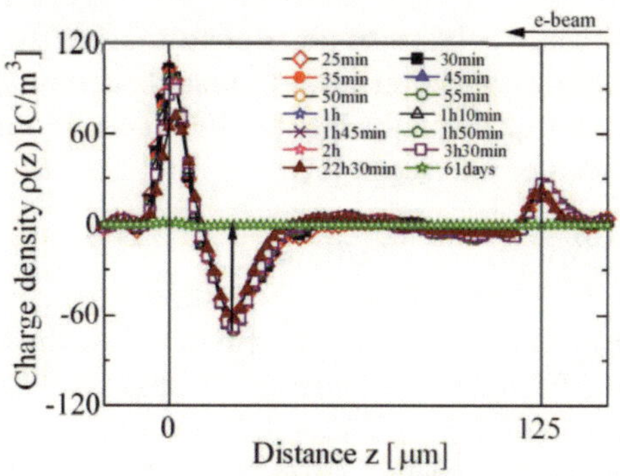

Figure 5.
Space charge distribution recorded by PEA on a previously electron irradiated Kapton®-H film. Irradiation conditions: 100 keV with a flux of 1 nA/cm² for 20 min [52].

slow and takes more than 24 h [52] in air. In this case, there was no spreading nor migration of the charge with time as it could be observed in other materials. As mentioned earlier, this type of information cannot be provided by surface potential data.

3.2 Effect of temperature and moisture on space charge distribution

As thermal effect is very important for materials used in space environment, some studies on the impact of temperature on irradiated PI have been realized. At first, a study on the effect of space charge formation under polarization by PEA on Kapton®-H using high electric field (from 110 up to 150 kV/mm) shows that heterocharge accumulates slowly until the breakdown occurs. This effect seems to be accelerated by the temperature when it is increased from 30 to 80°C [53]. Even long short-circuit period could not help to recover initial condition. After an annealing treatment (of at least 150 min in silicon oil at 100°C), it was concluded that the presence of moisture could favor the heterocharge accumulation and breakdown. That is why it is always recommended to take into account the humidity in the atmosphere to characterize PI samples.

X-rays photoelectron spectroscopy (XPS) studies on PI reveals that after a fast thermal cycling (based on IEC-60068-2-14:2009 method), the amount of C—N and C=O bonds increases so as the crystallinity, whereas the amount of C—C and C—O bonds decreases [54]. Due to these modifications, an increase in shallow traps (in the range of activation energy 0.04–0.05 eV) and in the deep traps (in the range of activation energy 0.95–1.31 eV) has been predicted. These alterations are directly linked with the increase of space charge amount detected by PEA during the 60 min polarization under –5 kV. The increase in shallow energy level traps act in favor of charge injection and recombination. In addition, the deep energy level traps make the charges stay into the material after the end of the polarization as observed on the PEA signal.

Another analysis of PEA measurements on a PI irradiated under a 140 keV electron beam at room temperature for various period of time (from 10 to 1800s) then heated up to 135°C has been reported [55]. The buildup of a negative peak into the bulk due to the accumulation of electrons provided by the beam after 10 s of irradiation was clearly observed. An additional negative peak was detected close to the irradiated surface but its origin was not clearly identified. After longer irradiation time, a migration of the negative charges toward the non-irradiated grounded surface was observed. The different patterns of migration accelerated by the temperature are also easy to understand with the representation of the electric field into the bulk. The polarity of the electric field helps to define the direction in which charged particles tend to move. It is also possible to analyze the charge migration into the bulk in combination with thermo-stimulated current (TSC) measurements. A global picture of the charge behavior into the bulk even if only the net charges were observed by PEA can be extracted [56].

3.3 PEA measurements on proton irradiated PI

The PEA technique has been also quite useful to follow the behavior of PI under proton irradiation with different energy on Upilex® S 125 μm films [57]. The energy selected was 1 and 1.5 MeV, and the current density varied from 0.3 to 30 nA/cm^2 knowing that 0.3 nA/cm^2 corresponds to the real flux within the inner radiation belts when a solar flare occurs. The irradiation was performed for 10 min. As expected positive charges are detected into the bulk but remain close to the irradiated surface despite the energy as expected by the calculation of proton range and stopping power by PSTAR [51]. For 1 and 1.5 MeV, the proton penetration depth in PI should be of about 19 and 37 μm, respectively. It was noticed that the SC distribution is highly dependent on the flux. The amplitude of the peak was saturated quickly during irradiation and the relaxation was quite fast for flux higher than 3 nA/cm^2. This tends to show that some chemical modifications might have happen into the irradiated area. The analysis by XPS UV-vis confirms that molecular scission was produced. Depending on the PI selected, it has been noticed that usually the C–C bond at benzene group increases, whereas C–N bond at imide group and C=O bound at carbonyl of imide group decrease [58]. At last the recovery of the dielectric properties was obtained after a day in air.

4. Conclusions

Polyimide films are used for spatial applications as they offer excellent thermal properties. As they are submitted to various charged particles and radiation during their lifetime, the determination of the ageing effect on such material is really challenging. In laboratories, many irradiation chambers have been developed so as to reproduce and couple the ageing sources such as electron radiation, proton radiation, UV exposure, atomic oxygen surface erosion, and temperature variation. In this chapter, the classic experimental measurements used to study materials for space applications have been presented. It is obvious that the surface potential decay is quite convenient to study the PI conductivity variation under various external conditions. It is also important to be able to characterize the SEE or photoconduction or surface flashover processes in order to mitigate ESD. In addition, the chemical analysis and the use of other techniques such as the PEA method to follow the dynamic of charge buildup and release are important to get a better understanding of the dielectric properties of PI under such an aggressive environment that remains quite difficult to reproduce entirely in laboratories. As mentioned, PI is very sensitive to air exposure, humidity, or UV exposure, so all the experiments must be performed with care in order to provide reliable data for the model that tend to reproduce satellite in its whole structure taking into account the material arrangement so as the surrounding environment.

Author details

Virginie Griseri
LAPLACE, Toulouse University, CNRS, INPT, UPS, Toulouse, France

*Address all correspondence to: virginie.griseri@laplace.univ-tlse.fr

References

[1] Pisacane VL. The Space Environment and Its Effect on Space Systems. 2nd ed. American Institute of Aeronautics and Astronautics: Reston, VA, USA; 2016

[2] Tribble AC. The Space Environment. New Jersey, USA: Princeton University Press; 2003. pp. 1-26

[3] Baker DN. The occurrence of operational anomalies in spacecraft and their relationship to space weather. IEEE Transactions on Plasma Science. 2000;**28**(6):2007-2016. DOI: 10.1109/27.902228

[4] Gubby R, Evans J. Space environment effects and satellite design. Journal of Atmospheric and Solar - Terrestrial Physics. 2002;**64**:1723-1733. DOI: 10.1016/S1364-6826(02)00122-0

[5] Hastings D, Garrett H. Spacecraft Environment Interactions. Cambridge University Press: Cambridge; 1996. pp. 132-177

[6] Paulmier T, Dirassen B, Belhaj M, Inguimbert V, Payan D, Balcon N. Experimenttestfacilitiesforrepresentative characterization of space used materials. In: Proceedings of the 14th ESA/ESTEC SCTC, Noordwijk, the Netherlands, April 2016. pp. 4-8

[7] Toyoda K, Masui H, Muranaka T, Cho M, Urabe T, Miura T, et al. ESD ground test of solar array coupons for a greenhouse gases observing satellite in PEO. IEEE Transactions on Plasma Science. 2008;**36**(5):2013-2424

[8] Griseri V, Malaval P, Berquez L, Tung TA, LeRoy S, Boudou L, et al. Charge build-up and transport in electron beam irradiated polymers in a new irradiation chamber. Annual Report of IEEE International Conference on Electrical Insulation and Dielectric Phenomena. 2010. DOI: 10.1109/CEIDP.2010.5724030

[9] Liaw DJ, Wang KL, Huang YC, Lee KR, Lai JY, Ha CS. Advanced polyimide materials: Syntheses, physical properties and applications. Progress in Polymer Science. 2012;**37**:907-974

[10] Koons HC, Mazur JE, Selesnick BJF, Fennell JL, Roeder JL, Anderson PC. The impact of space environment on space systems. In: Proceedings of the 6th Spacecraft Charging Conference. November 2-6, 1998. MA, USA: AFRL Science Center, Hanscom AFB; 1998. pp. 7-11

[11] Ferguson DC, Denig WF, Rodriguez JV. Plasma conditions during the galaxy 15 anomaly and the possibility of ESD from subsurface charging. In: 49th AIAA Aerospace Sciences Meeting including the New Horizons Forum and Aerospace Exposition 4-7 January 2011, Orlando, Florida. 2011

[12] Iucci N, Levitin AE, Belov AV, Eroshenko EA, Ptitsyna NG, Villoresi G, et al. Space weather conditions and spacecraft anomalies in different orbits. Space Weather. 2005;**3**:S01001. DOI: 10.1029/2003SW000056

[13] Frederickson AR, Dennison JR. Measurement of conductivity and charge storage in insulators related to spacecraft charging. IEEE Transactions on Nuclear Science. 2003;**50**(Part 1): 2284-2291

[14] Dennison JR, Brunson J, Swaminathan P, Green NW, Frederickson AR. Method for high resistivity measurements related to spacecraft charging. IEEE Transactions on Plasma Science. 2006;**24**(Part 2): 2191-2203

[15] Min D, Cho M, Khan AR, Li S. Charge transport properties of dielectrics revealed by isothermal surface potential decay. IEEE

Transactions on Dielectrics and Electrical Insulation. 2012;**19**(4):1465-1473

[16] Nitta K, Takahashi M. Material properties measurements related to spacecraft charging/discharging: Current status and future plan. IEEJ Transactions on Fundamentals and Materials. 2009;**129**:739-745. DOI: 10.1541/ieejfms.129.739

[17] Noor DAM, Khan AR, Chiga M, Okumura T, Masui H, Iwata M, et al. Effect of atomic oxygen exposure on surface resistivity change of spacecraft insulator material. Transactions of the Japan Society for Aeronautical and Space Sciences, Aerospace Technology Japan. 2011;**9**:1-8

[18] Min D, Cho M, Khan AR, Li S. Surface and volume charge transport properties of polyimide revealed by surface potential decay with genetic algorithm. IEEE Transactions on Dielectrics and Electrical Insulation. 2012;**19**(2):600-608

[19] Raju GG, Shaikh R, Haq SU. Electrical conduction processes in polyimide films-I. IEEE Transactions on Dielectrics and Electrical Insulation. 2008;**15**(3):663-670. DOI: 10.1109/TDEI.2008.4543102

[20] Tian F, Bu W, Yang C, He L, Wang H, Wang X, et al. A new method for direct determination of trap level distribution from TSC measurement. In: Proceeding of 9th International Conference on Properties and Applications of Dielectric Materials (ICPADM). 2009. DOI: 10.1109/ICPADM.2009.5252276

[21] Min D, Li S, Cho M, Khan AR. Investigation into surface potential decay of polyimide by unipolar charge transport model. IEEE Transactions on Plasma Science. 2013;**41**(12):3349-3358

[22] Teyssedre G, Laurent C. Charge transport modeling in insulating polymers: From molecular to macroscopic scale. IEEE Transactions on Dielectrics and Electrical Insulation. 2008;**12**(5):857-875

[23] Diaham S, Locatelli ML. Space-charge-limited currents in polyimide films. Applied Physics Letters. 2012;**101**:242905. DOI: 10.1063/1.4771602

[24] Le Roy S, Segur P, Teyssedre G, Laurent C. Description of bipolar charge transport in polyethylene using a fluid model with a constant mobility: Model prediction. Journal of Physics D: Applied Physics. 2004;**37**:298-305. DOI: 10.1088/0022-3727/37/2/020

[25] Plis EA, Engelhart D, Cooper R, Johnston WR, Ferguson D. Review of radiation-induced conductivity effects in polyimide. Applied Sciences. 2019. DOI: 10.3390/app9101999

[26] Tyunev AP, Saenko VS, Poshidaev ED, Ikhsanov RS. Experimental and theoretical studies of radiation-induced conductivity in spacecraft polymers. IEEE Transactions on Plasma Science. 2015;**43**(9):2915-2924

[27] Tyunev AP, Saenko VS, Zhadov A, Poshidaev ED. Radiation-induced conductivity in Kapton-like polymers featuring conductivity rising with an accumulating dose. IEEE Transactions on Plasma Science. 2019;**47**(8):3739-3745

[28] Minow JI, Parker LN. Spacecraft Charging in Low Temperature Environments, 45th AIAA Aerospace Science Meeting. 2007

[29] Watanabe R, Hiroaki H, Okumura T, Takahashi M. Effect of temperature and electron energy on volume resistivity of a polyimide film in space environment. In: IEEE International Conference on Electrical Insulation and Dielectric Phenomena. 2012. DOI: 10.1109/CEIDP.2012.6378870

[30] Hoffmann R, Dennisson JR, Thompson CD, Albretsen J. Low-fluence electron yield of highly insulating materials. IEEE Transactions on Plasma Science. 2008;**36**(5):2238-2245

[31] Hoffmann R, Denisson JR. Measurement method of electron emission over a full range of sample charging. IEEE Transactions on Plasma Science. 2012;**40**(2):298-304

[32] Shibuya K, Nomura K, Miyake H, Tanaka Y, Ochira M, Okumuta T, et al. Development of measurement system for secondary electron emission yield of insulating materials for spacecraft materials. In: IEEE International Conference on Condition Monitoring and Diagnosis. 2012. pp. 1102-1105

[33] Kleiman J, Iskanderova Z, Gudimenko Y, Horodetsky S. Atomic oxygen beam sources: A critical overview. In: Proceedings of the 9th International Symposium on Materials in a Space Environment, Noordwijk, the Netherlands. 2003. pp. 313-324

[34] Nitta K, Miyazaki E, Takahashi M. Influence of atomic oxygen irradiation on secondary electron emission yield of polyimide films. In: XXIVth International Symposium on Discharges and Electrical Insulation in Vacuum. 2010

[35] Wu J, Miyahara A, Khan AR, Iwata M, Toyoda K, Cho M et al. Effect of ultraviolet irradiation and atomic oxygen on total electron emission yield of polyimide. IEEE Transactions on Plasma Science. 2014;**42**(1)191-198

[36] Dennison JR, Sim A, Thomson CD. Evolution of the electron yield curves of insulators as a function of impinging electron fluence and energy. IEEE Transactions on Plasma Science. 2006;**34**(5):2204-2218. DOI: 10.1109/TPS.2006.883398

[37] Balcon N, Payan D, Belhaj M, Tondu T, Inguimbert V. Secondary electron emission on space materials: evaluation of the total secondary electron yield from surface potential measurements. IEEE Transactions on Plasma Science. 2012;**40**(2):282-290. DOI: 10.1109/TPS.2011.2172636

[38] Evans A, Dennison JR. The effect of surface modification on spacecraft charging parameters. IEEE Transactions on Plasma Science. 2012;**40**(2):305-310

[39] Wu J, Miyahara A, Khan A, Iwata M, Toyoda K, Cho M, et al. Effects of space environmental exposure on photoemission yield of polyimide. IEEE Transactions on Dielectrics and Electrical Insulation. 2015;**22**(2):1204-1212

[40] Engelhart DP, Plis E, Humagain S, Greenbaum S, Ferguson D, Cooper R, et al. Chemical and electrical dynamics of polyimide film damaged by electron radiation. IEEE Transactions on Plasma Science. 2017;**45**(9):2573-2577

[41] Dever JA, Messer R, Powers C, Townsend J, Wooldridge E. Effects of vacuum ultraviolet radiation on thin polyimide films. High Performance Polymers. 2001;**13**:391-399. PII: S0954-0083(01)26273-4

[42] Peng G, Hao W, Yang D, He S. Degradation of polyimide film under vacuum ultraviolet irradiation. Journal of Applied Polymer Science. 2004;**94**:1370-1374

[43] Cho M. Charging and discharge in vacuum and space. IEEE International Vacuum Electronics Conference. 2007. DOI: 10.1109/IVELEC.2007.4283197

[44] Anderson RA, Brainard JP. Flashover mechanism of pulsed surface flashover involving electron-stimulated desorption. Journal of Applied Physics. 1980;**51**:1414. DOI: 10.1063/1.327839

[45] Pillai AS, Reuben H. Surface flashover of solid insulators in atmospheric air and in vacuum. Journal

of Applied Physiology. 1985;**58**:146. DOI: 10.1063/1.335700

[46] Li G, Li S, Pan S, Min D. Effect of electron irradiation on dc surface flashover of polyimide in vacuum. IEEE Transactions on Dielectrics and Electrical Insulation. 2016;**23**(3):1846-1853. DOI: 10.1109/TDEI.2016.005429

[47] Liu C, Zheng Y, Yang P, Zheng X, Li Y, Bian E. The DC surface flashover performance research of polyimide under low-energy electron irradiation environment. IEEE Transactions on Plasma Science. 2016;**44**(1):85-92

[48] Wu J, Zhang Z, Ahang B, Ahang L, Zjeng X. Effects of degassing on DC surface flashover property of polyimide. In: IEEE 11th International Conference on the Properties and Applications of Dielectric Materials (ICPADM); 2015

[49] Griseri V, Levy L, Payan P, Maeno T, Fukunaga K, Laurent C. Space charge behaviour in electron irradiated polymers. In: Annual Report of IEEE International Conference on Electrical Insulation and Dielectric Phenomena. 2002. pp. 922-925. DOI: 10.1109/CEIDP.2002.1048946

[50] Griseri V, Fukunaga K, Maeno T, Laurent C, Payan D, Levy L. Assessment of measuring conditions with the pulse electro-acoustic system adapted to work under electronic irradiation. In: Annual Report of IEEE International Conference on Electrical Insulation and Dielectric Phenomena. 2003. pp. 20-23

[51] ESTAR&PESTAR Program Online. Available from: https://physics.nist.gov/PhysRefData/Star

[52] Horiguchi K, Kikuchi Y, Griseri V, Miyake H, Tanaka Y, Berquez L, et al. The relationship between charge decay process and current density differential polyimide and fluoride films irradiated by electrons. In: International Symposium on Electrical Insulation Materials. 2014. pp. 381-384

[53] Kishi Y, Miyake H, Tanaka Y, Takada T. Relationship between breakdown and space charge formation in polyimide film under dc high stress. In: Annual Report of IEEE International Conference on Electrical Insulation and Dielectric Phenomena. 2009. pp. 146-149

[54] Qin S, Tu Y, Wang S, Cheng Y, Chen B, Wang C, et al. Accelerated aging of fast thermal cycle effects on the behavior of space charge in polyimide. IEEE Transactions on Dielectrics and Electrical Insulation. 2017;**24**(52017):3182-3190. DOI: 10.1109/TDEI.2017.006483

[55] Sato S, Yanagisawa S, Tanaka Y, Watanabe R, Tomita N. Investigation of space charge behavior in polyimide film during elevating temperature. In: Annual Report of IEEE International Conference on Electrical Insulation and Dielectric Phenomena. 2007. pp. 57-60

[56] Takada T, Kitajima H, Kodaka M, Tanaka Y. Analysis of conduction current in electron-beam irradiated PMMA by simultaneous measurement of thermally stimulated current and space charge distribution. In: Annual Report of IEEE International Conference on Electrical Insulation and Dielectric Phenomena. 1997. pp. 463-466

[57] Numat S, Miyake H, Tanaka Y, Takada T. Dielectric characteristic evaluation of proton beam irradiated polyimide films. In: Annual Report of IEEE International Conference on Electrical Insulation and Dielectric Phenomena. 2010

[58] Miyake H, Uchiyama R, Tanaka Y. The relationship between charge accumulation and scission of molecular chain in the proton irradiated PI. In: IEEE International Conference on Solid Dielectrics. 2016. DOI: 10.1109/ICD.2016.7547562

Section 4

Polyimide in Power Conversion Applications

Chapter 6

Polyimide Films for Digital Isolators

Baoxing Chen and Sombel Diaham

Abstract

Digital isolators provide compelling benefits over legacy opto-couplers in terms of high speed, low power consumption, high reliability, small size, high integration, and ease of use. Billions of digital isolators using micro-transformers have been widely adopted in many markets including automotive, industry automation, medical, and energy. What are essential for the high voltage performance for these digital isolators are polyimide films deposited in between the top spiral winding and bottom spiral winding for the stacked winding transformers. In this chapter, digital isolator construction using polyimide films as isolation layers will be reviewed. To meet various safety standards such as UL and VDE, digital isolators need to satisfy various high voltage performances, such as short duration withstand voltage, surge voltage, and working voltage. Polyimide aging behavior under various high voltage waveforms such as AC or DC was studied, and isolator's working voltage is extrapolated through a polyimide lifetime model. Structural improvements to improve polyimide high voltage lifetime will also be discussed.

Keywords: polyimide films, digital isolators, high voltage, lifetime model, charge injection, barrier effect

1. Introduction

Isolation between circuit components is typically required for safety and/or data integrity considerations. For example, isolation protects sensitive circuit components and human interface on the system side from dangerous voltage levels present on the field side, where more robust components such as sensors and actuators reside. Isolation can also eliminate common-mode noises or ground loops that affect data acquisition accuracy. While opto-couplers are choices of isolation for many decades, they present significant limitations in terms of low speed, high power consumption, and limited reliability. Its low bandwidth and long propagation delay presented significant challenges in meeting the ever-increasing speed requirements for many isolated field bus communications such as RS485 in industry automation systems. Its high power consumption due to the need to lighting up LED puts significant constraint on overall system power budget in power limited industry systems such as process control 4–20 mA systems. As current transfer ratio for the opto-couplers degrades over time, especially at high temperatures, they fail to meet reliability for demanding applications such as automotive.

Digital isolators remove penalties associated with isolation, provide compelling advantages over opto-couplers in terms of high speed, low power consumption, high reliability, small size, high integration, and ease of use. Digital isolators using micro-transformers [1, 2] allow the integration of multiple transformers and with other necessary circuit functions. These stacked spirals used in digital isolators provide tight magnetic coupling between the top coil and bottom coil and very little coupling between spirals side by side. This enables multiple channel integration with little interference between the channels. The magnetic coupling between the top spiral and bottom spiral depends only on the size and separation, unlike the current transfer ratio for the opto-couplers, and does not degrade over time, which leads to the high reliability for these digital isolators based on transformers. These transformers have self-resonant frequency from a few hundred MHz to a few GHz and can be easily used to realize digital isolators from 150 to 600 Mbps. With high-quality factor well over 10 for these transformers, the power consumption for these digital isolators is orders of magnitude lower than those of opto-couplers.

Opto-couplers as shown in **Figure 1** rely on a few mm thick molding compound between the LED die and photodiode die to achieve isolation. For transformer-based digital isolators as shown in **Figure 2**, isolation performance is mainly limited by 20 to 40 μm thick polyimide layers sandwiched between the top and bottom coils of the chip-scale micro-transformers. In this chapter, we will review detailed construction of these isolators, the deposition methods for these polyimide films, characterization of the polyimide films, the high voltage performance, and the aging behavior for the digital isolators.

Polyimide was chosen as the insulating material for many reasons, including excellent breakdown strength, thermal and mechanical stability, chemical resistance, ESD performance, and relatively low permittivity. Besides good high voltage performance, polyimide has an excellent ESD performance and is capable of handling EOS and ESD events exceeding 15 kV [3]. During energy-limited ESD events, the polyimide polymer absorbs some of the charges to form stable radicals that interrupt the avalanche process and bleeds away some of the charge. Other dielectric materials such as oxides typically do not have this ESD tolerant characteristic and may go into avalanche once the ESD level exceeds the dielectric strength, even if the ESD energy is low. The polyimide also has high thermal stability, with a weight loss temperature over 500°C and a glass transition temperature above 260°C. The polyimide also has high mechanical stability with a tensile strength over 120 MPa and a high elastic elongation over 30%. In spite of its high elongation, polyimide does not deform easily, because the Young's Modulus is about 3.3 Gpa.

The polyimide has excellent chemical resistance, which is one reason it has been widely used for insulation coatings for high voltage cables. High chemical resistance also helps to facilitate IC processing on top of polyimide layers, such as the Au plating used to create *i*Coupler transformer coils. Lastly, the thick polyimide layers, with a dielectric constant of 3.3, work well with the small diameter Au transformer coils to minimize capacitance across the isolation barrier. Most Coupler products exhibit

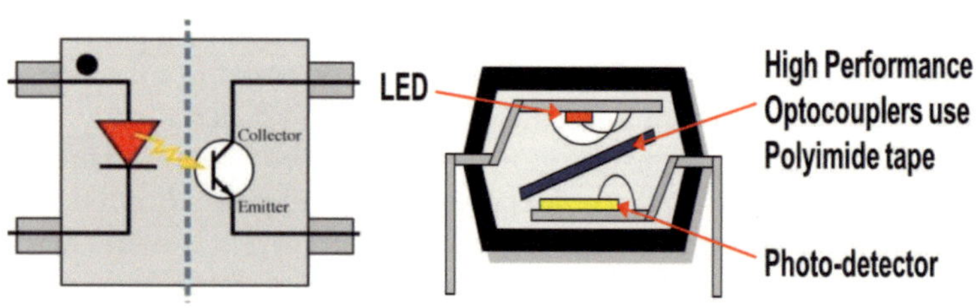

(a) (b)

Figure 1.
(a) Opto-coupler schematic. (b) Opto-coupler package cross section.

(a) (b)

Figure 2.
(a) Digital isolator in a plastic package. (b) Transformer cross section.

less than 2.5 pF capacitance between the input and output. Because of these characteristics, polyimide is increasingly used in microelectronics applications, and it is an excellent choice as insulating material for the *i*Coupler high voltage digital isolators.

2. Digital isolator construction and fabrication

There are three major components for a digital isolator, isolation barrier coupling element, insulation material, and signaling schemes through the isolation barrier. Insulation material is used for the isolation barrier to achieve certain isolation rating, and the isolation rating mainly depends on the dielectric strength and its thickness. There are two main types of dielectric materials, organic such as polyimide and inorganic such as silicon dioxide or silicon nitride. Oxide or nitride has an excellent dielectric strength of 700–1000 V/μm; however, it has an inherent high stress to prevent film thicker than 15–20 μm to be formed reliably on a large-scale modern IC wafer. The other limitation to organic films is that they are susceptible to ESD, and a tiny energy of voltage overstress will lead to catastrophic avalanche breakdown. Organic films such as polyimides consist of long C-H chains, and a small ESD event with limited energy may break some local C-H links without compromising material structural integrity, and they tend to be much more ESD tolerant. Polyimide does not compare favorably to oxide or nitride in terms of dielectric strength, around 400–700 V/μm; however, with inherent low film stress, much thicker polyimide layers as much as 40–60 μm can be formed economically. Thus, 30 μm polyimide films provide withstand voltages of 12–21 kV, comparable to 20 μm oxide with withstand voltages of 14–20 kV. For applications with robust ESD performance and high voltage withstand capability against impulse voltages, such as those present during lightning strikes, polyimide-based isolators provide the most robust choice.

Commercial polyimide films are available in photoresist forms that are deposited on wafers with well-controlled thicknesses and then easily patterned with standard photolithography processes. Here is the process flow as shown in **Figure 3** for the isolation transformers used for the digital isolators. A CMOS wafer with its top metal layer forming the bottom coil is spin-coated with the first photosensitive polyimide, and the polyimide layer is patterned through photolithography. The polyimide is then thermally cured to achieve high structural quality. Top coil layer is plated after which a second polyimide layer is coated, patterned, and cured to form the encapsulation for the top coil. Because deposited polyimide films are free of voids as shown in **Figure 4** and do not suffer from corona discharge, the transformer devices also exhibit good aging behavior and work well under continuous AC voltages and DC voltages.

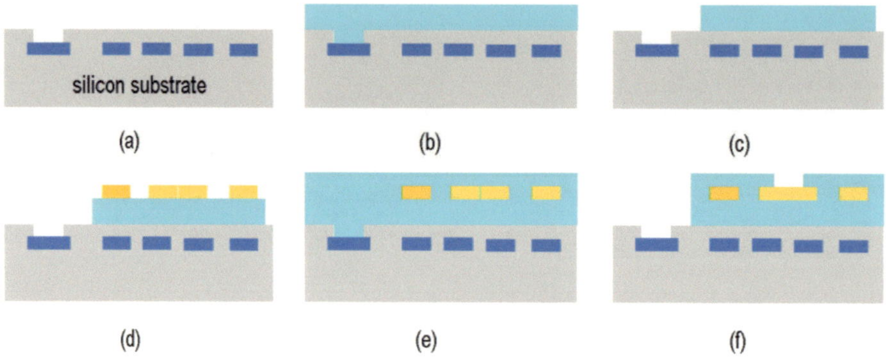

Figure 3.
Isolation transformer process flow. (a) CMOS substrate with top metal, (b) polyimide layer spin coated, (c) polyimide layer patterned & cured, (d) top coil plated, (e) 2nd polyimide layer spin coated and (f) 2nd polyimide layer patterned & cured.

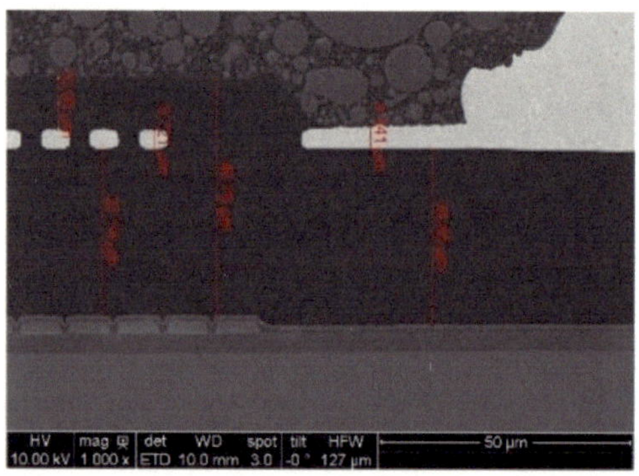

Figure 4.
Cross section for the fabricated isolation transformer.

3. High voltage performances

Isolation rating is defined by the maximum withstand voltage with 1 min duration per UL1577. For production test, the digital isolators were tested for 1 s at 120% of rated voltage. For example, for 2.5 kV_{rms}, 1 min-rated digital isolators, the production test is 3 kV_{rms} for 1 s. For practical applications, there are two important high voltage performance parameters; one is the maximum working voltage where the insulation needs to be intact over the lifetime of continuous operation, AC or DC. For example, per VDE0884-11, the lifetime for isolators with reinforced isolation at 120% of the rated voltage needs to be greater than 37.5 years at 1 ppm failure rate. As an example, if the rated working voltage for a reinforced digital isolator is 1 kV_{rms}, its lifetime at 1.2 kV_{rms} needs to be greater than 37.5 years at 1 ppm failure rate. Similarly, the lifetime for isolators with basic insulation at 120% of the rated voltage needs to be better than 26 years with 1000 ppm failure. The other important application specification is the maximum transient isolation voltage where the part needs to survive. Transient test waveforms may vary, and an example waveform per EN 60747-5-5 or IEC 61010-1 is shown in **Figure 5**. It rise time from 10–90% is about 1.2 μS, while the falling time from peak to 50% is 50 μS. This intends to simulate the lightening condition, so it is important for isolators to have robust surge performance to be robust in the field. ESD tolerance is an important attribute for semiconductor devices, and high surge performance also implies excellent ESD performance.

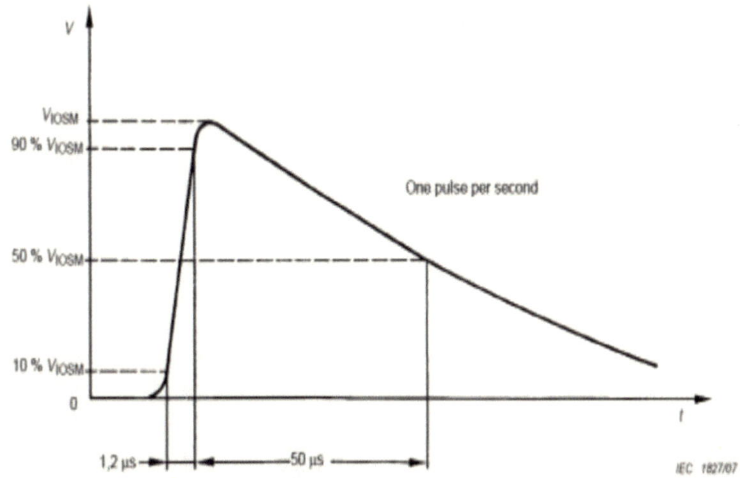

Figure 5.
IEC61010-1 Surge test waveform.

4. Polyimide films' characterization

Figure 6 shows the main intrinsic electrical properties of spin-coated polyimide films measured at wafer level. On the one hand, the DC bulk conductivity of polyimide shows very low values around 10^{-16} S/m over an applied electric field range up to 40 V/μm (i.e., ohmic range), but remaining quite low at least up to 150 V/μm. On

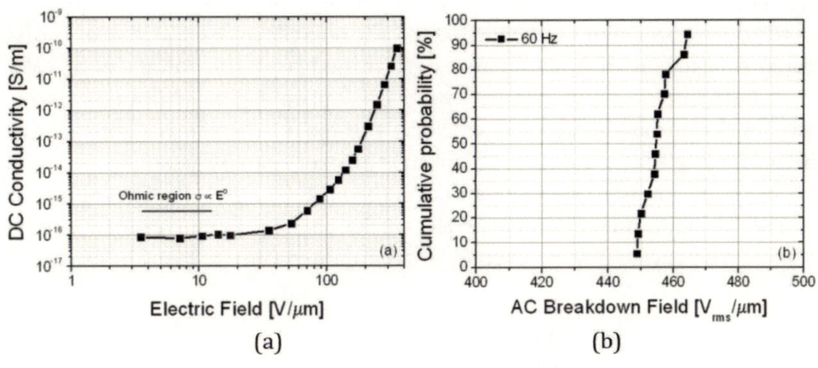

Figure 6.
Main intrinsic electrical properties of spin-coated polyimide films measured at wafer level: (a) DC conductivity versus electric field and (b) AC breakdown field distribution.

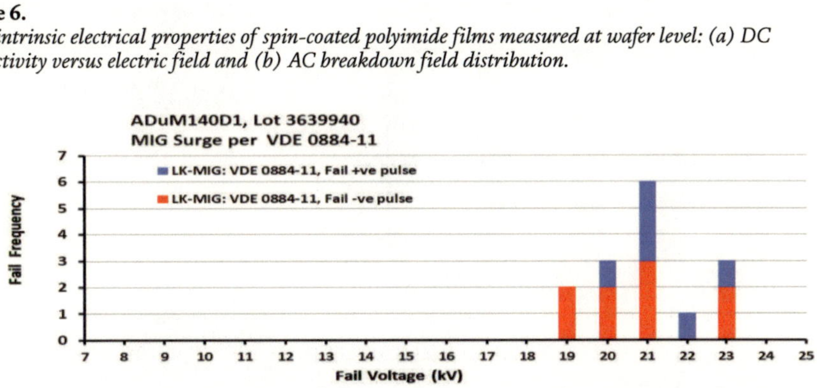

Figure 7.
Surge performance for isolators with polyimide films 30 μm thick.

the other hand, the AC breakdown field of the polyimide films exhibits a minimum value of 450 $V_{rms}/\mu m$ at 60 Hz. All these make spin-coated polyimide films very good insulating materials for reliable digital isolator application.

Figure 7 shows the surge performance for isolators with 30-μm thick polyimide films. As it can be seen, these isolators will pass surge tests up to 18 kV, and the first failure voltage is 19 kV for negative pulse and the first failure voltage is 20 kV for positive pulse.

5. Polyimide aging

Polyimide lifetime is studied through high voltage endurance test. Any insulator, given sufficient time and voltage, will break down. An example setup is shown in **Figure 8**. Multiple parts are connected electrically in parallel, and multiple groups of parts are stressed in different high voltages from high voltage power supplies, and a switch/measurement unit such as Agilent 34,980 together with a PC can be used to monitor the time the number of units have broken down. This can be a time-consuming process where it can take days to months for the units to break down.

The distributions for the time to failure can be analyzed through Weibull plots as shown in **Figure 9**. Groups of 16 parts were stressed at six different voltages, where each group forms a fairly decent Weibull distribution. Through Weibull plots, mean time to failure (MTTF) or time to failure at certain failure rates such as 1 ppm can be estimated. Obviously, time to failure at high voltages takes much less time compared to that at low voltages. Per VDE0884-11, the smallest to the largest MTTFs need to span at least two orders of magnitude and at the lowest test voltage, the 63% time to failure needs to be longer than 10^7 s or about 116 days. As it can be seen from **Figure 9**, the data sets generated at these six voltages meet these requirements.

To extrapolate working voltage, time to failure is plotted against stress voltages. For basic insulation, working voltage is determined from the voltage with 20% derating where time to failure or lifetime at 1000 ppm is greater than 24 years. Similarly for reinforced insulation, working voltage is determined from the voltage with 20% derating where lifetime at 1 ppm is greater than 30 years.

The dominant breakdown mechanism is through charge injections as a result of the direct electron impact from the electrodes to the polyimide surface regions. The breakdown process begins as charges are injected into polyimide surface under

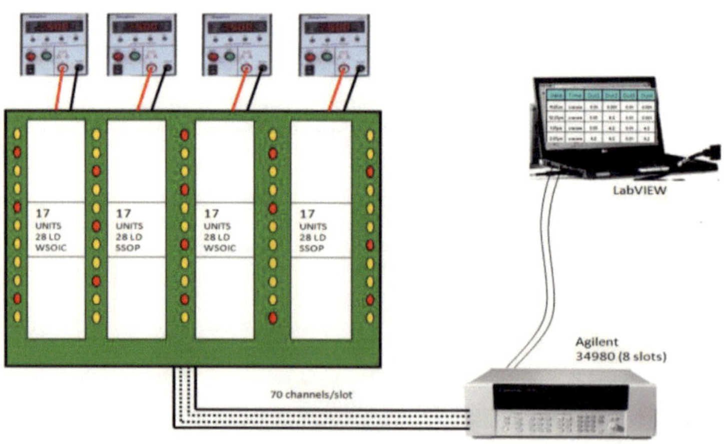

Figure 8.
Experimental setup for high voltage endurance test.

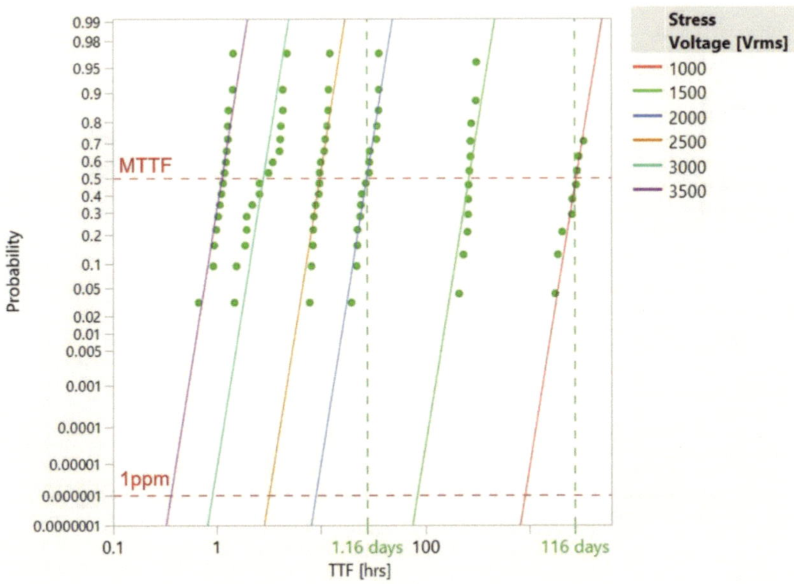

Figure 9.
Weibull distribution for isolators with 20-μm thick polyimide.

HV_{ac} conditions. The charges can become trapped in some local trapping sites at the surface. Once trapped, energy will be released, which will cause local mechanical tension because of stored electrostatic energy. Through quantum activation process, this tension will eventually cause local free volumes, voids or micro-cracks, which act as more local trapping sites. If the HV_{ac} remains long enough, this process will lead to the continued degradation of insulation and eventually electrical punch-through.

Through thermodynamic analysis, the lifetime, L [4], can be expressed as Eq. (1),

$$L \sim \frac{e^{-(E-E_t)^n}}{(E-E_t)^m} \tag{1}$$

where E_t is the threshold field where no charge injection will happen, and m, n are scaling constants.

The HV_{ac} endurance data of *i*Coupler devices were analyzed according to the procedure specified by ANSI/IEEE Std 930-1987, the "IEEE Guide for the Statistical Analysis of Electrical Insulation Voltage Endurance Data," and they are observed to follow:

$$L \sim e^{V^{-n}} \tag{2}$$

This phenomenological fit as shown in Eq. (2) was used to get worst-case lifetime because it assumes no threshold field as specified by the thermodynamic model. The duration of the HV test becomes prohibitively long if we try to measure the threshold field. Eq. (2) was used to model the time to failure for **Figure 10**. As you can see, the model fits the data rather well.

We also observed that the lifetime of *i*Coupler devices under DC or unipolar AC is much longer compared to that under bipolar AC; it is at least two orders of magnitude higher. For unipolar waveforms, the trapped charges tend to form an

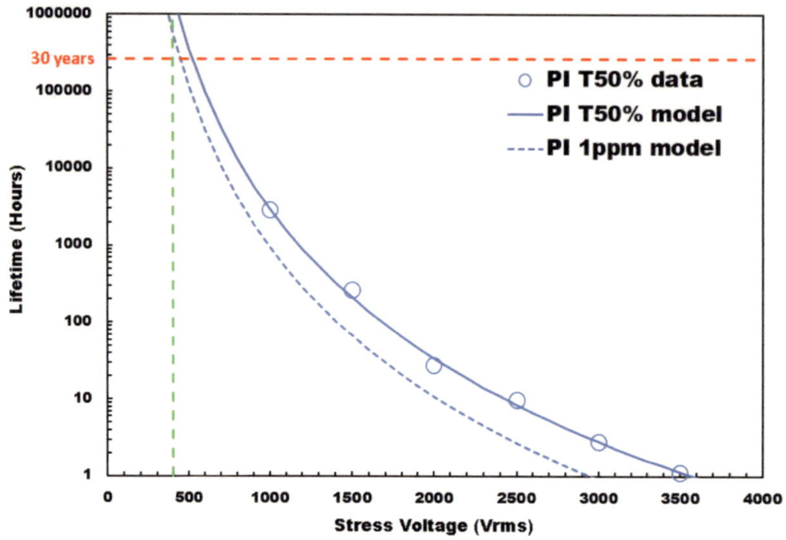

Figure 10.
Time-to-failure plot for isolators with 20-μm thick polyimide.

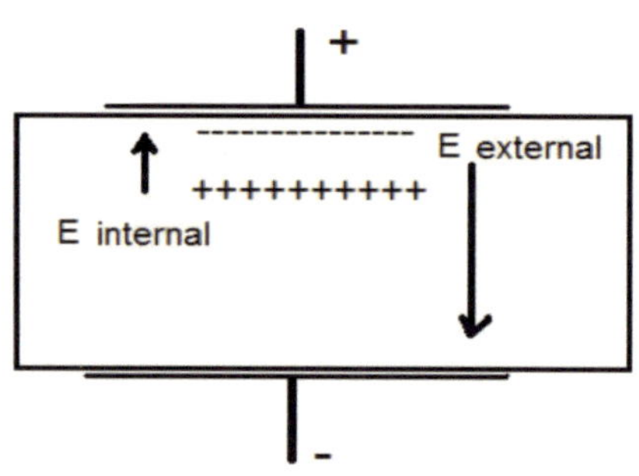

Figure 11.
Field barrier region with zero net e-field formed by the trapped charges.

internal field barrier region (i.e., homocharge) around the electrodes that prevents further injection of charge into the polyimide as shown in **Figure 11**. With a bipolar AC waveform, the reverse of field will prevent formation of this steady field barrier, and the trapped regions will keep progressing into the polyimide and eventually lead to the electrical breakdown. SiO_2, on the other hand, tends to give worse lifetime for DC or unipolar AC, especially for thick films [5].

The lifetime as shown in **Figure 10** is based on worst-case bipolar AC waveforms. HV lifetime is even greater for unipolar AC or DC waveforms. It should be noted that the models described in this chapter relate to polyimide insulation and have no bearing on isolators that use SiO_2 insulators as the primary means for isolation. Likewise, models that predict the HV lifetime of SiO_2-based digital isolators have no bearing on polyimide-based isolation systems.

Figure 12 shows how lifetime for unipolar is compared to that of bipolar for polyimide films. As it can be seen, the peak stress voltage for unipolar is about twice that of the peak stress voltage for AC bipolar for the same time to failure. In essence, the lifetime is dependent on peak to peak rather than the peak stress voltage for the polyimide films.

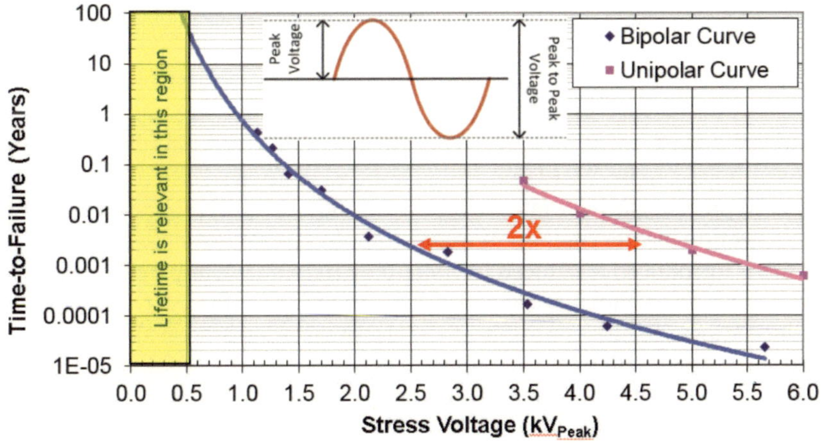

Figure 12.
Time-to-failure comparisons for AC bipolar versus unipolar.

6. Polyimide structural improvements

To improve high voltage endurance for the polyimide, a charge injection barrier can be used as shown in **Figure 13** [6, 7]. The charge injection barrier is preferred to be oxide or nitride with a large bandgap and high dielectric constant. High dielectric constant will help to reduce the electric field close to the electrode, while the large bandgap raises the energy barrier for charge injection.

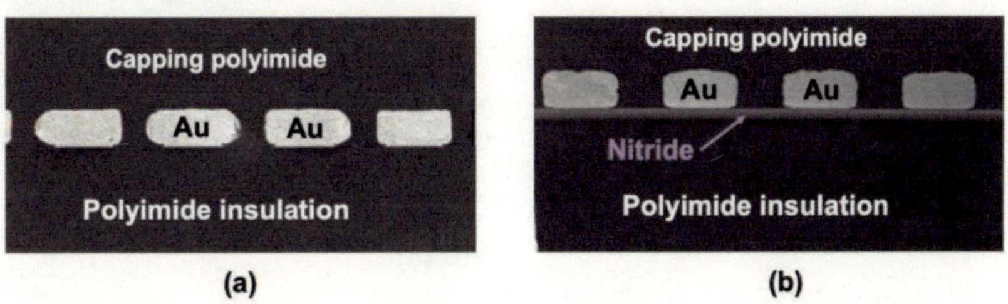

Figure 13.
Isolation transformer without (a) and with SiN charge injection barrier (b).

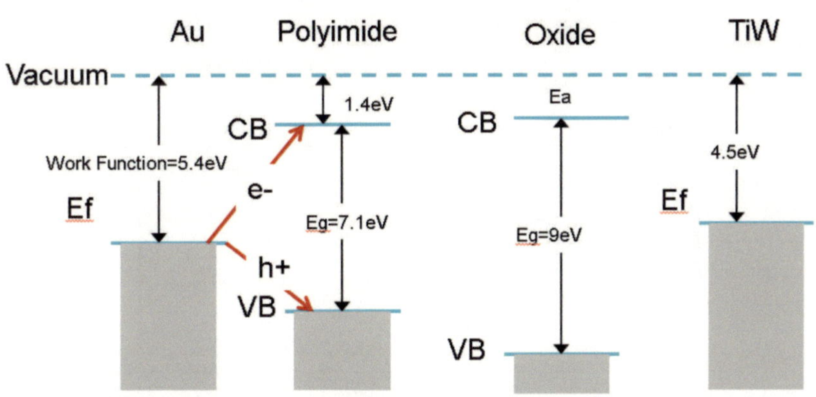

Figure 14.
Band diagram for charge injection.

To analyze the charge injection for a given isolation system, a band diagram can be drawn as shown in **Figure 14**. Four key materials in the isolation system shown in **Figure 12** are Au, the top coil material; polyimide, the isolation material between the top coil and bottom coil; oxide, the charge injection barrier; and TiW, the seed layer under the Au. Charge injection from Au or TiW into polyimide or oxide for electrons or holes can be calculated from the band diagram.

Figure 15 presents the charging currents over time for polyimide and polyimide with SiN injection barriers measured under 1000 V. The steady-state current when the SiN barrier is introduced is reduced by more than five times compared to that of polyimide only. This highlights a significant reduction of the charge injection processes that are well known to be responsible of the electrical aging at high electric field.

Figure 16 presents the time-to-failure (HVE tests) versus AC applied voltage from 1 kV_{rms} up to 3.5 kV_{rms} at 60 Hz for isolators with polyimide and polyimide/SiN barriers single die configurations. The lifetime at 50% and the extrapolation at 1 ppm of the data set are presented. Moreover, for both cases, the extrapolated working voltages at 30-year lifetime are reported. Digital isolator devices with polyimide insulation exhibit a 400 V_{rms} working voltage, while the improved design involving SiN injection barriers shows >900 V_{rms} working volatge at 1 ppm (750 V_{rms} after

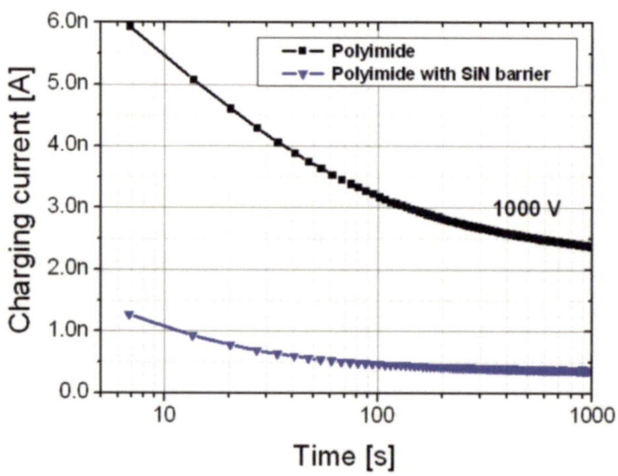

Figure 15.
Charging currents comparison for polyimide and polyimide with SiN injection barrier under 1 kV.

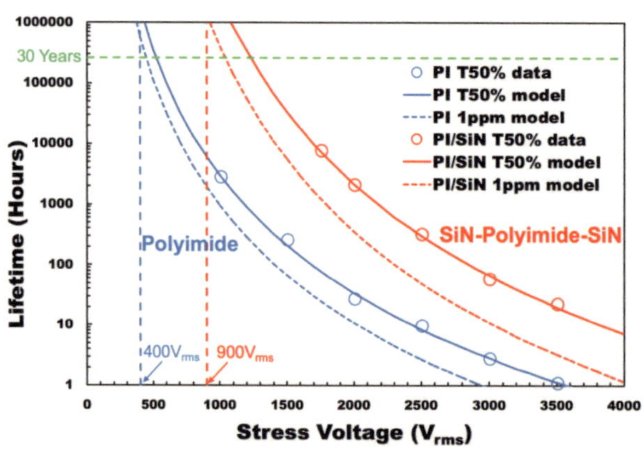

Figure 16.
Time-to-failure comparison for polyimide isolators with and without SiN charge injection barrier.

20% voltage derating). Based on wafer-level analysis comparison, it is reasonable to attribute the lifetime and working voltage improvements to the SiN injection barriers between polyimide and metallic coils. These SiN thin layers, by mitigating bipolar charge injection at the onset of space charge formation, reduce the electrical current, the related thermal effects, and, very likely, extend the lifetime for a given voltage.

7. Conclusion

Polyimide films have excellent high voltage performance from surge voltage to its high voltage endurance. These films have been characterized, and the aging behavior can be further enhanced through a charge injection barrier with large dielectric constant and large bandgap. These polyimide materials are excellent candidates as isolation barriers for digital isolators.

Acknowledgements

The authors would like to thank members of the isolation team at Analog Devices for contribution and the European Union through the Marie Skłodowska-Curie Action (MSCA-IF, H2020 program) for financial funding participation in the frame of the PRISME project (grant N°846455, 2019-2021).

Author details

Baoxing Chen[1*] and Sombel Diaham[2,3]

1 Analog Devices Inc., Wilmington, MA, USA

2 University of Toulouse, LAPLACE Research Institute, CNRS, Toulouse, France

3 Analog Devices International, Limerick, Ireland

*Address all correspondence to: baoxing.chen@analog.com

References

[1] Chen B. iCoupler® products with isoPower™ technology: Signal and power transfer across isolation barrier using microtransformers. Technial Paper. 2006. Available from: www.analog.com

[2] Chen B, Wynne J, Kliger R. High speed digital isolators using microscale on-chip transformers. Elektronik Magazine; 2003. Available from: www.elektroniknet.de

[3] Analog Devices. Surging across the barrier: Digital isolators set the standard for reinforced insulation. Technical Article MS-2341. 2012. Available from: www.analog.com

[4] Dissado LA, Mazzanti G, Montanari G. The incorporation of space charge degradation in the life model for electrical insulating materials. IEEE Transactions on Dielectrics and Electrical Insulation. 1995;**2**(6):1147-1158

[5] Chaparala P, Suehle J, Messick C, Roush M. Electric field dependent dielectric breakdown of intrinsic SiO_2 films under dynamic stress. In: IEEE 1995 International Integrated Reliability Workshop Final Report. 1995

[6] McLoughlin C et al. Isolator and method of forming an isolator. US Patent, No. 9,941,565. 2018. Available from: www.uspto.gov

[7] Diaham S, O'Sullivan L, Ceccarelli E, Lambkin P, O'Malley J, Fitzgibbon J, et al. Improving polyimide isolation performance by tailoring interfaces with nitride layers for digital isolator application. In: IEEE ICD Conference Proceedings; 2020. Available from: www.ieeexplore.ieee.org

Chapter 7

Behavior of Space Charge in Polyimide and the Influence on Power Semiconductor Device Reliability

Kunihiko Tajiri, Hirotaka Muto, Didier Marty-Dessus, Laurent Berquez, Gilbert Teyssedre, Marie-Laure Locatelli, Sombel Diaham, Virginie Griseri and Flora Carrasco

Abstract

Polyimide is widely used in film form as a passivation material for power semiconductor devices such as Si, SiC, and GaN. The magnitude of the electric field at the edge termination area of these semiconductor devices is becoming higher due to the increase of operational voltage and/or demand for shrinking the edge termination area to increase device active area. Hence, it is concerned that the accumulation of space charge in the encapsulation and passivation material may affect the insulation performance of these devices, for example, the degradation of withstand voltage due to distortion of the internal electric field caused by space charge accumulation. To design space charge resistance of semiconductor devices, it is important to understand the space charge behavior in polyimide films with a thickness of several to several tens of micrometers. This chapter addresses practical implementation, specifications, and issues on space charge in polyimide insulation on power semiconductor devices focusing on the space charge measurements in thin polyimide films using the latest developed LIMM method and DC conductivity measurements.

Keywords: power electronics and devices, power modules, power semiconductor device, passivation material, space charge, LIMM, DC conductivity

1. Introduction

In recent years, high voltage electronics and power electronics applications have emerged needing the use of power semiconductor devices with widely used Si and wide bandgap materials such as SiC and GaN. In these devices, thin polymer material has been widely used as passivation coating to protect device surfaces. Particularly, polyimide (PI) is of great interest due to its excellent thermal and electrical properties and its easy processing. Some of the most important applications of these material films are as inter-level dielectric insulators and as electronic device surface passivation [1]. The typical image of PI layer at the edge surface of semiconductor chip is shown in **Figure 1**.

IntechOpen

(a)

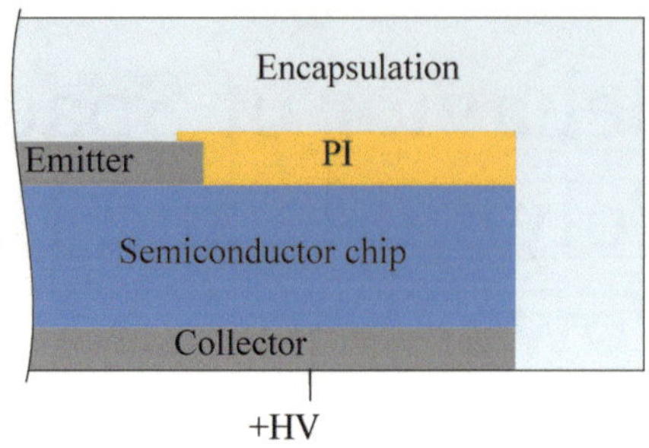

(b)

Figure 1.
Power module and edge termination structure of semiconductor chip. (a) HV IGBT power module. (b) Edge termination structure of semiconductor chip

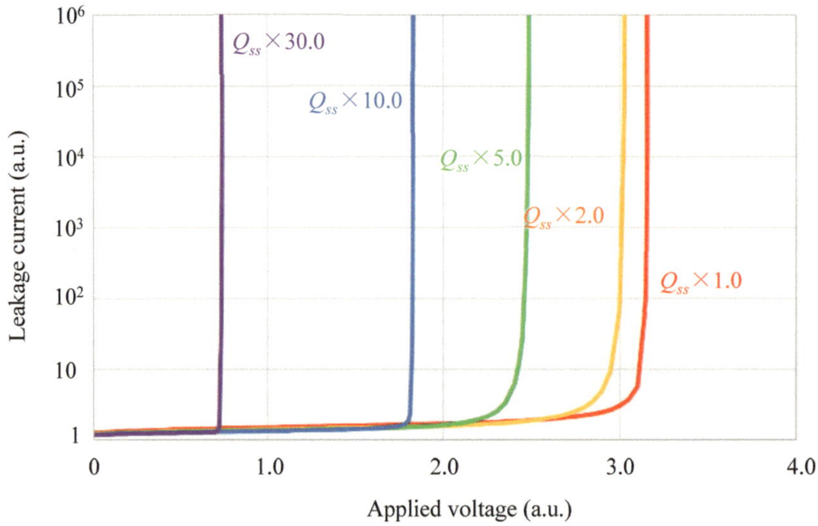

Figure 2.
Avalanche voltage simulation results of HVIGBT chip for different Q_{ss}.

Space charge is generally reported as a triggering mechanism for the degradation of insulators [2] and considered as a cause of the decrease of electrical breakdown voltage of those polymer films. Furthermore, the space charge could degrade semiconductor leakage current characteristics by strengthening the electric field in semiconductor substrate region. The influence of space charge transportation and accumulation on surface of semiconductor chip has been discussed in [3–6] using TCAD simulation. Reliability tests of IGBT devices under accelerated conditions were reported in [7, 8], showing that the degradation (increase) of leakage current of the device is caused possibly by the formation of space charge on the surface at semiconductor edge termination area. The decrease of withstand voltage of HV IGBT due to accumulated space charge (Q_{ss}) on the surface of device edge termination area has been demonstrated by TCAD simulation as shown in **Figure 2**. Attempts have also been made to evaluate the space charge resistance of a real chip by evaluating the withstand voltage of a semiconductor chip using a guard probe electrode to simulate an external electric field due to space charge as shown in **Figure 3** [9].

To evaluate the actual influence by the space charge accumulation, it is necessary to clarify the space charge distribution around the edge termination area of

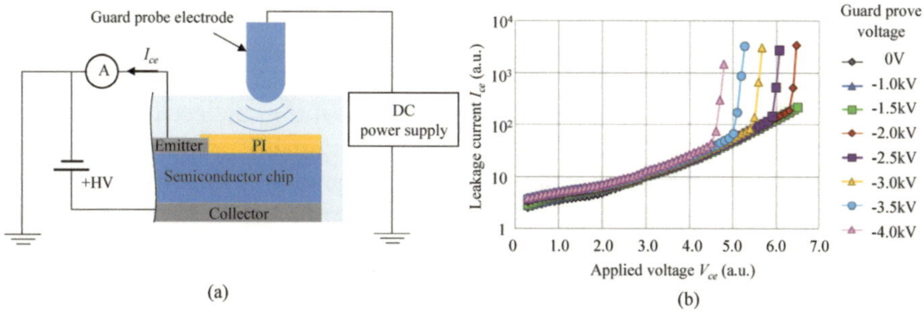

Figure 3.
(a) Schematic view of guard probe (GP) method and (b) Leakage current dependent on the guard prove voltage.

semiconductor chip. Many trials have been performed to measure the space charge characteristics in polyimide and encapsulation material such as silicone gel as described below.

Earlier reports featuring space charge in PI [10–12] reported the processes indirectly, for example, through charging current measurements or directly as with pulsed electroacoustic (PEA) method with relatively thick films (125 μm) [10]. In [10], the effect of humidity in air and the difference due to electrode material (Al and Au) on space charge formation with time evolution are discussed. Recently, the Laser Intensity Modulation Method (LIMM) [13] has been performed to investigate space charge characteristics in thin PI films with few micrometers in thickness under the DC field up to 125 kV/mm, close to breakdown voltage [14–17]. In [17], the space charge characteristics were reported under DC voltage of from low field of 2.5 to 125 kV/mm with the correlation with DC conductivity trends. Considering the encapsulation material, some studies have been performed through charge accumulation measurement in silicone gel by using PEA [18, 19] and LIPP (Laser-Induced Pressure Pulse) [20] methods, and space charge characteristics of only surface information and total amount of charge have been measured in [18–20], respectively.

As described, many attempts have been made to evaluate the influence of space charge, and quantitative evaluation has been under development, especially in thin PI. However, the space charge distribution in the PI film, which is important for providing space charge tolerance of semiconductor chip, is gradually becoming apparent along with the influence of the operating environment. In this chapter, the latest developments with space charge measurements in thin polyimide films using the LIMM method, with the focus on local field strengthening and correlation with conductivity measurements, are discussed.

2. Space charge evaluation methods

2.1 PEA method

Since the 1980s, the PEA method has been used to measure the space charge of sheet or film samples in order to investigate space charge phenomena and evaluate the aging of dielectrics. A space charge distribution by converting a pressure wave generated by applying a pulse electric field to a sample under high voltage application and converting the pressure wave into an electric signal using a piezoelectric element. The thickness of the sample that can be evaluated by PEA is more than 100 μm, and its spatial resolution is about several tens of micrometers.

2.2 (F)LIMM method

The (Focused) Laser Intensity Modulation Method or (F)LIMM is a thermal wave method dedicated to the space charge analysis of thin dielectric films (with a thickness from 5 to 50 μm) [21, 22]. This method is originally proposed by Lang [23] in the 1980s, and the characterization of space charge distribution had been performed under volt-off after external DC voltage application. Recent development has been reported that LIMM measurement is carried out under volt-on [24]. The LIMM is much suitable to clarify the space charge characteristics in a polymer film with only several micrometer thickness, which is a typical thickness of passivation layer on a semiconductor chip. **Figure 4** shows the schematic diagram of online LIMM. A DC potential (V_{ht}) is applied to the top electrode of the measuring cell through a low-pass filter. (F)LIMM currents are recorded after pre-amplification and extracted from noise by a lock-in amplifier. To protect a damage from an eventual breakdown of the sample, an electrical protection device was introduced at the output of the measuring cell.

The thermal gradient created by the laser beam induces periodical and local expansions that lead to relative charge displacement regarding to the electrodes within the irradiated volume. Varying the laser beam modulation frequency, one can control the depth of thermal diffusion and then calculate its effect on the total current signal. Finally, a mathematical treatment allows charge density profile reconstruction in the direction of the sample thickness. A modulated laser beam of a frequency f heating a surface S of a top electrode sputtered on a dielectric nonpolar sample of thickness L, to which an additional DC voltage V_{ht} is applied, the fundamental (F)LIMM equation for the complex current $I(f)$ can be expressed by [24]:

$$I(f) = -j\frac{2\pi f}{L}(\alpha_z - \alpha_\varepsilon)\varepsilon S\left[\int_0^L E_i(z)T(z,f)dz + \frac{V_{ht}}{L}\int_0^L T(z,f)dz\right] \quad (1)$$

where z is the direction normal to the sample, α_z is the coefficient of thermal expansion (K^{-1}), αε is the coefficient of thermal dependence of the dielectric permittivity (K^{-1}), ε is the permittivity, $E_i(z)$ is the internal electric field along z-axis (V.m^{-1}), and $T(z,f)$ is the simulated spatial variation of the temperature versus frequency.

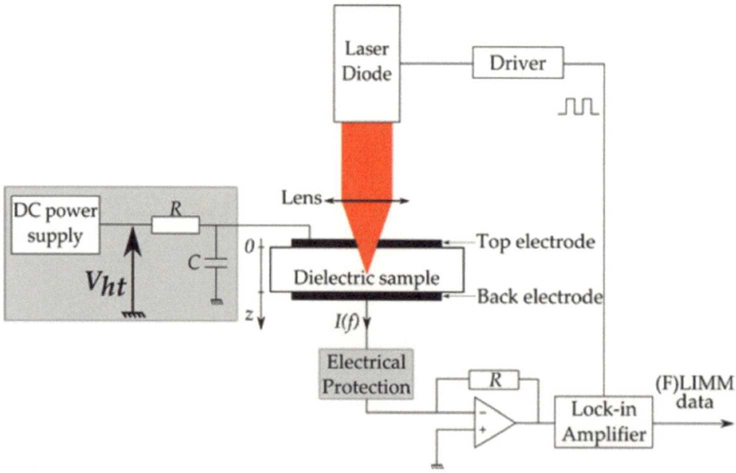

Figure 4.
Schematic diagram of (F)LIMM [24].

3. Space charge behavior in polyimide film

3.1 Space charge distribution in polyimide film

In this subsection, the LIMM measurement results of PI film in a typical case using cumulative DC voltage application. **Figure 5** shows the sample structure and the example of LIMM measurement protocol. The thickness of PI film is 20 μm. The polarization steps consist in applying a DC electric field from 25 to 125 kV/mm with positive or negative polarity. The corresponding positive or negative DC voltages are applied on the top electrode of the sample in air at room temperature, whereas the Si substrate is grounded. After each polarization step, a depolarization process is performed by short-circuiting the sample. A cumulative protocol is applied for successive voltage steps on the sample as shown in the right side of **Figure 5**. Volt-on and volt-off steps of each 60 min are applied on the sample from 25 to 125 kV/mm, with the steps of 25 kV/mm. LIMM measurements are performed throughout the whole protocol, with every 10 min scanning of 63 frequency points from 10 Hz to 10 kHz.

Figure 6 shows LIMM current waveform under the field of 25, 50, 75, 100, and 125 kV/mm, positive polarity. The space charge and internal electric field

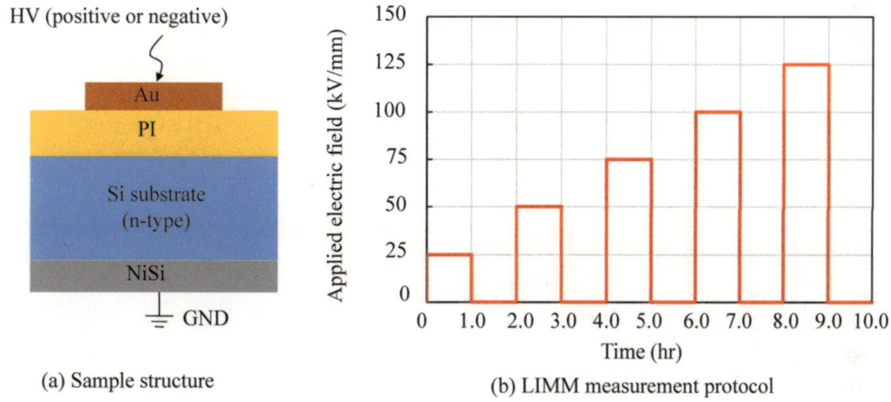

Figure 5.
Typical sample structure and LIMM measurement protocol (cumulative voltage application of 25–125 kV/mm, positive polarity).

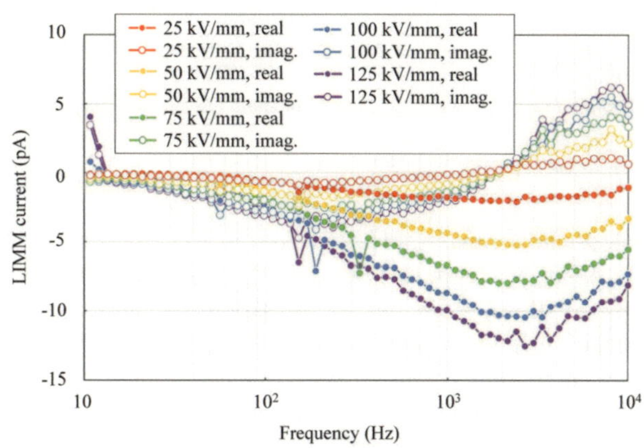

Figure 6.
Example of LIMM current waveform (25–125 kV/mm, positive polarity).

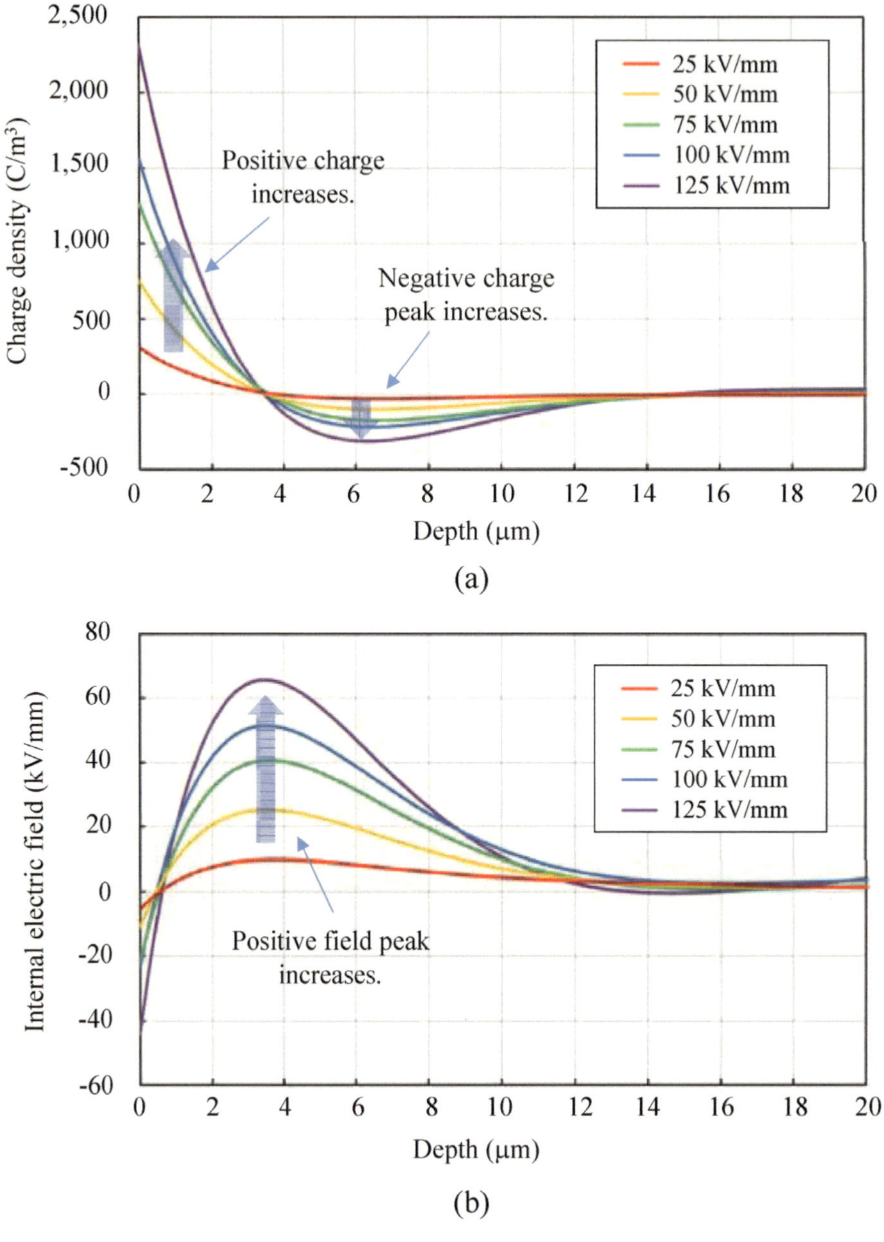

Figure 7.
Example of the depth profiles of (a) space charge and (b) internal electric field (25–125 kV/mm, positive polarity).

distributions are deduced from Eq. (1) as shown in **Figure 7(a)** and **(b)**, respectively. In this case, positive and negative space charges are confirmed close to and at the depth of from 4 to 10 μm from top Au electrode, respectively, which causes the enhancement of positive field strengthening at 2–6 μm depth. The maximum values of these are confirmed to increase along with the applied DC field. Time evolution of maximum peaks of enhanced electric field in both polarities is shown in **Figure 8**. As mentioned so far in this subsection, the space charge and enhanced electric field characteristics in polyimide under external DC voltage on and off can be evaluated by using LIMM method including both depth-profile and time evolution.

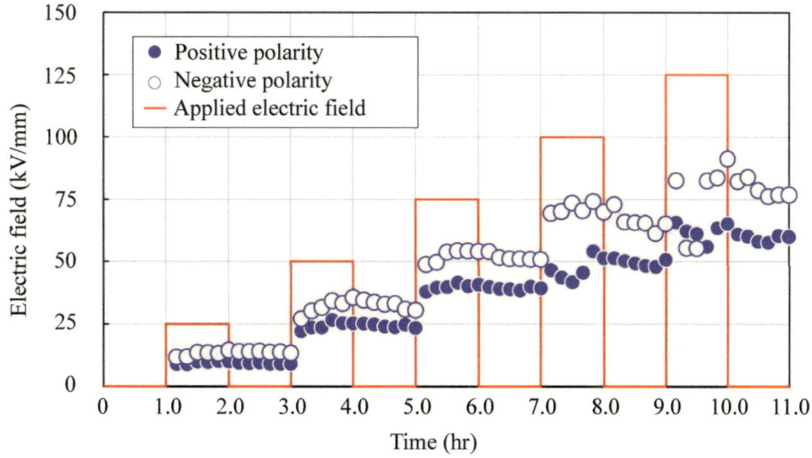

Figure 8.
Magnitude of positive peaks of internal electric field.

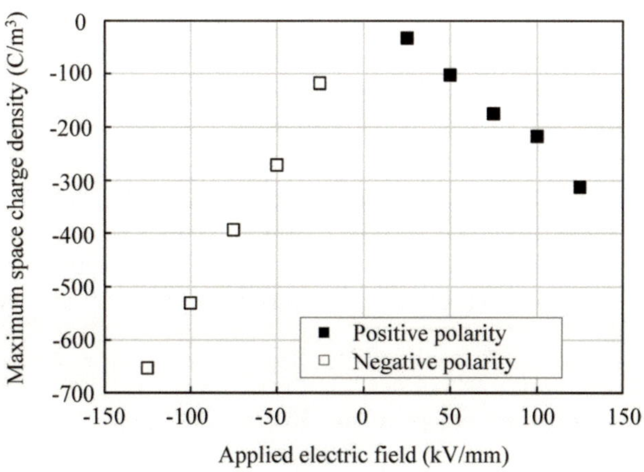

Figure 9.
Maximum negative space charge density.

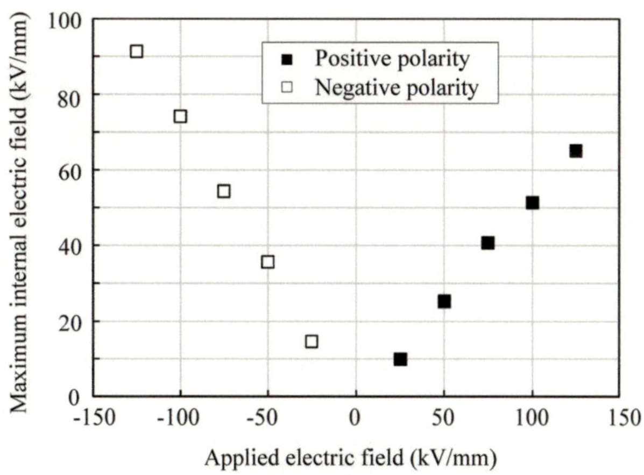

Figure 10.
Maximum positive internal electric field.

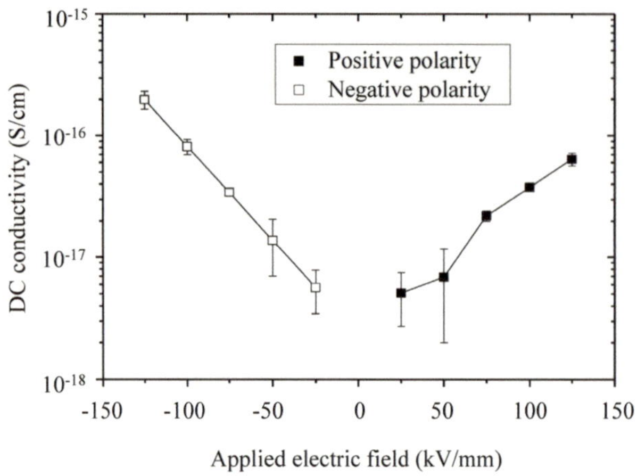

Figure 11.
DC conductivity.

3.2 DC conductivity

To clarify the influence of space charge accumulation to the conductivity behavior, space charge density and internal electric field are compared with DC conductivity. **Figures 9–11** show the maximum space charge density of negative charge, maximum positive internal electric field, and DC conductivity, respectively, under the applied field from 25 to 125 kV/mm for both polarities. To compare these figures, no remarkable influence on the conductivity by space charge accumulation inside polyimide at this level works. To establish a direct link between space charge dynamics and conduction phenomena, further evaluations are still necessary.

4. Effect of space charge on semiconductor devices

As described in Section 1, the degradation (increase) of leakage current of the device is caused possibly by the formation of space charge on the surface at

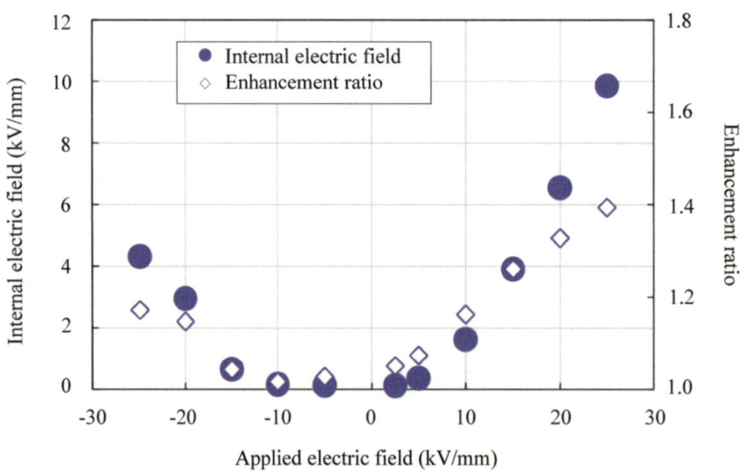

Figure 12.
Peak value of internal electric field linked to field strengthening (filled circle) and enhancement ratio of those peaks (open square) versus applied electric field.

semiconductor edge termination area. **Figure 12** shows that the electric field enhancement ratio defined in Eq. (2) confirmed in the LIMM measurement depending on the applied DC field of from −25 to 25 kV/mm. It is confirmed that the enhancement ratio in PI film is negligibly small (less than 1.1) for applied fields less than 10 kV/mm. However, it can reach much higher like 1.4 at higher applied field like 25 kV/mm.

$$Enhancement\ ratio = \frac{Applied\ electric\ field + Measured\ electric\ field}{Applied\ electric\ field} \quad (2)$$

This result implies that electric field at edge termination area should be carefully designed under such an operational electric field as is the case of SiC devices to prevent the influence of space charge accumulation. It should be noted that not only inside polyimide, but also space charge accumulations in encapsulation material and the interface between polyimide and encapsulation as its structure shown in **Figure 1** are also needed to be clarified. Moreover, the influence of environmental conditions such as temperature and humidity on the space charge behavior should be taken into account.

5. Conclusion

It has been presented that the space charge distribution can be clarified in thin polyimide film using LIMM method from relatively low field of 2.5 kV/mm up to 125 kV/mm, which is close to breakdown field. The electric field enhancement due to space charge accumulation is shown to be 1.4 times at 25 kV/mm under room temperature condition. To design much reliable structure against space charge accumulation in polyimide and encapsulation material, its distributions are needed to be clarified by the further study.

Acknowledgements

The authors would like to express our gratitude to all the people who have involved in this work, especially, Dr. Thierry Lebey, who is former Director of LAPLACE for his tremendous support during the collaborative research about LIMM evaluation.

Author details

Kunihiko Tajiri[1]*, Hirotaka Muto[1], Didier Marty-Dessus[2], Laurent Berquez[2], Gilbert Teyssedre[2], Marie-Laure Locatelli[2], Sombel Diaham[2], Virginie Griseri[2] and Flora Carrasco[2]

1 Mitsubishi Electric Corporation, Amagasaki, Hyogo, Japan

2 LAPLACE, Université de Toulouse, CNRS, INPT, UPS, Toulouse, France

*Address all correspondence to: tajiri.kunihiko@ab.mitsubishielectric.co.jp

References

[1] Zelmat S, Locatelli M-L, Lebey T, Diaham S. Investigations on high temperature polyimide potentialities for silicon carbide power device passivation. Microelectronic Engineering. 2006;**83**:51-54. DOI: 10.1016/j.mee.2005.10.050

[2] Mazzanti G, Montanari GC, Dissado LA. Electrical aging and life models: The role of space charge. IEEE Transactions on Dielectrics and Electrical Insulation. 2005;**12**:876-890. DOI: 10.1109/TDEI.2005.1522183

[3] Rahimo M, Richter F, Fischer F, Vemulapati U, Kopta A, Corvasce C, et al. The impact on power semiconductor device operation due to local electric field alterations in the planer junction termination. Microelectronics Reliability. 2016;**58**:51-57. DOI: 10.1016/j.microrel.2015.12.021

[4] Imperiale I, Reggiani S, Gnani E, Gnudi A, Baccarani G, Nguyen L, Denison M. TCAD modeling of charge transport in HV-IC encapsulation materials. In: Proceedings of the 26th International Symposium on Power Semiconductor Devices & IC's; Waikoloa, Hawaii, USA; 15–19 June, 2014. DOI: 10.1109/ISPSD.2014.6856073

[5] Imperiale I, Reggiani S, Gnani E, Gnudi A, Baccarani G, Nguyen L, Denison M. TCAD modeling of encapsulation layer in high-voltage, high-temperature operation regime. In: Proceedings of 44th European Solid State Device Research Conference (ESSDERC 2014); Venice, Italy; 22–26 September 2014. DOI: 10.1109/ESSDERC.2014.6948826

[6] Imperiale I, Imperiale I, Reggiani S, Gnani E, Gnudi A, Baccarani G, et al. Role of encapsulation formulation on charge transport phenomena and HV device instability. In: Proceedings of IEEE 65th Electronic Components and Technology Conference (ECTC 2015); San Diego, CA, USA; 26–29 May 2015. DOI: 10.1109/ECTC.2015.7159586

[7] Tanaka N, Ota K, Iura S, Kusakabe T, Nakamura K, Wiesner E, et al. Robust HVIGBT module design against high humidity In: Proceedings of Power Conversion and Intelligent Motion (PCIM) Europe 2015; Nuremberg, Germany; 19–21 May 2015. pp. 368-373. ISBN: 978-3-8007-3924-0

[8] Kitajima Y, Hatori K, Iura S, Nakamura K, Kusakabe Y, Kurachi K, et al. Lifetime estimation model of HVIGBT considering humidity. In: Proceedings of Power Conversion and Intelligent Motion (PCIM) Europe 2017; Nuremberg, Germany; 2017. pp. 353-358. ISBN: 978-3-8007-4424-4

[9] Yamatake A, Shiota H, Muto H, Tsurimoto T. New technology for evaluating insulation reliability of termination of power semiconductor chip. In: Proceedings of the IEEJ Transactions on Electrical and Electronic Engineering 2017; Muroran, Hokkaido, Japan; 19–20 September 2017. p. 157

[10] Kaneko K, Shiomi R, Suzuoki Y, Mizutani T, Shiota H, Muto H. Effects of electrode material on injection and space charge formation in polyimide. In: Proceedings of IEEE International Conference on Solid Dielectrics; 2007. pp. 518-521. DOI: 10.1109/ICSD.2007.4290865

[11] Sessler GM, Hahn B, Yoon DY. Electrical conduction in polyimide films. Journal of Applied Physics. 1986;**60**: 318-326. DOI: 10.1063/1.337646

[12] Liang T, Makita Y, Kimura S. Effect of film thickness on the electrical properties of polyimide thin films. Polymer. 2001;**42**:4867-4872. DOI: 10.1016/S0032-3861(00)00881-8

[13] Petre A, Marty-Dessus D, Berquez L, Franiceschi J-L. A comparison of different mathematical treatments for solving the inverse problem in focused laser intensity modulation method. Japanese Journal of Applied Physics. 2004;**43**:2572-2579. DOI: 10.1143/JJAP.43.2572

[14] Locatelli M-L, Pham CD, Diaham S, Berquez L, Marty-Dessus D, Teyssedre G. Space charge formation in polyimide films and polyimide/SiO_2 double-layer measured by LIMM. IEEE Transactions on Dielectrics and Electrical Insulation. 2017;**24**(2): 1220-1228. DOI: 10.1109/TDEI.2017.006172

[15] Carrasco F, Velazquez-Salazar A, Berquez L, Diaham S, Griseri V, Lebey T, et al. Assessment of the high field behaviour of polyimide films: Experimental methods and impact of electrode material. In: Proceedings of the International Symposium on Electrical Insulating Materials 2017 (ISEIM 2017); Toyohashi, Aichi, Japan; 11–15 September 2017. pp. 122-125. DOI: 10.23919/ISEIM.2017.8088704

[16] Carrasco F, Berquez L, Marty-Dessus D, Locatelli M-L, Lebey T, Teyssedre G, et al. Effect of film thickness and electrode material on space charge formation and conductivity in polyimide films. In: Proceedings of the IEEE International Workshop on Integrated Power Packaging 2019 (IWIPP 2019); Toulouse, Haute-Garonne, France; 24–26 April 2019

[17] Tajiri K, Muto H, Carrasco F, Berquez L, Marty-Dessus D, Locatelli ML, et al. Electric field enhancements due to space charge in thin polyimide film. In: Proceedings of the Conference on Electrical Insulation and Dielectric Phenomena 2019 (CEIDP 2019); Richland, Washington, USA; 20–23 October 2019

[18] Shiota H, Muto H. Space charge behavior and electrical conduction of silicone gel. In: Symposium on Electrical and Electronic Insulating Materials and Applications in Systems; Chiba, Chiba, Japan; 12–13 October 2006. pp. 49-52

[19] Shiota H, Muto H. Effect of space charge behavior in silicone gel on electrical insulation characteristics. In: IEEJ Transactions on Electrical and Electronic Engineering; Tokushima, Tokushima, Japan; 17–19 March 2005

[20] Breit H, Malec D, Lebey T. Investigations on DC conductivity and space charge in silicone gel. In: Annual Report Conference on Electrical Insulation and Dielectric Phenomena, 2002 (CEIDP 2002); Cancun, Quintana Roo, Mexico; 20-24 October 2002. DOI: 10.1109/CEIDP.2002.1048733

[21] Singh R. A review of developments in thermal techniques for charge profile measurements in polymer electrets. Journal of Electrostatics. 2014;**72**: 322-329. DOI: 10.1016/j.elstat.2014.05.007

[22] Collins RE. Practical application of the thermal pulsing technique to the study of electrets. Journal of Applied Physics. 1980;**51**:2973-2986. DOI: 10.1063/1.328109

[23] Lang SB, Das-Gupta DK. A technique for determining the polarization distribution in thin polymer electrets using periodic heating. Ferroelectrics. 1981;**39**:1249-1252. DOI: 10.1080/00150198108219626

[24] Velazquez-Salazar A, Berquez L, Marty-Dessus D. Thermal modeling and calibration in (F)LIMM using an external bias field: Theory and experiment. IEEE Transactions on Dielectrics and Electrical Insulation. 2018;**25**(3):783-790. DOI: 10.1109/TDEI.2017.007029

Section 5

Polyimide for Energy Storage

Chapter 8

Polyimides as High Temperature Capacitor Dielectrics

Janet Ho and Marshall Schroeder

Abstract

Nearly five decades of effort has focused on identifying and developing new polymer capacitor films for higher-than-ambient temperature applications, but simultaneous demands of processability, dielectric permittivity, thermal conductivity, dielectric breakdown strength, and self-clearing capability limit the number of available materials. Demands on these criteria are even more stringent in growing numbers of applications demanding high power performance. Aromatic polyimides, though not a panacea, are a class of heat-resistant polymers of great interest to researchers as capacitor dielectrics because of good thermal and mechanical stability. In this chapter, the key aspects and advantages of metallized polymer film capacitors are compared to analogous alternative technologies (polymer-film-metal-foil, ceramic, and electrolytic capacitors), followed by a comprehensive review of commercial resin development leading up to recent research on polyimides targeted for operating temperature above 150°C. Finally, this chapter provides a brief discussion on the recent effort on combining computation and synthesis to design polymers with desirable dielectric properties.

Keywords: Kapton, Ultem, polyimides, high temperature, capacitor dielectrics, thermal conductivity, self-clearing, metallized film capacitors, power electronics, DC link capacitors

1. Introduction

Capacitors are one of the primary components in power electronic devices, some of which are required to operate in hostile environments across a variety of consumer, industrial, and military sectors. For example, the sensors in "down hole" electronics for characterizing oil, gas, and geothermal wells can experience temperatures exceeding 200°C depending on the well depth. In the aircraft industry, new engine control systems require placement of sensor/actuator and signal conditioning electronics in or near aircraft engines where temperature can be in range of 200–300°C. Similar demands exist in the automobile industry, with power electronics located in the engine compartment and near the wheels of hybrid and electric vehicles where temperatures can reach 150°C [1–5].

For high voltage (>500 V [6]) applications, metallized polymer film capacitors are generally selected over polymer-film-metal-foil, ceramic, or electrolytic capacitors because of the enhanced volumetric efficiency and improved safety. Failure of a charged high energy capacitor (>10 kJ) is equivalent to a bomb, which requires design engineering with an appropriate failsafe mechanism such that energy is released gradually as the capacitor fails. Presently, the only compatible technology is based on polymer film capacitors with thin metallization as electrodes, for which a thin layer of vacuum-deposited metallization (usually 20–100 nm of aluminum, zinc, or alloy [7]) functions as a fuse. When a localized breakdown of the film occurs during operation, (i) the current flowing through the breakdown site is limited by the metallization resistivity and (ii) the energy dissipated in the breakdown is sufficient to vaporize/oxidize the metallization near the breakdown, isolating the breakdown site. This results in a small decrease in capacitance but continued operation of the capacitor at the rated voltage. This "graceful" recovery mechanism is known as "self-clearing" and a photograph of a breakdown ("clearing") site in a metallized polymer film is shown in **Figure 1**. In contrast, polymer film capacitors with metal-foil electrodes (5–10 μm thick [9]) and ceramic capacitors, for which the electrode is a thick metal coating, often fail catastrophically if shorted [10, 11].

With power system designers striving for miniaturization and reliability at high temperatures and operating voltages, they must turn to specialized components.

Figure 1.
Photograph of a breakdown site in a metallized polymer film. Figure reproduced from Figure 3 of [8] with permission from IEEE.

These applications are enabled, in part, by wide bandgap semiconductors (e.g., silicon carbide), which support operation at temperatures well above 150°C [3, 12]. However, these types of environments are too aggressive for conventional polymer capacitor dielectrics unless the voltage is derated, or an active cooling mechanism is implemented, introducing additional cost and complexity while reducing energy efficiency. High temperature polymer film capacitors offer a promising solution for these issues due to reduced thermal management requirements and elimination of the voltage derating due to improved stability of the breakdown strength at high temperatures. Aromatic polyimides are one specific class of high temperature polymers which have been commercially available since the early 1960s [13], but the form in which these materials are manufactured generally does not meet the specifications required for capacitor films. One of the main requirements is the processability of the polymer into a continuous thin film (<12 μm thickness), since the capacitance scales inversely with film thickness [14]. This limitation precludes the use of Kapton® polyimide as a capacitor dielectric [15], even though it has been used extensively as wire and cable insulation for aircraft with a continuous operating temperature of 300–350°C since the early 1980s [16–18].

One major impediment to the development and integration of new capacitor dielectrics is that specialty film chemistries optimized specifically for high performance polymer capacitors represent relatively small markets (i.e. military, aerospace, and down hole exploration [3, 12]) compared to those necessary for profitable commercial production of a polymer resin. Other than biaxially oriented polypropylene (BOPP) and polyethylene terephthalate (BOPET), which are commodity films in a variety of commercial applications such as packaging [19, 20], nearly all commercial polymer capacitor films are specialty polymers synthesized for other applications. For example, poly (phenylene sulfide) (PPS) and poly(ethylene 2,6-naphthalate) (PEN) are available as premium capacitor dielectrics, but the majority of their use is in automotive, household, and food packaging applications [21–24].

This chapter discusses the important criteria for high temperature polymer capacitor dielectrics and presents a comprehensive review on commercial resin development up to recent research progress on polyimide (PI) targeted for operating temperature above 150°C. While many review articles on various aspects of polymeric capacitor dielectrics are available [25–32], this chapter has a specific focus on polyimides for high temperature applications.

2. Relevant polymer properties for capacitors dielectrics

The deliverable energy density of a capacitor scales linearly with the dielectric constant and quadratically with breakdown strength. As such, engineers and scientists often focus on these two properties while neglecting other relevant characteristics, such as thermal conductivity and the implications of chemical composition of a polymer on self-clearing capability. These properties will be discussed in the following subsections.

2.1 Thermal conductivity

For reliable operation at elevated temperatures, efficient heat removal from the interior of capacitors is essential but often challenging, especially in metallized polymer film designs due to the low thermal conductivity of both the polymer dielectric and the thin metallization [33]. As electrical insulators, polymer dielectrics are poor thermal conductors with thermal conductivities ranging from

0.1 W/(m·K) for amorphous polymers [34] to about 0.6 W/(m·K) for the present state-of-the-art BOPP capacitor films [35]. As for the electrodes, while the thermal conductivity of metals are typically three orders of magnitude higher than those of polymers, the metallization is also ~1000 times thinner than the dielectric and may not have the thermal conductivity of the bulk metal from which it is deposited. One approach to increase thermal conductivity of the polymer is by blending in ceramic nano-fillers. Researchers have shown that mixing 10 vol% boron-nitride nano-sheets (BNNS) with an amorphous polymer resulted in a six-fold increase in thermal conductivity from 0.3 to 1.8 W/(K·m) [36]. The resulting BNNS/polymer nanocomposite also exhibited an increase in dielectric constant, breakdown strength, and charge-discharge efficiency, the latter of which was attributed to reduction of the conduction current. A similar trend has also been observed in other polymer systems including polyimides [37, 38].

2.2 Self-clearing capability: implications of chemical composition

For a self-clearing event to occur successfully, the metallization needs to be thin enough to be vaporized/oxidized by the energy dissipated during the dielectric breakdown. In addition, the polymer also needs to be oxidized completely, leaving no conductive paths from free carbon in the region of cleared metallization around the breakdown site, as shown in **Figure 1** [8]. Based on an in-depth study of the physics and chemistry of clearing phenomena, polymers with low ratios of carbon to (hydrogen + oxygen) in the repeating units, such as cellulose, often exhibit excellent self-clearing. Conversely, polymers with high carbon to (hydrogen + oxygen) ratios, like polystyrene, tend to form greater amounts of free carbon for a given clearing energy and clear poorly [7]. This trend is illustrated in **Table 1**,

Polymer	Repeating unit	C	O	H	N	S	$\frac{(C+N+S)}{(H+O)}$ [a]	Carbonization [b] (Calc.) %	Carbonization [b] (Meas.) %	Self Clearing [c]
Cellulose	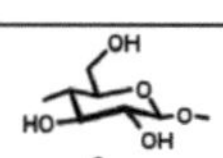	6	5	10	0	0	0.40	2	2.2	excellent
Cellulose Acetate	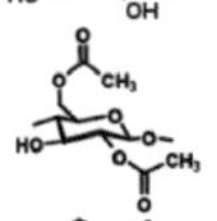	10	7	14	0	0	0.48	5	3	excellent
Polypropylene	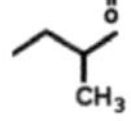	3	0	6	0	0	0.5	54	50.5	good
Polyethylene terephthalate	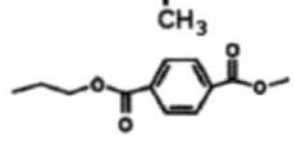	10	4	8	0	0	0.83	41	37.5	medium
Polycarbonate	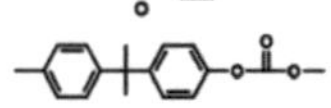	16	3	14	0	0	0.94	60	58.5	--
Polyethylene naphthalate	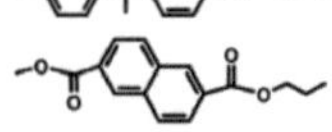	14	4	10	0	0	1.0	N/A	N/A	medium-low
Polystyrene	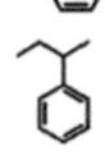	8	0	8	0	0	1.0	76	75.5	low
Polyphenylene sulfide		6	0	5	0	1	1.4	N/A	N/A	low
Kapton polyimide	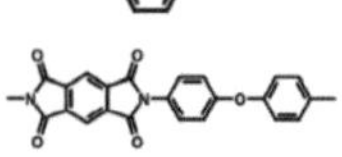	22	5	10	2	0	1.6	N/A	N/A	--

[a] *Adapted from [5].*
[b] *Data taken from [7].*
[c] *Observation taken from [5, 7].*

Table 1.
Chemical composition of various polymer dielectrics including ratios of (carbon + nitrogen + sulfur) to (hydrogen + oxygen), amount of residual carbon, and the observation of self-clearing behavior.

in which the chemical composition of various polymer dielectrics is correlated with the amount of carbon residue, both measured and calculated, based on comparison between amount of carbon in the gaseous by-products and composition of the dielectrics [5, 7]. Kapton® polyimide, also shown in **Table 1**, has a high degree of aromaticity in the structure and has a tendency toward carbonization (arc-tracking), as observed in wire insulation in commercial aircraft before fluoropolymers were applied as thin coatings to resist arc-tracking [39, 40]. Poor self-clearing can be mitigated to some degree by applying a thin coating of acrylate over the dielectric to increase the oxygen content. The clearing energy can also be reduced by increasing the metallization resistance in order to avoid damaging adjacent layers of dielectric during clearing (**Figure 2**), given that a metallized film capacitor is typically wound from two layers of single-side-metallized dielectric film as illustrated in **Figure 3** [8]. However, the trade-offs for increasing metallization resistance include higher equivalent series resistance (ESR), lower thermal conductivity, and lower ripple current handling capability. Thus, achieving an optimum balance between capacitor ESR and clearing energy depends on the application requirements.

2.3 Tensile strength

In order to withstand the tension applied by winding machines during capacitor manufacturing, a polymer dielectric must have sufficient tensile strength. A typical value of tension used by capacitor manufacturers is about 10 MPa while the typical tensile strength for various commercially available capacitor films is in the range of 160–200 MPa [41]. In comparison, Kapton® polyimide has a tensile strength of 72 MPa and poly(ether imide) 97 MPa [31]. Based on these values, the tensile strengths of the polymers discussed herein are sufficient for capacitor winding.

2.4 Temperature dependence of dielectric constant, dissipation factor, and breakdown strength

Polar polymers often exhibit a substantial increase in dielectric constant when the temperature passes through the glass transition temperature (T_g) of

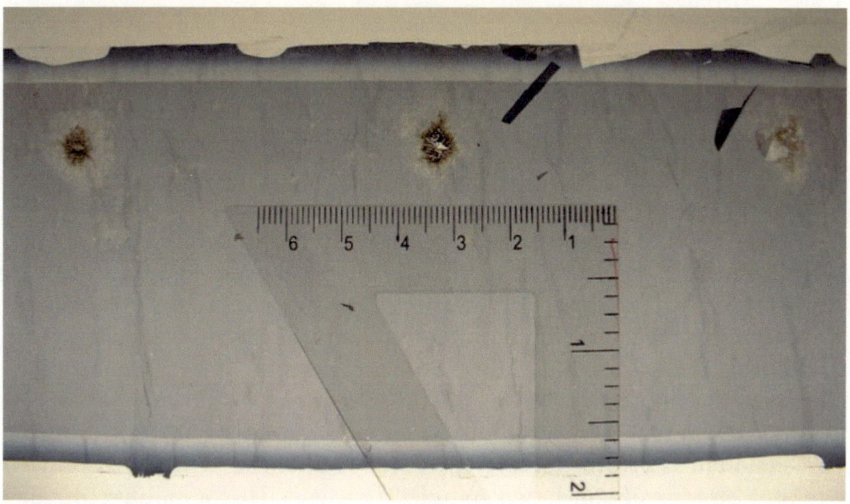

Figure 2.
Photograph of an unwound metallized film capacitor having poor self-clearing behavior with damage involving adjacent layers.

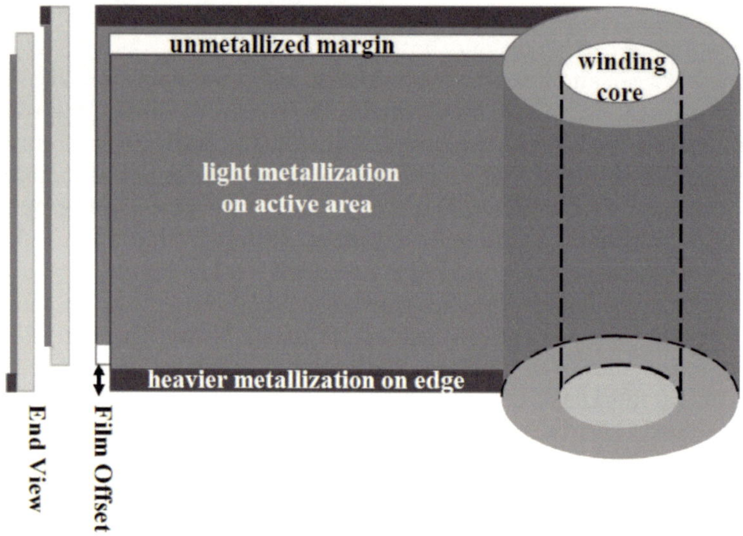

Figure 3.
Schematic of the construction of a metallized polymer film capacitor winding without the end connections.

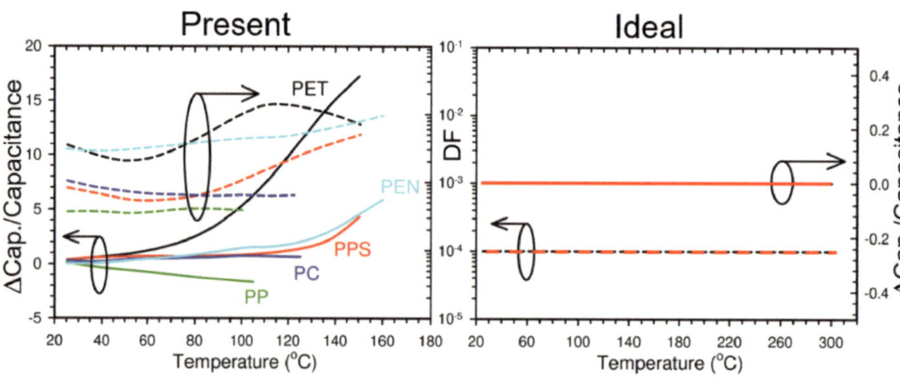

Figure 4.
Typical change in capacitance and dissipation factor (DF) of various polymer film capacitors as a function of temperature at 1 kHz. PET: poly(ethylene terephthalate), PEN: poly(ethylene 2,6-naphthalate), PPS: poly(phenylene sulfide), PC: poly(carbonate), PP: poly(propylene). Data adapted from [42, 43].

the polymer; however, non-polar polymers, which have only electronic polarization, are less temperature-dependent (**Figure 4**). The temperature dependence of dielectric constant for polar polymers is the result of increased polymer chain mobility as temperature increases through the T_g. This leads to greater freedom in accessible molecular dipole orientations at frequencies below 10^{12} Hz, while the electronic and atomic contributions tend to occur at around 10^{12} Hz and above [44]. Furthermore, polymers tend to have a broad distribution of response times as a result of interconnectivity and steric-hindrance imposed by neighboring molecules [45]. Substantial increases in dielectric constant near T_g limit capacitor operation for polar dielectrics to somewhat below T_g for many applications requiring a stable capacitance, including power conditioning/conversion in electronic circuitry, where the operating frequency ranges from tens of kHz for silicon-based switches to a few GHz for gallium nitride transistors [46].

Dissipation factor (DF) of a dielectric is a measure of electrical energy dissipated, usually in the form of heat, when an oscillating electric field is applied.

Similar to the dielectric constant, the DF is temperature and frequency-dependent and is more pronounced for polar polymers (**Figure 4**). In a capacitor, the total DF is a combination of contributions from the electrode and the polymer. For the electrode, metallization has a characteristic DF of ~0.1%, while the DF of metal foils is negligible. For polymers, DF can range from 0.01% for nonpolar polymers like PP, to as high as 1% for polar polymers such as PET, as shown in **Figure 4**. Depending on the capacitor applications, such as in a snubber or DC link, the requirement for dissipation factor ranges from ~0.01% for the former, to 0.1% for the latter [47–50]. Thus, for a snubber to meet the low DF requirement, metal foil electrodes with a BOPP polymer film is a common capacitor configuration.

The breakdown strength of a dielectric is defined as the maximum electric field that a dielectric can sustain for a given electrode configuration, test area, [51, 52] and voltage waveform (e.g. linear voltage ramp rate for DC breakdown [53]). The breakdown field is a statistical parameter, typically characterized by a Weibull distribution, although sometimes the log-Normal distribution provides a better fit [54]. It is usually determined by extrinsic factors such as weak points or defects such as embedded foreign particles in the dielectric [55]. Therefore, when reporting the breakdown field at the film level (as opposed to a wound capacitor), measurement conditions such as the electrode configuration (ball-plane or parallel plates) and test area should be included. For comparison purposes, all breakdown field data contained in this chapter included a description of the measurement conditions or control measurements with other capacitor-grade polymer films. Presently, BOPP capacitor films have the highest breakdown field of ~700 MV/m (at the 63% Weibull cumulative probability for a test area of ~2 cm^2 and 300 V/s linear ramp voltage) among all commercial polymer capacitor films [56]. The breakdown field of polymers decreases as temperature approaches T_g of amorphous (e.g. poly-ether-imide or PEI) or T_m of semi-crystalline polymers, [such as BOPP and BO poly-phenylene-sulfide

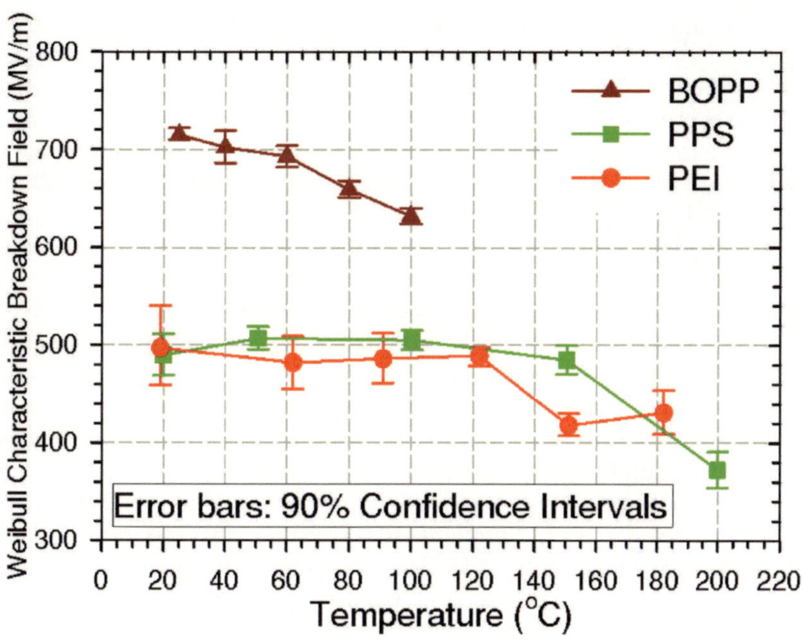

Figure 5.
Weibull characteristic breakdown field as a function of temperature for various polymers. Electrode area: ~2 cm^2. BOPP: biaxially oriented poly(propylene), PPS: poly(phenylene sulfide), PEI: poly(ether imide). Figure taken from [57].

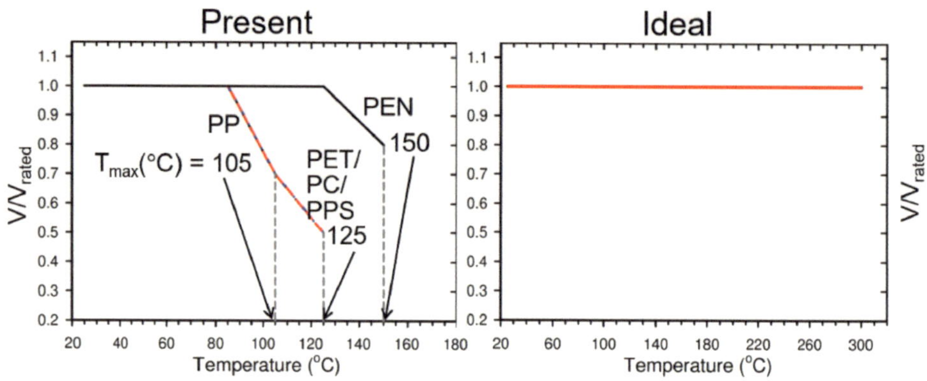

Figure 6.
Typical derating of operating voltage for various polymer film capacitors as a function of temperature. PET: poly(ethylene terephthalate), PEN: poly(ethylene 2,6-naphthalate), PPS: poly(phenylene sulfide), PC: poly(carbonate), PP: poly(propylene). Data adapted from [42, 43].

(PPS)], as shown in **Figure 5** [57]. Hence, the operating voltage of a capacitor is usually derated at elevated temperatures to protect against failure and prolong operational lifetime (**Figure 6**). Although many breakdown mechanisms have been proposed [58–61], extrinsic breakdown in solids is generally driven by power dissipation and eventually "thermal runaway", which leads to breakdown [62, 63].

3. Commercial polyimides evaluated for capacitor applications

In the early 1990s, the lack of suitable high temperature polymer film capacitors prompted a concerted effort funded by NASA and the U.S. Air Force to develop new capacitor dielectrics based on commercially available heat-resistant polymers with operating temperatures above 200°C. While Kapton® polyimide has been used extensively since the early 1980s as wire and cable insulation for aircraft (continuous operating temperature of 300–350°C [16–18]), it has never been used as a capacitor dielectric due in part to its previous inability to be manufactured in thin films. Nevertheless, it is a common benchmark for development of new dielectrics. One study reported that the dissipation factor of Kapton® at the film level increased from 0.1% at 25°C and 1 kHz to 6% at 300°C, while that the dielectric constant decreased from 3.1 at 25°C and 1 kHz to 2.8 at 300°C [64, 65]. At 10 kHz, Kapton® showed a similar decreasing trend with increasing temperature for the dielectric constant while the dissipation factor remained constant at 0.1% at temperature up to 300°C, as shown in **Figure 7** [36]. Another recent study showed that Kapton® at 1 kHz in a film-foil capacitor configuration exhibited a gradual decrease in capacitance from 25 up to 225°C, followed by an abrupt increase at higher temperature, as illustrated in **Figure 8**. In this configuration, the dissipation factor of Kapton® remained constant at 0.1% up to 200°C and then increased sharply as temperature was further increased [66].

Another aromatic polyimide is SIXEF-44™ (**Figure 9**), which is manufactured by Hoechst Celanese based on 2,2-bis(3,4dicarboxyphenyl) hexafluoropropane dianhydride (6FDA) and 2,2-bis(4-aminophenyl) hexafluoropropane diamine (4,4′-6F diamine). This fluorinated polyimide has a glass-transition temperature (T_g) of 323°C [67] and a dielectric constant of 2.8 at 1 kHz with <10% change across temperature range from −55 to 300°C. The dissipation factor is around 0.1% over the temperature range of 25–250°C from 100 Hz to 10 kHz, as illustrated in

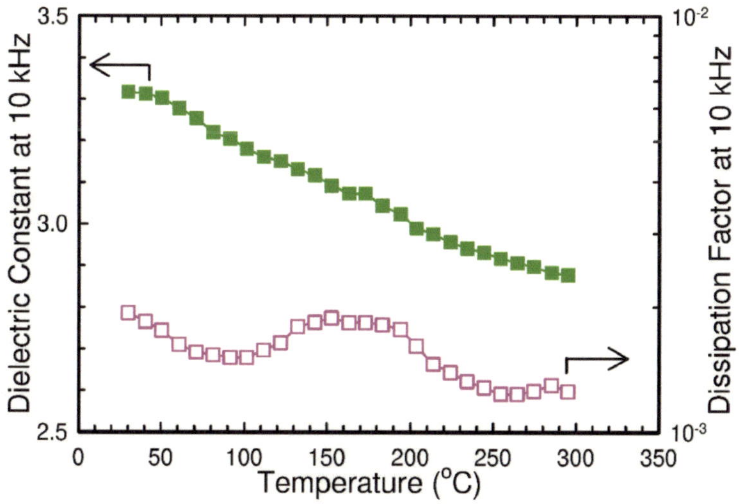

Figure 7.
Dielectric constant and dissipation factor at 10 kHz for Kapton® film as a function of temperature. Data adapted from [36].

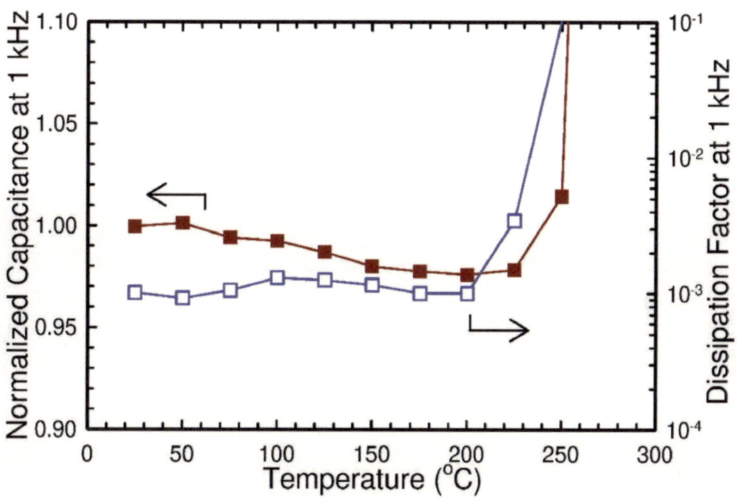

Figure 8.
Normalized capacitance and dissipation factor at 1 kHz as a function of temperature for an 8-μm Kapton® HN-30 film capacitor wound with 12-μm copper foil. Data adapted from [66].

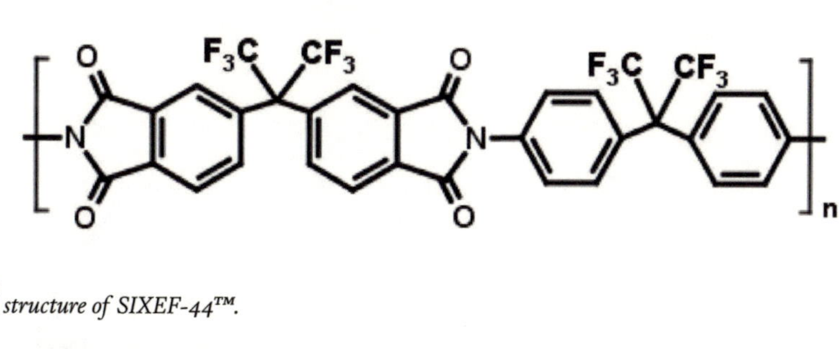

Figure 9.
Chemical structure of SIXEF-44™.

Figure 10 [68]. The reduced dielectric constant in SIXEF-44™ relative to nonfluorinated counterparts was attributed to the symmetry of the fluorine atoms on the polymer backbone [67, 69, 70].

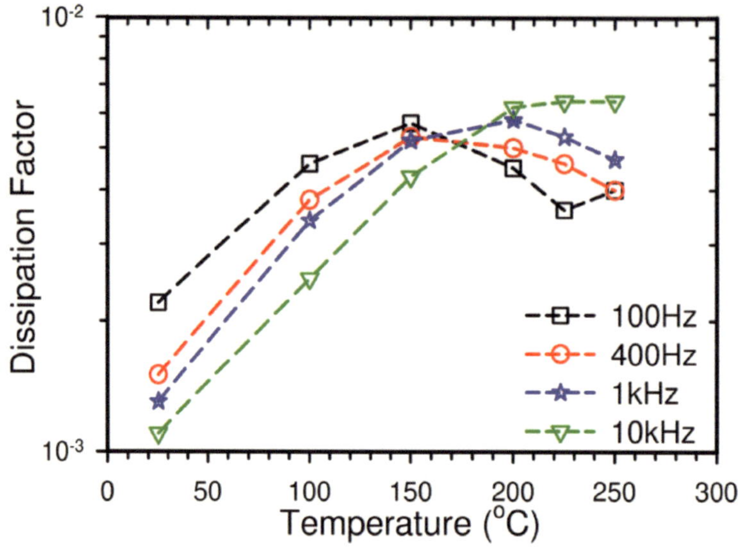

Figure 10.
Dissipation factor of SIXEF-44™ as a function of temperature at various frequencies. Data adapted from [68].

Other aromatic polyimides investigated include perfluoropolyimide (PFPI) (**Figure 11**) and Upilex-S® (**Figure 12**). PFPI was developed by TRW, Inc., and is a perfluoroisopropylidene diamine of 2,2-bis(4,4-aminophenoxy)-phenyl-hexafluoropropane (4-BDAF) and pyromellitic dianhydride (PMDA) [71]. Upilex-S® was originally synthesized by ICI Americas, Inc. from monomers of 3,3′,4,4′-biphenyl tetracarboxylic dianhydride (BPDA) and p-phenylene diamine (p-PDA). The T_g for PFPI is above 390°C [72] while that of Upilex-S® is 355°C [73]. The dielectric constant of PFPI is 3.1 at 25°C but decreases to 2.9 at 300°C [64, 65], while that of Upilex-S® is 3.3 for 1 kHz and temperatures from 25 to 300°C. The dissipation factor for both PFPI and Upilex-S® at 1 kHz is about 0.1% at both 25 and 300°C.

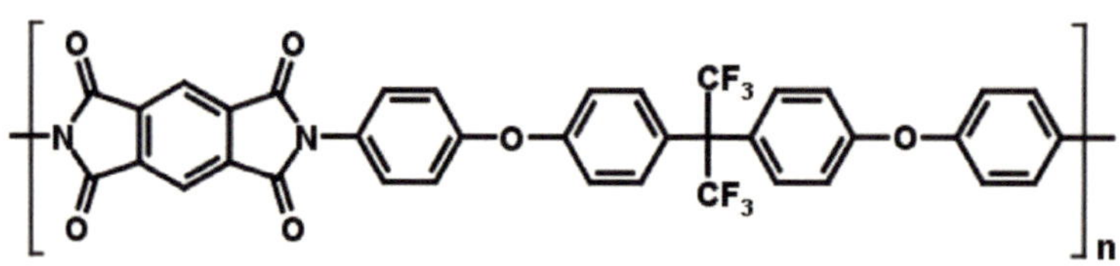

Figure 11.
Chemical structure of PFPI.

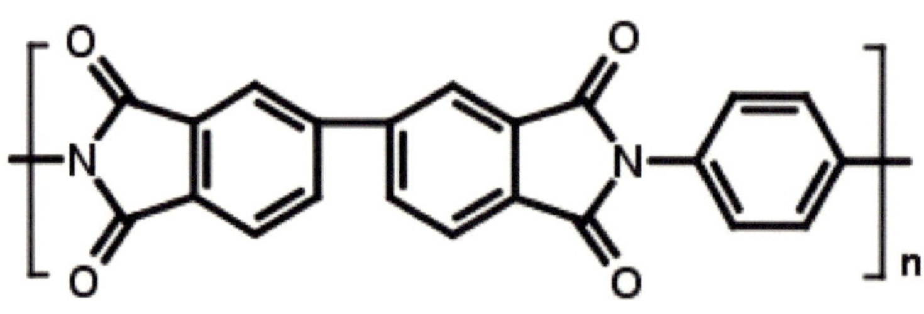

Figure 12.
Chemical structure of Upilex-S®.

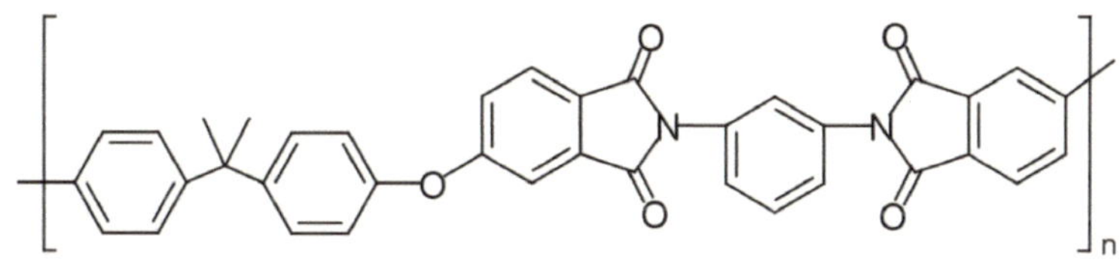

Figure 13.
Chemical structure of PEI.

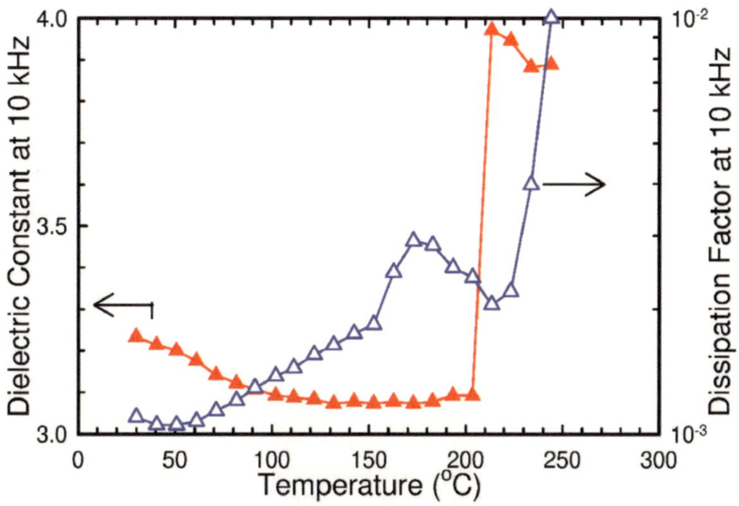

Figure 14.
Dielectric constant and dissipation factor at 10 kHz for PEI film as a function of temperature. Data adapted from [36, 57].

Most aromatic polyimides suffer from the same processing issues as a result of the high degree of aromaticity, which led researchers to develop modified systems with flexible moieties such as ether linkages and alkyl groups in the polymer backbone. One example is Ultem™, which is poly(ether imide) (PEI) (**Figure 13**) that is synthesized from the disodium salt of bisphenol A and 1,3-bis(4-nitrophthalimido)benzene [74]. Subsequent development efforts focused on melt-extrusion and stretching of PEI films have enabled film thicknesses as thin as 5 μm [31]. The T_g of PEI is ~215°C, which is considerably lower than many of the wholly aromatic polyimides due to the increase in flexibility imparted by the ether linkages and alkyl groups [17]. The dielectric constant and dissipation factor of PEI are about 3.1 and 0.2%, respectively, at frequencies from 100 Hz to 10 kHz and temperatures from 25 to 200°C; however, they increase sharply as temperature approaches the T_g, as shown in **Figure 14** [36, 57]. Such temperature dependence for the dielectric constant and dissipation factor is a characteristic for polymers with molecular dipoles as discussed in Section 2.4.

4. Recently developed polyimides

Over the past 10 years, research efforts have continued to develop new polyimides for capacitor dielectrics. The thermal stability of these new polymers is primarily derived from a high degree of aromaticity and fused-ring heterocyclic rigid structures. As energy density of a capacitor scales with the dielectric constant, some researchers have specifically focused on designing new aromatic polyimides

(PIs) that have a higher dielectric constant than the typical value of 3 while preserving the essential thermal/mechanical properties.

One common approach to increasing the dielectric constant is by incorporating polar moieties into the backbone of a polymer chain to enhance the dipole moment. The nitrile group (—CN) is one of the polar moieties that was explored by Wang et al. [75, 76] and Treufeld et al. [77], in which the number of —CN dipoles on the diamine was varied between 0, 1, and 3. The nitrile-containing diamines synthesized in the study were aminophenoxy-benzonitrile (APBN)-based with the two amino groups varying between three isomeric positions (m,m, m,p, and p,p) (**Figure 15**) to explore the effect of isomeric position on the 6FDA-based PIs. A nitrile-free analog, LaRC-CP2™ [78, 79], was used as a control.

Among the 6FDA-based PIs, the addition of one nitrile group in the diamine increased the room temperature dielectric constant from 2.9 (independent of frequency between 0.1 Hz and 1 MHz for the nitrile-free PI) to between 3.1 and 3.5 depending on the frequency and isomeric positions of the diamine linkage. In general, the m,m- and m,p-positions exhibited higher values (3.25–3.5) as compared to the p,p-position (**Figure 16**). Further increasing the nitrile content to three nitrile groups per diamine monomer resulted in an additional increase of the room temperature dielectric constant to between 3.5 and 3.7 (frequency dependent) for the 6FDA-based PIs with p,p-linkage. Interestingly, substitution of the other two isomeric positions had no significant effect (**Figure 17**). The investigators explained that the greater flexibility of the three —CN dipoles in the p,p-linkage as compared to the m,m- or m,p-linkage in the diamine resulted in greater dipolar polarization. An increase in dielectric constant was also observed in the OPDA-based PIs with m,m diamine linkage with increasing content of —CN dipoles.

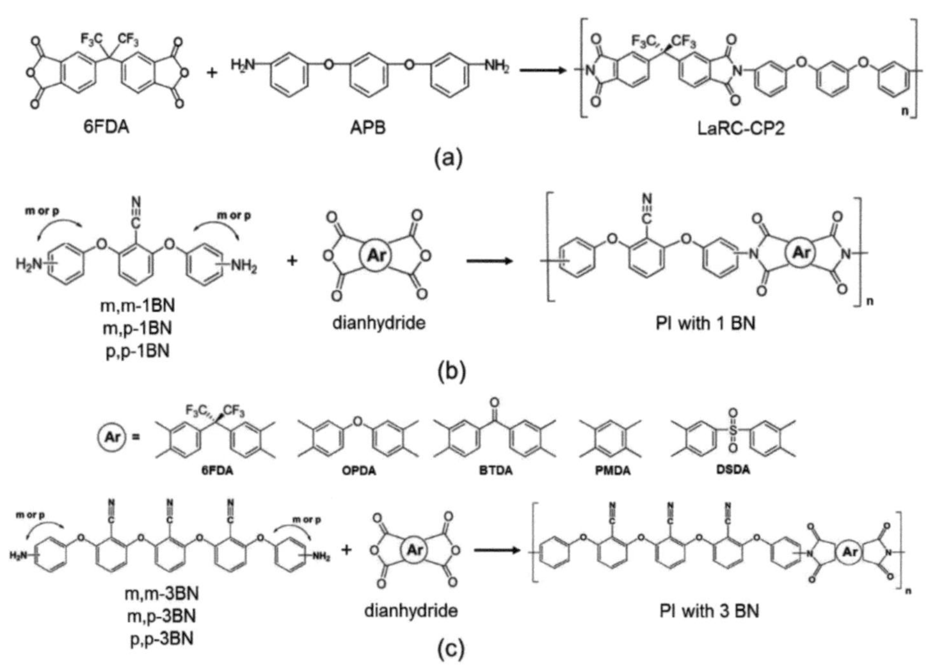

Figure 15.
Reaction schemes of various polyimides: (a) LaRC-CP2 from 2,2-bis(phthalic anhydride)-1,1,1,3,3,3-hexafluoroisopropane (6FDA) and the 1,3-bis(3-aminophenoxy) benzene (APB) diamine, (b) with one benzonitrile (BN) group, (c) with three benzonitrile groups. Reaction schemes adapted from [75, 76].

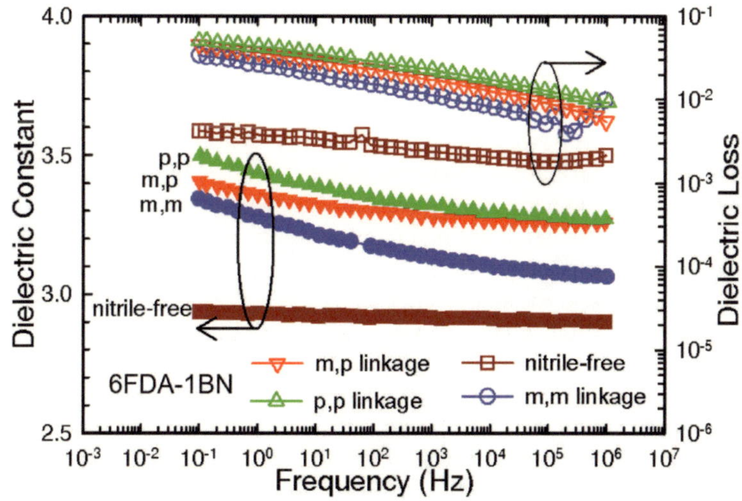

Figure 16.
Dielectric constant and dielectric loss as a function of frequency at ambient conditions for 6FDA-APB nitrile free PI and the three analogs containing one **—***CN dipole per diamine at different isomeric positions. Data adapted from [75].*

The key drawback of the addition of nitrile groups was the increase in temperature dependence in both the dielectric constant and dissipation factor, as illustrated in **Figure 17**. For example, the dielectric constant of the nitrile-free 6FDA-based PI at 10 kHz increased by only 1.4% over a temperature range of −150 to 190°C, whereas that of the three nitrile-containing analogs with the m,m-, p,p-, and m,p-linkages rose 6.3, 8.3, and 15%, respectively. With respect to the dissipation factor, the nitrile-free PI remained below 0.5% over the tested temperature/frequency

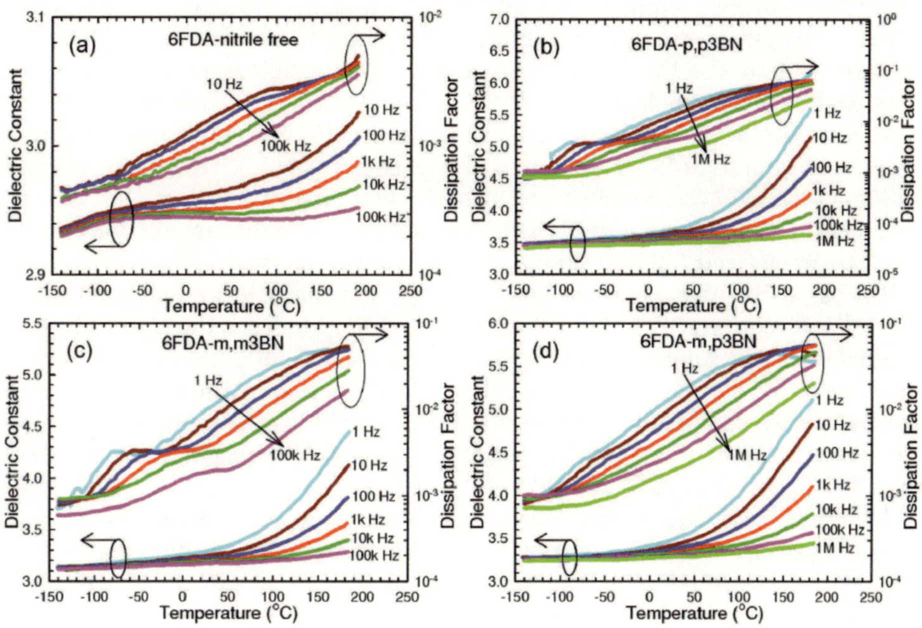

Figure 17.
Dielectric constant and dissipation factor as a function of temperature at various frequencies for the 6FDA-based PIs containing three nitrile groups in different isomeric positions (a–c) and the nitrile-free analog, LaRC-CP2 (d). Data taken from [77].

range, while the same nitrile-containing PIs increased from below 0.1% at −150°C to above 1% at 190°C between 10 Hz and 100 kHz. The temperature scan was limited to 190°C to avoid deforming the sample significantly. The investigators attributed the increase in dielectric constant with temperature at high frequencies to increased short-range segmental motion as the T_g was approached, which resulted in stronger dipolar polarization. However, since the —CN dipoles were attached to the polymer backbone in a 90° configuration, the segmental motion was hindered, causing friction with the neighboring chains and leading to high dissipation factor at high temperatures and frequencies. At low frequencies (<10 Hz), the increase in dielectric constant and dissipation factor with temperature was attributed to ions present from residual solvent and unreacted poly(amic acid) precursors in the 35–50 μm thick samples.

The investigators also compared the structural effect of four dianhydrides on the dielectric properties of PIs with three —CN dipoles in the diamine unit. The four dianhydride monomers under study were 2,2-bis(phthalic anhydride)-1,1,1,3,3,3-hexafluoroisopropane (6FDA), 4,4′-oxydiphthalic dianhydride (OPDA), 4,4′-benzophenonetetracarboxylic dianhydride (BTDA), and pyromellitic dianhydride (PMDA). The chemical structures of these dianhydrides are shown in **Figure 15**. A comparison of the increase in dielectric constant (Δε) from −150 to +190°C shows that for the p,p-linkage, PMDA exhibited the largest increase (e.g. ~25% at 10 kHz from ~3.4 at −150°C), followed by OPDA (23% from ~3.3), 6FDA (16% from ~3.4), and finally BTDA with the lowest increase (10% from ~3.6 at 10 kHz at −150°C) (**Figure 18a, c,** and **e**). The values of Δε decreased with increasing frequency for all four PIs and followed a decreasing order PMDA > OPDA > 6FDA > BTDA, which agreed with the trends observed in the polarization from dipole orientation determined experimentally and predicted based on a freely rotating single-dipole model. This model assumes an anti-parallel configuration for the dianhydride and diamine dipoles, such that the net dipole moment of a repeat unit is the difference between the dianhydride and diamine dipole moments without an external electric field. PMDA, being a symmetrical dianhydride, has no net dipole moment, whereas BTDA has the largest dipole moment (2.96 Debye) as a result of the benzophenone functional group. The 1.14 D dipole moment of OPDA results from the diphenyl ether, and the 2.0 D dipole moment of 6FDA from the 1,1,1,3,3,3-hexafluoropropane. The diamine portion has a dipole moment of 17.1 D which is the sum of three benzonitrile groups (4.18 D) and four diphenyl ethers (1.14 D). The largest Δε value for PMDA-based PI was accompanied by the highest dissipation factor relative to the other three PIs, but all four PIs had dissipation factor in the range of 3–8% at 190°C between 1 and 100 kHz (**Figure 18b, d,** and **f**). The results suggest that the PMDA-based PI with p,p diamine linkage had greater chain flexibility, which led to larger dipole motion at 190°C. In comparison to the p,p-linkage, the m,m-linkage resulted in a smaller increase in dielectric constant and slightly lower dissipation factor at 190°C, although the trends with respect to the dianhydrides and frequency dependence were similar. In addition, the 6FDA-based PI with m,m-linkage shows a dielectric constant of 3.1, which is noticeably lower than the p,p-linkage (ε = 3.4) and other PIs (ε ~ 3.3 to 3.6) at −150°C.

In terms of discharge energy density and efficiency, as characterized by measuring electric displacement as a function of electric field, the nitrile-containing PIs delivered 25–40% greater discharge energy density than the nitrile-free analog at an applied electric field of 100 MV/m at 190°C and frequencies of 10 Hz and 1 kHz (**Figure 19a**). For comparison, the investigators also evaluated Kapton® PI and Ultem® PEI, both of which gave similar discharge energy density to the nitrile-free PI. While the nitrile-containing PIs delivered more energy than the nitrile-free counterparts, their discharge efficiency was generally lower, as

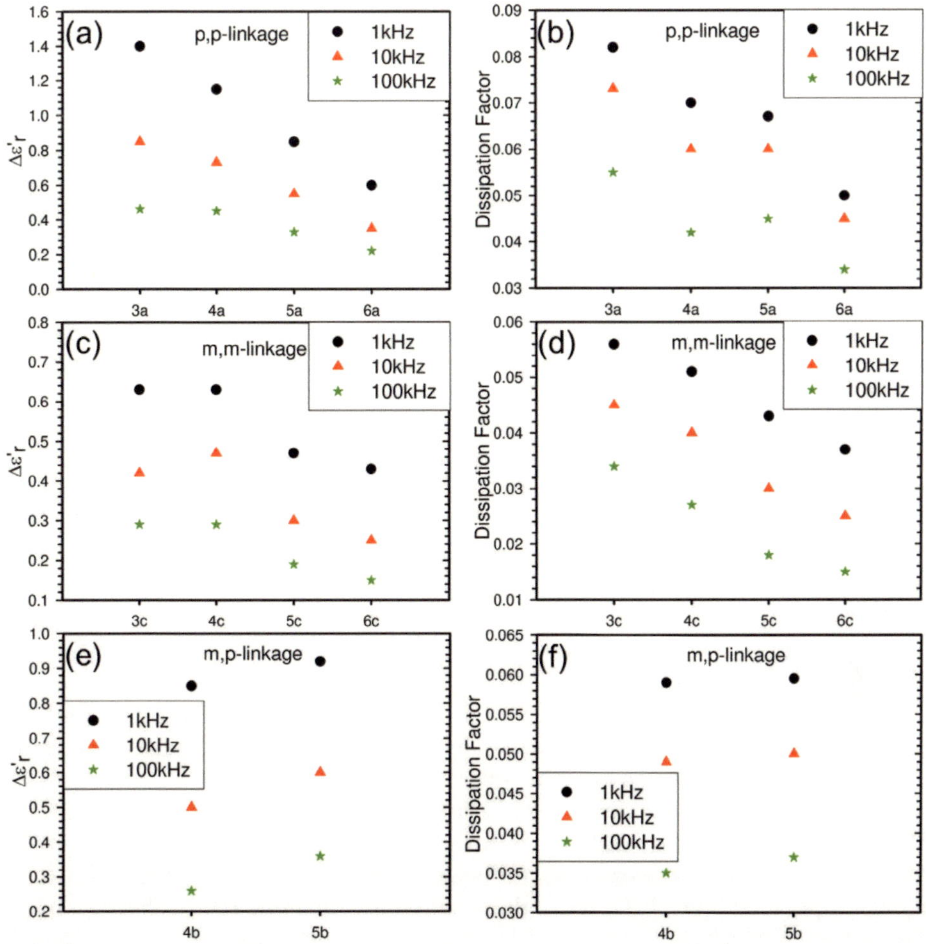

Figure 18.
(a, c, and e) increase in dielectric constant (Δε) from −150 to +190°C and (b, d, and f) dissipation factor at 190°C at various frequencies for PIs containing three nitrile groups per diamine with various dianhydride (labeled as 3–6) and different isomeric diamine linkage (denoted as a: p,p; b: m,m; c: m,p). 3–6 represent PMDA-based, OPDA-based, 6FDA-based, and BTDA-based, respectively. Data adapted from [77].

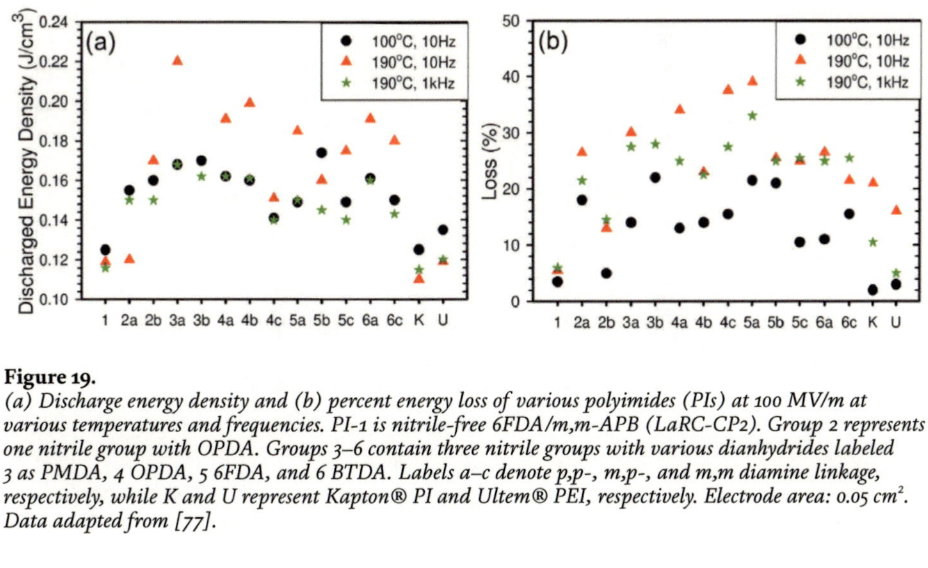

Figure 19.
(a) Discharge energy density and (b) percent energy loss of various polyimides (PIs) at 100 MV/m at various temperatures and frequencies. PI-1 is nitrile-free 6FDA/m,m-APB (LaRC-CP2). Group 2 represents one nitrile group with OPDA. Groups 3–6 contain three nitrile groups with various dianhydrides labeled 3 as PMDA, 4 OPDA, 5 6FDA, and 6 BTDA. Labels a–c denote p,p-, m,p-, and m,m diamine linkage, respectively, while K and U represent Kapton® PI and Ultem® PEI, respectively. Electrode area: 0.05 cm².
Data adapted from [77].

shown in **Figure 19b**. The discharge energy losses for the nitrile-containing PIs were 15–40% compared to ~5% for the nitrile-free counterpart at 190°C at 10 Hz. Kapton® PI and Ultem® PEI exhibited 15–20% loss at that temperature and frequency.

With respect to the thermal properties, based on the limited glass transition temperature (T_g) data provided by the investigators, the addition of nitrile groups in the polymer backbone appears to increase T_g, but the enhancement decreased with increasing nitrile content, as indicated by the comparison of T_g for the seven 6FDA-based PIs (**Figure 20**). The structure of the dianhydride also affects the T_g. The four PMDA-based nitrile-containing PIs appear to have the highest T_g values (300–350°C with high crystallinity), whereas the other three dianhydride families of 6FDA, BTDA, and OPDA were mostly amorphous with T_g values ranging from 220 to 300°C. By comparison, these values were at least 20°C above that of the nitrile-free LaRC-CP2 (T_g = 200°C). Additionally, the PIs with m,m diamine linkage appeared to have lower T_g than those with a p,p linkage. This effect was attributed to the greater free volume created between polymer chains by the two bent amino groups in the m,m-linkage.

An ether (**—O—**) linkage is another polar moiety explored for the potential enhancement in dielectric constant of PIs [80, 81]. Jeffamines® EDR-104, D230 and HK511 (**Figure 21**), which are commercial polyether aliphatic diamines, were used to introduce ether linkages into PIs. The rationale behind the use of linear alkyl diamines was to impart close-packing for reducing the overall free volume while maintaining a high density of the imide functional groups in the polymer backbone. For comparison, 1,3-diaminopropane (1,3-DAP) and 1,6-diaminohexane (1,6-DAH), were synthesized as ether-free PIs with commercial linear alkyl diamines. The dianhydrides used in the study were PDMA, BTDA, OPDA, and 6FDA. Depending on the dianhydride, the PIs with diamine D230 exhibited dielectric constants of 2.5 for 6FDA and ~4.5 for the other three dianhyrides at room temperature and 1 kHz, while those containing diamine HK511, because of the greater ether content, produced higher dielectric constants (~5.3 for PMDA and 6FDA, 6.2 for OPDA, and 7.8 for BTDA) (**Figure 22a**). Hence, the general trend for dielectric constant of the ether-containing PIs with respect to the dianhydrides followed the order BTDA ≥ OPDA > PMDA > 6FDA. The lowest dielectric constant of 2.5 observed in the PI with 6FDA/D230, was attributed to the weak electronic interaction between chains caused by the bulky CF_3 groups, which disrupted close

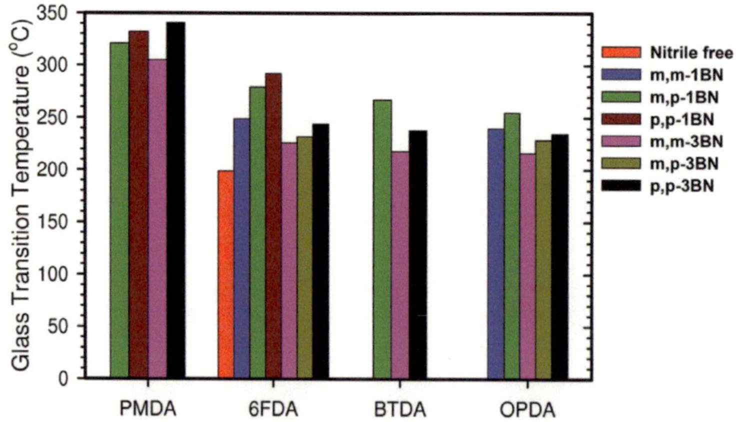

Figure 20.
Glass transition temperatures of PIs prepared from various dianhydrides: PMDA, 6FDA, BTDA, OPDA, and APB diamine isomers containing 0, 1 or 3 benzonitrile (BN) groups. Data adapted from [75–77].

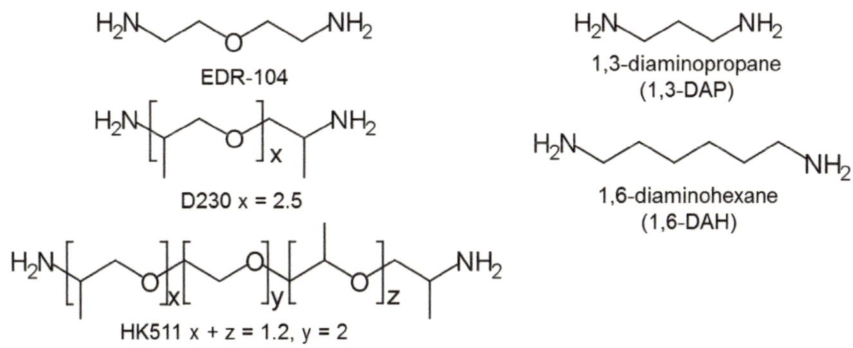

Figure 21.
Chemical structures of polyether aliphatic diamines: Jeffamines EDR-104, D230, and HK511, and linear alkyl diamines of 1,3-DAP and 1,6-DAH.

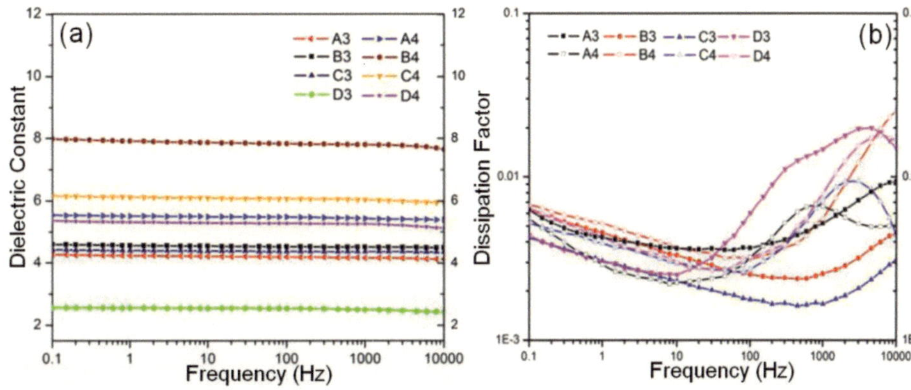

Figure 22.
(a) Dielectric constant and (b) dissipation factor at room temperature as a function of frequency for various ether-containing PIs with A–D represent PMDA, BTDA, OPDA, and 6FDA, respectively, while 3 denotes D230 polyether diamine and 4 is HK511. Figure reproduced from Figure 2 of [80] with permission from American Chemical Society.

packing. In comparison, the ether-free PIs, namely PMDA/1,3-DAP, BTDA/1,3-DAP, and BTDA/1,6-DAH exhibited dielectric constants of 5.6, 4, and 3.6, respectively. While the dielectric constants at room temperature were frequency independent in the range from 0.1 Hz to 10 kHz; the dissipation factor of the ether-containing PIs showed a strong frequency dependence with dielectric relaxation peaks of 1–2% occurring at 1 kHz and above, except for BTDA/D230 and OPDA/D230, which remained below 1% at 10 kHz (**Figure 22b**). In terms of temperature dependence, the ether-containing PIs showed almost an order of magnitude increase in dissipation factor up to 1% at frequencies of 1 kHz and 10 kHz as the temperature approached their corresponding T_g (50–100°C) (**Figure 23**). The low T_g was attributed to free volume created by the methyl side groups in the two diamines, D230 and HK511. In comparison, the two ether-free PIs, BTDA/1,3-DAP and BTDA/1,6-DAH, which were prepared from linear alkyl diamines, showed T_g of 175°C and 150°C, with crystal melting temperature of 271 and 234°C, respectively.

Another group of researchers utilized bipyridines and bipyrimidines, which are electron-rich diamines, to enhance the dielectric constant of PIs [82, 83]. The compounds of interest included 5,5′-bis(4-aminophenoxy)-2,2′-bipyridine (BPBPA) and 5,5′-bis(4-aminophenoxy)-2,2′-bipyrimidine (BAPBP) (**Figure 24**). The dianhydrides used in the study were BTDA, OPDA, PMDA, and BPDA

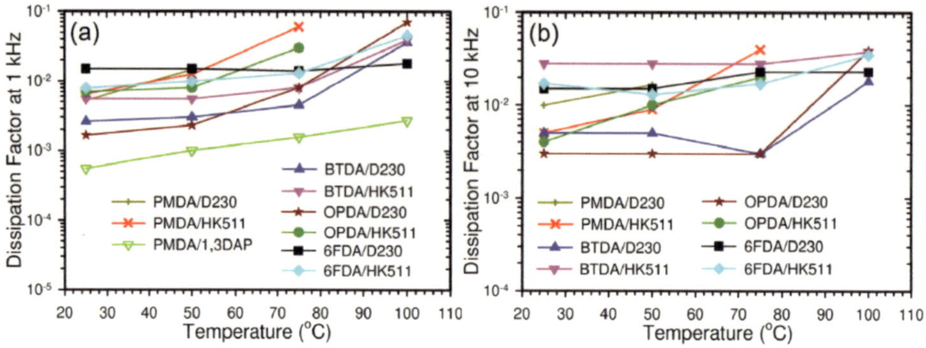

Figure 23.
Dissipation factor at (a) 1 kHz and (b) 10 kHz as a function of temperature for various ether-containing PIs (solid symbols) and an ether-free analog (inverted green triangle). Data adapted from [80, 81].

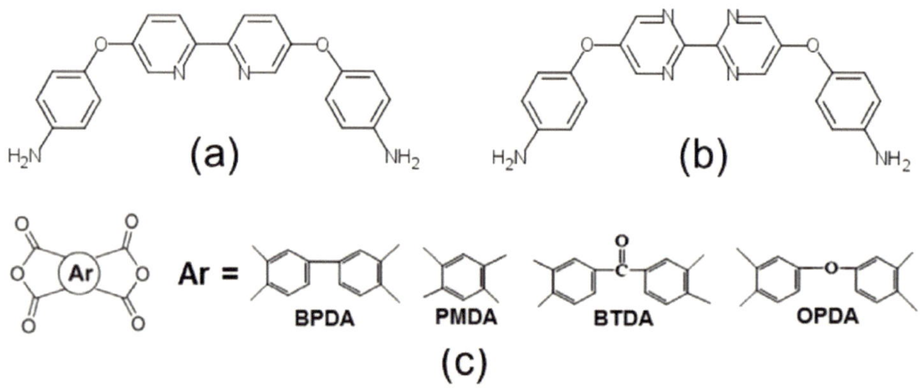

Figure 24.
Chemical structures of (a) 5,5′-bis(4-aminophenoxy)-2,2′-bipyridine (BPBPA) and (b) 5,5′-bis (4-aminophenoxy)-2,2′-bipyrimidine (BAPBP), and (c) various dianhydrides.

(3,3′,4,4′-biphenyltetracarboxylic dianhydride). The resulting PIs with BPBPA diamine possessed relatively high T_g, which differed depending on the dianhydride unit, giving a decreasing order PMDA (320°C) > BTDA (296°C) > BPDA (285°C) > OPDA (275°C). Analogously, the PI containing BAPBP diamine and BPDA dianhydride showed a T_g of 291°C. The dielectric constants for the BPBPA-based PIs ranged from ~5.5 for the PMDA-based PI to ~6.9 for the BTDA-based at room temperature and frequencies from 1 to 100 kHz, following a decreasing order of BTDA > BPDA > OPDA > PMDA (**Figure 25**). At 220°C, the dielectric constants decreased about 4% with the same decreasing trend with respect to the dianhydrides over the same frequency range. The dissipation factor for all the BPBPA-based PIs was below 4% at both room temperature and 220°C from 100 Hz to 100 kHz. For the BAPBP/BPDA PI, the dielectric response at room temperature was similar to that of the BPBPA analog.

The demand for even higher operating temperatures (~350°C) for avionics prompted researchers to develop new polymers with increased degrees of aromaticity and heterocyclic rings in the polymer backbone. In the work by Venkat et al. [84], one of the polymers synthesized was a fluorinated polyimide based on 6FDA and a diamine of 2,2-bis(4-aminophenyl) hexafluoropropane grafted with adamantane (ADE) ester pendant groups (**Figure 26**). The T_g of the PI-ADE is 305°C with a dielectric constant of 2.85–2.91 for the temperature range of 25–250°C at 10 kHz while the dissipation factor increased from 0.6% at 25°C to 0.8% at 250°C (**Figure 27**).

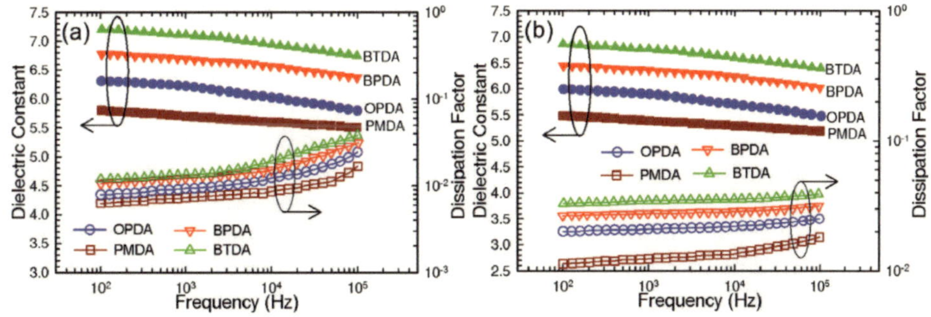

Figure 25.
Dielectric constant and dissipation factor as a function of frequency for the BPBPA-based PIs with various dianhydrides at (a) room temperature and (b) 220°C. data adapted from [82] and supporting information.

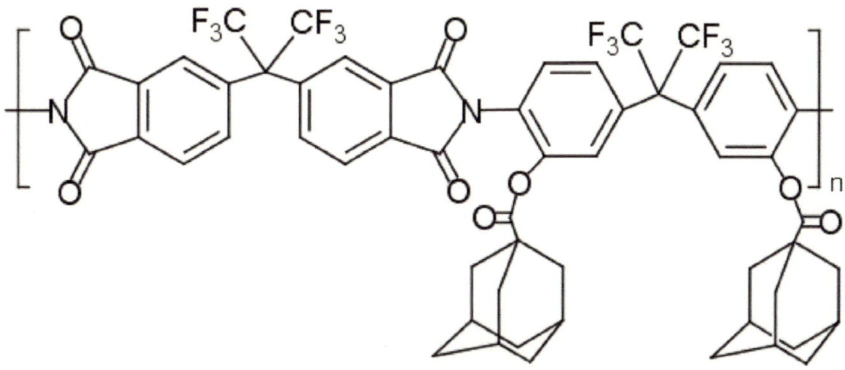

Figure 26.
Chemical structure of PI-ADE. Structures adapted from [84].

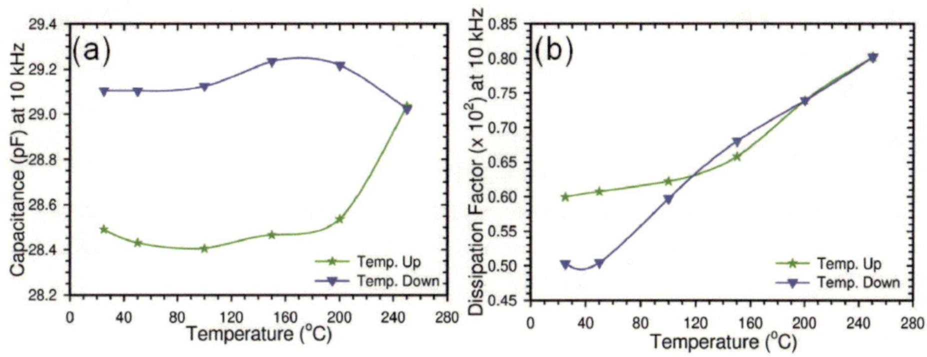

Figure 27.
Capacitance (a) and dissipation factor (b) of PI-ADE as a function of temperature at 10 kHz. Data adapted from [84].

5. Outlook

Researchers have achieved some progress in developing polyimides for capacitor dielectrics targeted for operating temperature above 150°C, but there is still major room for improvement. There is a promising new strategy of combining polymer synthesis and computational techniques (e.g., computational quantum mechanics and molecular dynamics) to synergistically search the material space for polymers with desirable material properties (dielectric constant, glass transition temperature, etc.) for capacitor applications. The ultimate goal is to develop synthesis pathways for polymers that are as close as possible to those "discovered" through computational methods [85–89]. The synergy comes in (i) directing the focus of synthesis and (ii) providing data which simplifies characterization of synthesized polymers through computation of likely crystalline structures (with relative energy differences), IR spectra, and NMR spectra. Work in this area is still relatively limited in that some critical properties cannot be estimated through computational methods, including thermal conductivity, dielectric breakdown field (other than intrinsic), and dielectric loss at frequencies of practical interest (in the kHz region). These deficiencies in the present state of the art result in the approach being somewhat "hit or miss," but with greater guidance than conventional trial and error approaches. For example, a computational approach uncovered a new chemistry based on organotin esters that was subsequently synthesized and shown to provide greater dielectric constant than is normally available from organic polymers while maintaining a reasonable band gap [90]. This interplay between predictive modeling and synthetic chemistry will become increasingly productive with integration of additional material properties into the explored materials space, and polyimides will certainly be one material class of continued interest to researchers.

6. Concluding remarks

The use of power electronics in a growing list of high temperature and high voltage (>500 V) applications currently requires voltage derating and/or active cooling of capacitors with state of the art polymer dielectrics (such as BOPP). High temperature polymers such as polyimides offer promising advantages over the status quo by eliminating this need for voltage derating, easing thermal management requirements, and providing a failsafe for HV applications. However, the lack of commercial interest in processing some polyimides or other specialty polymers as thin films for these comparatively niche polymer applications currently limits implementation of high temperature polymers.

Acknowledgements

This chapter is dedicated in memory to Dr. Steven Boggs who was a rigorous and passionate researcher, a selfless mentor, an intellectual sounding board, a great inspiration, and a truly amazing friend.

Conflict of interest

The authors declare no conflict of interest.

Author details

Janet Ho* and Marshall Schroeder
US Army Research Laboratory, Adelphi, MD, USA

*Address all correspondence to: janet.s.ho.civ@mail.mil

References

[1] Demcko RS, editor. Evolution of high-temperature capacitors. In: 38th IEEE Electronics Components Conference, Los Angeles, CA, USA; 9-11 May 1988

[2] Dreike PL, Fleetwood DM, King DB, Sprauer DC, Zipperian TE. An overview of high-temperature electronic device technologies and potential applications. IEEE Transactions on Components, Packaging, and Manufacturing Technology-Part A. 1994;**17**(4):594-609

[3] Buttay C, Planson D, Allard B, Bergogne D, Bevilacqua P, Joubert C, et al. State of the art of high temperature power electronics. Materials Science and Engineering B. 2011;**176**(4):283-288

[4] Watson J, Castro G. High-temperature electronics pose design and reliability challenges. Analog Dialogue. 2012;**46**(7):3-9

[5] Caliari L, Bettacchi P, Boni E, Montanari D, Gamberini A, Barbieri L, et al. KEMET film capacitors for high temperature, high voltage and high current. In: CARTS International. Houston, TX, USA; 25-28 March 2013

[6] Prevallet M, Bagdy S, Prymak J, Randall M. High voltage considerations with MLCs. In: IEEE International Power Modulator Symposium and High Voltage Workshop. San Francisco, CA, USA; 23-26 May 2004

[7] Reed CW, Cichanowski SW. The fundamentals of aging in HV polymer-film capacitors. IEEE Transactions on Dielectrics and Electrical Insulation. 1994;**1**(5):904-922

[8] Boggs SA, Ho J, Jow TR. Overview of laminar dielectric capacitors. IEEE Electrical Insulation Magazine. 2010;**26**(2):7-13

[9] Gebbia M. Introduction to Film Capacitors. 2009. Available from: https://www.illinoiscapacitor.com/pdf/Papers/introduction_to_film.pdf

[10] Weachock RJ, Liu DD. Failure analysis of dielectric breakdowns in base-metal electrode multilayer ceramic capacitors. In: CARTS International. Houston, TX, USA; 25-28 March 2013

[11] Pan M-J, Randall CA. A brief introduction to ceramic capacitors. IEEE Electrical Insulation Magazine. 2010;**26**(3):44-50

[12] Watson J, Castro G. A review of high-temperature electronics technology and applications. Journal of Materials Science: Materials in Electronics. 2015;**26**(12):9226-9235

[13] Sroog CE. History of the invention and development of the polyimides. In: Ghosh MK, Mittal KL, editors. Polyimides: Fundamentals and Applications. New York, USA: Marcel Dekker, Inc.; 1996. pp. 1-6

[14] Cozens JH. Development of plastic dielectric capacitors. IRE Transactions on Component Parts. 1959;**6**(2):114-118

[15] Carter MA. Is There a Substitute for Polycarbonate Film Capacitors? 2002. Available from: http://powerelectronics.com/site-files/powerelectronics.com/files/archive/powerelectronics.com/mag/Carter%20April%202002.pdf

[16] Critchley JP, Knight GJ, Wright WW. Polymers with heterocyclic rings in the chain: Polyimides. In: Heat-Resistant Polymers: Technologically Useful Materials. New York, USA: Plenum Press; 1983. pp. 186-258

[17] Odian G. Principles of Polymerization. 3rd ed. New York, USA: John Wiley & Sons, Inc.; 1991. p. 313

[18] Feger C, Franke H. Polyimides in high-performance electronics packaging

and optoelectronic applications. In: Ghosh MK, Mittal KL, editors. Polyimides: Fundamentals and Applications. New York, USA: Marcel Dekker, Inc.; 1996. pp. 759-814

[19] Pasquini N. Polypropylene—The business. In: Pasquini N, editor. Polypropylene Handbook. 2nd ed. Munich, Germany: Hanser; 2005. pp. 489-571

[20] DeMeuse MT. Other polymers used for biaxial films. In: DeMeuse MT, editor. Biaxial Stretching of Film: Principles and Applications. Cambridge, UK: Woodhead Publishing Limited; 2011. pp. 47-59

[21] Solvay. Ryton PPS Applications. 2020. Available from: http://www.solvay.com/en/markets-and-products/featured-products/Ryton-Applications.html

[22] Dupont Teijin Films, Mylar PET Polyester Film and Teonex PEN Polyester Film for Use as Capacitor Dielectric. 2017. Available from: http://usa.dupontteijinfilms.com/marketspaces/electricalcomponents/capacitors.aspx

[23] Tekra. PEN Film Teonex Product Outline. 2020. Available from: https://www.tekra.com/products/brands/dupont-teijin-films/teonex

[24] Odian G. Aromatic Polysulfides. In: Principles of Polymerization. 3rd ed. New York, USA: John Wiley & Sons, Inc.; 1991. p. 158

[25] Barber P, Balasubramanian S, Anguchamy Y, Gong S, Wibowo A, Gao H, et al. Polymer composite and nanocomposite dielectric materials for pulse power energy storage. Materials. 2009;**2**(4):1697-1733

[26] Hao X. A review on the dielectric materials for high energy-storage application. Journal of Advanced Dielectrics. 2013;**3**(1):1330001

[27] Qi L, Petersson L, Liu T. Review of recent activities on dielectric films for capacitor applications. Journal of International Council on Electrical Engineering. 2014;**4**(1):1-6

[28] Chen Q, Shen Y, Zhang S, Zhang QM. Polymer-based dielectrics with high energy storage density. Annual Review of Materials Research. 2015;**45**:433-458

[29] Huan TD, Boggs S, Teyssedre G, Laurent C, Cakmak M, Kumar S, et al. Advanced polymeric dielectrics for high energy density applications. Progress in Materials Science. 2016;**83**:236-269

[30] Prateek, Thakur VK, Gupta RK. Recent progress on ferroelectric polymer-based nanocomposites for high energy density capacitors: Synthesis, dielectric properties, and future aspects. Chemical Reviews. 2016;**116**(7):4260-4317

[31] Tan D, Zhang L, Chen Q, Irwin P. High temperature capacitor polymer films. Journal of Electronic Materials. 2014;**43**(12):4569-4575

[32] Ho J, Greenbaum SG. Polymer capacitor dielectrics for high temperature applications. ACS Applied Materials & Interfaces. 2018;**10**:29189-29218

[33] Ennis JB, MacDougall FW, Yang XH, Bushnell AH, Cooper RA, Gilbert JE. High-specific-power capacitors. In: IEEE International Power Modulators and High Voltage Conference. Las Vegas, NV, USA; 27-31 May 2008

[34] Yang Y. Thermal conductivity. In: Mark JE, editor. Physical Properties of Polymers Handbook. 2nd ed. New York, USA: Springer Science+Business Media, LLC; 2007. pp. 155-164

[35] Qin S, Ho J, Rabuffi M, Borelli G, Jow TR. Implications of the anisotropic thermal conductivity of capacitor

windings. IEEE Electrical Insulation Magazine. 2011;**27**(1):7-13

[36] Li Q, Chen L, Gadinski MR, Zhang S, Zhang G, Li H, et al. Flexible high-temperature dielectric materials from polymer nanocomposites. Nature. 2015;**523**(7562):576-580

[37] Diaham S, Saysouk F, Locatelli M-L, Lebey T. Huge nanodielectric effects in polyimide/boron nitride nanocomposites revealed by the nanofiller size. Journal of Physics D: Applied Physics. 2015;**48**(38):385301

[38] Diaham S, Saysouk F, Locatelli M-L, Lebey T. Huge improvements of electrical conduction and dielectric breakdown in polyimide/BN nanocomposites. IEEE Transactions on Dielectrics and Electrical Insulation. 2016;**23**(5):2795-2803

[39] Cahill PL, Dailey JH. Aircraft Electrical Wet-Wire Arc Tracking. New Jersey, USA: US Department of Transportation/Federal Aviation Administration; August 1988. Report No.: DOT/FAA/CT-88/4

[40] Kurek J, Bernstein R, Etheridge M, LaSalle G, McMahon R, Meiner J, et al. Aircraft Wiring Degradation Study. Washington, DC: US Department of Transportation/Federal Aviation Adminstration: Raytheon Technical Services Company LLC/Federal Aviation Administration; January 2008. Report No.: DOT/FAA/AR-08/2

[41] Steinerfilm. Steinerfilm Dielectrics for Film Capacitors. 2020. Available from: http://www.steinerfilm.de/en/our-products/

[42] Smith DH, Simpson RJ, Geoghegan EDA. Final Report on Manufacturing Methods for Metallized Teflon Capacitors (Subminiature 200°C). Ohio, USA: Aeronautical Systems Division Air Force Systems Command: Dearborn Electronic Laboratories, Inc.; 1963 April. Report No.: Technical Documentary Report Nr. ASD-TDR-63-308

[43] TDK. Film Capacitors: General Technical Information. 2018. Available from: https://en.tdk.eu/download/530754/bb7f3c742f09af6f8ef473fd34f6000e/pdf-generaltechnicalinformation.pdf

[44] Blythe AR, Bloor D. Dielectric relaxation. In: Electrical Properties of Polymers. 2nd ed. Cambridge, UK: Cambridge University Press; 2005. pp. 58-110

[45] McCrum NG, Read BE, Williams G. Phenomenological theories of mechanical and dielectric relaxation. In: Anelastic and Dielectric Effects in Polymeric Solids. New York, USA: Dover Publications, Inc.; 1991. pp. 102-140

[46] Lidow A, Strydom J, Rooij M, Reusch D. Replacing silicon power MOSFETs. In: GaN Transistors for Efficient Power Conversion. 2nd ed. West Sussex, UK: John Wiley & Sons Ltd.; 2015. pp. 232-239

[47] Cornell Dubilier Capacitors. Application Guide Snubber Capacitors. 2016. Available from: http://www.cde.com/resources/catalogs/igbtAPPguide.pdf

[48] Severns R. Design of Snubbers for Power Circuits. 2016. Available from: http://www.cde.com/resources/technical-papers/design.pdf

[49] Electrocube. High Current DC Link Film Capacitors. 2020. Available from: https://www.electrocube.com/pages/958a-series-metallized-polypropylene-dc-link-capacitors-data-sheet?_pos=10&_sid=9c2533969&_ss=r

[50] Colella T. How to Select a DC Link Capacitor. 2020. Available from: https://www.electrocube.com/pages/how-to-select-dc-link-capacitor-data-sheet

[51] Laihonen SJ, Gafvert U, Schutte T, Gedde UW. DC breakdown strength of polypropylene films: Area dependence and statistical behavior. IEEE Transactions on Dielectrics and Electrical Insulation. 2007;**14**(2):275-286

[52] Xu C, Ho J, Boggs SA. Automatic breakdown voltage measurement of polymer films. IEEE Electrical Insulation Magazine. 2008;**24**(6):30-34

[53] Schneider MA, MacDonald JR, Schalnat MC, Ennis JB. Electrical breakdown in capacitor dielectric films: Scaling laws and the role of self-healing. In: IEEE International Power Modulator and High Voltage Conference. San Diego, CA, USA; 3-7 June 2012

[54] Dissado LA, Fothergill JC. Statistical features of breakdown. In: Stevens GC, editor. Electrical Degradation and Breakdown in Polymers. London, UK: Peter Peregrinus Ltd.; 1992. pp. 319-355

[55] Blythe AR, Bloor D. Dielectric breakdown. In: Electrical Properties of Polymers. 2nd ed. Cambridge, UK: Cambridge University Press; 2005. pp. 186-216

[56] Ho J, Ramprasad R, Boggs S. Effect of alteration of antioxidant by UV treatment on the dielectric strength of BOPP capacitor film. IEEE Transactions on Dielectrics and Electrical Insulation. 2007;**14**(5):1295-1301

[57] Ho J, Jow R. Characterization of High Temperature Polymer Thin Films for Power Conditioning Capacitors. Adelphi, MD, USA: US Army Research Laboratory; 2009 July. Report No.: ARL-TR-4880

[58] Frohlich H. The theory of dielectric breakdown in solids. Proceedings of the Royal Society A. 1947;**188**(1015):521-532

[59] Simmons JG. Poole-Frenkel effect and Schottky effect in metal-insulator-metal systems. Physical Review. 1967;**155**(3):657-660

[60] Lampert MA. Current Injection in Solids. New York, USA: Academic Press; 1970

[61] Fukuma M, Nagao M, Kosaki M. Numerical analysis of dielectric breakdown in polypropylene film based on thermal and electronic composite breakdown model. In: Conference on Properties and Applications of Dielectric Materials. Tokyo, Japan; 8-12 July 1991

[62] O'Dwyer JJ, Beers BL. Thermal breakdown in dielectrics. In: Conference on Electrical Insulation and Dielectric Phenomena. Whitehaven, PA, USA; 26-28 October 1981

[63] Dissado LA, Fothergill JC. Thermal breakdown. In: Stevens GC, editor. Electrical Degradation and Breakdown in Polymers. London, UK: Peter Peregrinus Ltd.; 1992. pp. 242-262

[64] Jones RJ, Wright WF. High Temperature Polymer Dielectric Film Insulation. Aero Propulsion and Power Directorate Wright Laboratory Air Force Systems Command Wright-Patterson Air Force Base: TRW Space and Defense. Ohio, USA; 1992 February. Report No.: WL-TR-91-2105. Available from: http://www.dtic.mil/dtic/tr/fulltext/u2/a255243.pdf

[65] Jones RJ. High temperature polymer dielectric film insulation. In: 2nd NASA Workshop on Wiring for Space Applications. Cleveland, OH, USA; 6-7 October 1993

[66] Donhowe M, Lawler J, Souffie S, Lee Stein Jr E. 250°C operating temperature dielectric film capacitors. In: International Microelectronics Assembly and Packaging Society, High Temperature Electronics Network. Oxford, UK; 18-20 July 2011

[67] Vora RH, Krishnan PSG, Goh SH, Chung T-S. Synthesis and properties of designed low-k fluoro-copolyetherimides. Part I. Advanced Functional Materials. 2001;**11**(5):361-373

[68] Mandelcorn L, Miller RL. High temperature, >200 deg. C, polymer film capacitors. In: IEEE 35th International Power Sources Symposium, Cherry Hill, NJ, USA; 1992

[69] Sasaki S, Nishi S. Synthesis of fluorinated polyimides. In: Ghosh MK, Mittal KL, editors. Polyimides: Fundamentals and Applications. New York, USA: Marcel Dekker, Inc.; 1996. pp. 71-120

[70] Ghosh A, Mistri EA, Banerjee S. Fluorinated polyimides: Synthesis, properties, and applications. In: Banerjee S, editor. Handbook of Specialty Fluorinated Polymers. Massachusetts, USA: Elsevier; 2015. pp. 97-185

[71] Jones RJ, O'Rell MK, Hom JM, inventors. Polyimides Prepared from Perfluoroisopropylidene Diamine patent US 4111906; 5 September 1978

[72] Jones RJ, Chang GE, Powell SH, Green HE. Polyimide Matrix Resins for up to 700 degree F Service. Ohio, USA: TRW Engery Development Group; 1985. Report No.: N86-I1280-NTRS-NASA

[73] Kochi M, Yonezawa T, Yokota R, Mita I. Monoaxial drawing techniques for high modulus/high strength aromatic polyimide films. In: Feger C, Khojasteh MM, Htoo MS, editors. Advances in Polyimide Science and Technology. Pennsylvania, USA: Technomic Publishing Company, Inc.; 1993. p. 376

[74] Clagett DC. Engineering plastics. In: Mark HF, Bikales NM, Overberger CG, Menges G, editors. Encyclopedia of Polymer Science and Engineering. Vol. 6. New York, USA: Wiley-Interscience; 1986. pp. 94-131

[75] Wang DH, Riley JK, Fillery SP, Durstock MF, Vaia RA, Tan L-S. Synthesis and characterization of unsymmetrical benzonitrile-containing polyimides: Viscosity-lowering effect and dielectric properties. Journal of Polymer Science Part A: Polymer Chemistry. 2013;**51**:4998-5011

[76] Wang DH, Kurish BA, Treufeld I, Zhu L, Tan L-S. Synthesis and characterization of high nitrile content polyimides as dielectric films for electrical energy storage. Journal of Polymer Science Part A: Polymer Chemistry. 2015;**53**(3):422-436

[77] Treufeld I, Wang DH, Kurish BA, Tan L-S, Zhu L. Enhancing electrical energy storage using polar polyimides with nitrile groups directly attached to the main chain. Journal of Materials Chemistry A. 2014;**2**:20683-20696

[78] Clair AKS, Clair TLS, Shevket KI. Synthesis and characterization of essentially colorless polyimide films. Polymeric Materials Science and Engineering. 1984;**51**:62-66

[79] Jacobs JD, Arlen MJ, Wang DH, Ounaies Z, Berry R, Tan L-S, et al. Dielectric characteristics of polyimide CP2. Polymer. 2010;**51**:3139-3146

[80] Ma R, Baldwin AF, Wang C, Offenbach I, Cakmak M, Ramprasad R, et al. Rationally designed polyimides for high-energy density capacitor applications. ACS Applied Materials & Interfaces. 2014;**6**(13):10445-10451

[81] Baldwin AF, Ma R, Wang C, Ramprasad R, Sotzing GA. Structure-property relationship of polyimides based on pyromellitic dianhydride and short-chain aliphatic diamines for dielectric material applications. Journal of Applied Polymer Science. 2013;**130**(2):1276-1280

[82] Peng X, Wu Q, Jiang S, Hanif M, Chen S, Hou H. High dielectric constant polyimide derived from 5,5'-bis[(4-amino) phenoxy]-2,2'-bipyrimidine. Journal of Applied Polymer Science. 2014;**131**(24):40828

[83] Peng X, Xu W, Chen L, Ding Y, Xiong T, Chen S, et al. Development of high dielectric polyimides containing bipyridine units for polymer film capacitor. Reactive and Functional Polymers. 2016;**106**:93-98

[84] Venkat N, Dang TD, Bai Z, McNier VK, DeCerbo JN, Tsao B-H, et al. High temperature polymer film dielectrics for aerospace power conditioning capacitor applications. Journal of Materials Science and Engineering B. 2010;**168**(1-3):16-21

[85] Sukumar N, Krein M, Luo Q, Breneman C. MQSPR modeling in materials informatics: A way to shorten design cycles? Journal of Materials Science. 2012;**47**:7703-7715

[86] Pilania G, Wang CC, Jiang X, Rajasekaran S, Ramprasad R. Accelerating materials property predictions using machine learning. Scientific Reports. 2013;**3**

[87] Sharma V, Wang CC, Lorenzini RG, Ma R, Zhu Q, Sinkovits DW, et al. Rational design of all organic polymer dielectrics. Nature Communications. 2014;**5**:1-8

[88] Huan TD, Mannodi-Kanakkithodi A, Ramprasad R. Accelerated materials property predictions and design using motif-based fingerprints. Physical Review B. 2015;**92**:014106

[89] Wu K, Sukumar N, Lanzillo NA, Wang C, Ramprasad R, Ma R, et al. Prediction of polymer properties using infinite chain descriptors (ICD) and machine learning: Toward optimized dielectric polymeric materials. Journal of Polymer Science Part B: Polymer Physics. 2016;**54**:2082-2091

[90] Baldwin AF, Huan TD, Ma R, Mannodi-Kanakkithodi A, Tefferi M, Katz N, et al. Rational design of organotin polyesters. Macromolecules. 2015;**48**:2422-2428

Chapter 9

High-Temperature Polyimide Dielectric Materials for Energy Storage

Jun-Wei Zha, Xue-Jie Liu, Yaya Tian, Zhi-Min Dang and George Chen

Abstract

The availability of high-temperature dielectrics is key to develop advanced electronics and power systems that operate under extreme environmental conditions. In the past few years, many improvements have been made and many exciting developments have taken place. However, currently available candidate materials and methods still do not meet the applicable standards. Polyimide (PI) was found to be the preferred choice for high-temperature dielectric films development due to its thermal stability, dielectric properties, and flexibility. However, it has disadvantages such as a relatively low dielectric permittivity. This chapter presents an overview of recent progress on PI dielectric materials for high-temperature capacitive energy storage applications. In this way, a new molecular design of the skeleton structure of PI should be performed to balance size and thermal stability and to optimize energy storage property for high-temperature application. The improved performance can be generated via incorporation of inorganic units into polymers to form organic-inorganic hybrid and composite structures.

Keywords: polyimide, thermal property, dielectric property, energy storage, nanocomposites

1. Introduction

With the rapid development of the global economy and a rising population, the search for efficient and clean energy and energy storage technologies has become a priority worldwide. Because of its exceptionally fast energy conversion rate, long life, and environmental friendliness, dielectric energy storage technology has been used in applications for the electronics and power industries such as wearable electronic devices, hybrid vehicles, and weapon systems [1]. As the trend toward high-performance miniaturized electronic devices continues, the demand for dielectric materials with high energy storage density (U_e) is increasing. U_e is an important parameter to measure the energy storage performance of dielectric materials:

$$U_e = 1/2\varepsilon_r\varepsilon_o{E_b}^2 \qquad (1)$$

where ε_r is the permittivity of material and ε_o is the permittivity of free space (8.85×10^{-12} F m^{-1}) [2]. This requires that the dielectric material has a high ε_r while having a low dielectric loss and a high breakdown strength.

Commonly known high-energy storage dielectric materials are mainly biaxially oriented polypropylene (BOPP), polyester, polycarbonate (PC), polyphenylene disulfide, polyurea, polyurethane, and polyvinylidene fluoride [3]. Among many polymers, polyimide (PI) is a type of polymer containing an imide ring on the main chain [4]. PI is widely used in packaging materials, insulation layers, circuit boards, and interlayer dielectrics due to its high tensile strength, excellent mechanical properties, high glass transition temperature (T_g), and good solvent resistance and thermal stability [5]. However, the ε_r of polyimide is not sufficiently high (usually less than 10) to meet the requirements of the applications of high-energy density film capacitors.

The chemical groups of a dielectric medium contribute to its molar polarization; as the molar polarization increases, the ε_r increases. The dielectric properties (including ε_r and dielectric loss) of polymers are mainly related to molecular polarization, which includes electron polarization, vibration polarization (or atomic polarization), orientation polarization (or dipole polarization), ion polarization, and interfacial polarization. However, low-quality/purity polar molecules can reduce the dielectric properties of PI materials [6].

Ma et al. [7] used high-throughput density functional theory (DFT) to rationally design high ε_r and band gaps and linked experimental and theoretical results to changes in PI to demonstrate the relationship between chemical functionality and dielectric properties. Currently, researchers usually use two methods to prepare polyimide film capacitors with high ε_r, low dielectric loss, and high breakdown strength. One method is directly based on the molecular design of polyimide: Polar groups, conjugated components, or electron-rich groups are introduced into the main polymer chain to increase molecular polarizability, thereby increasing the ε_r [8]. The other method, which is currently the most studied, prepares a composite material by introducing high-ε_r ferroelectric materials such as TiO_2, $BaTiO_3$ (BT), Pb (Zr, Ti)O_3, etc. into the polymer matrix to significantly improve the dielectric properties [5].

2. Commercial PI

Kapton PI is an aromatic PI film that has been commercially available since the mid-1960s. Due to its continuous operating temperature of 300–350°C, it is widely used as a high-temperature wire and cable insulation material. At 25°C and 1 kHz, Kapton's ε_r is 3.1, but it drops to 2.8 at 300°C. Despite its good thermal stability, Kapton PI cannot be applied to capacitor films because it is difficult to manufacture films with a thickness of <12 μm, and problems of carbonization during breakdown. Consequently, research is required to find other PIs with superior dielectric properties. SIXEF-44 is a fluorinated PI (from Hoechst Celanese) prepared from 2,2-bis(3,4-dicarboxyphenyl)-1,1,1,3,3,3-hexafluoropropane dianhydride (6FDA) and 2,2-bis(4-aminophenyl)hexafluoropropane (4,4′-6F diamine). This fluorinated PI has a ε_r of 2.8 at 1 kHz, T_g of 323°C, and a change in ε_r of less than 10% over a temperature range of −55 to 300°C. Other aromatic PIs include: perfluoropolyimide (PFPI; developed by TRW), prepared from the perfluoroisopropylidene diamine of 2,2-bis[4-(4-aminophenoxy)phenyl]hexafluoropropane (4-BDAF) and pyromellitic dianhydride (PMDA); and Upilex-S (from ICI), prepared from 3,3′,4,4′-biphenyltetracarboxylic dianhydride (BPDA) and p-phenylenediamine

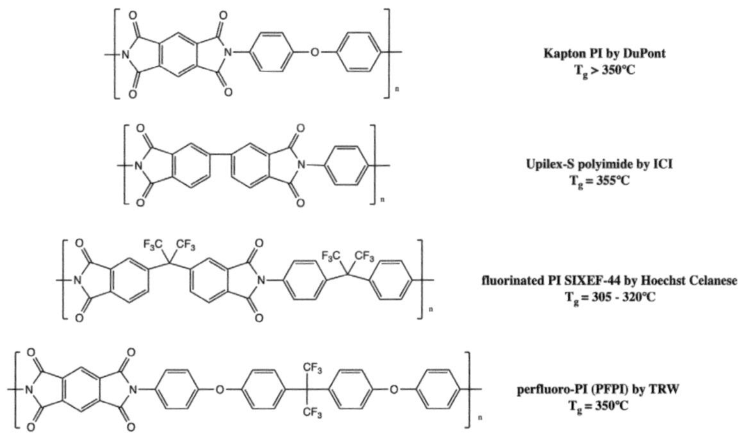

Figure 1.
Heat-resistant PI polymers used as capacitor dielectrics [9].

(p-PDA). The T_g of PFPI is >300°C, the ε_r is 3.1 at 25°C, and decreases to 2.9 at 300°C; Upilex-S has a T_g of 355°C, ε_r of 3.3 at 25°C and 1 kHz, and remains stable at 300°C. Most of these aromatic PIs affect practical applications due to processing difficulties. A polyetherimide (PEI) called Ultem is modified by the addition of flexible moieties such as ether bonds and alkyl groups in the polymer backbone; it is synthesized from the disodium salt of bisphenol A and 1,3-bis(4-nitrophthalimido)benzene. After development, the PEI film can attain a thickness of 5 μm by melt extrusion and stretching. In order to give PEI flexibility, ether bonds and alkyl groups are added, which results in the following changes (compared to PI): the T_g reduces to circa 215°C; the ε_r increases to 25 at 200°C, and the εr over 100 Hz–10 kHz is 3.1 [9]. Some of the heat-resistant PI polymers used as capacitor dielectrics are shown in **Figure 1**.

3. Structure modification of PI

The relationship between the dielectric properties of PI and molecular structure can be studied by changing the structure of the aromatic tetracarboxylic dianhydride and diamine monomers used to prepare PI. However, the preparation of the aromatic tetracarboxylic dianhydride is often complex and the yield is low while the synthetic method for phenyl-substituted aromatic diamine is relatively simple, diverse, and high yield. Consequently, modifying the structure of the aromatic diamine monomers has become the primary choice to improve the properties of PIs [10].

3.1 Introduction of bispyridine groups in PI chains

Peng et al. [3] used 5,5′-bis[(4-amino)phenoxy]-2,2′-bipyridine (BPBPA) diamine monomer (as shown in **Figure 2**) and different dianhydrides [BPDA, PMDA, 3,3′,4,4′-Benzophenonetetracarboxylic dianhydride (BTDA), 4,4′-Oxydiphthalic anhydride (OPDA)] to give a series of bispyridyl-containing PIs using a two-step synthesis. The bipyridyl unit enhanced the electronic polarization and coupling: The polarized PI had a ε_r of ≤7.2, the dielectric loss was ≥0.04, and the energy density was ≤2.77 J cm^{-3}. At the same time, it demonstrated good thermal and mechanical properties.

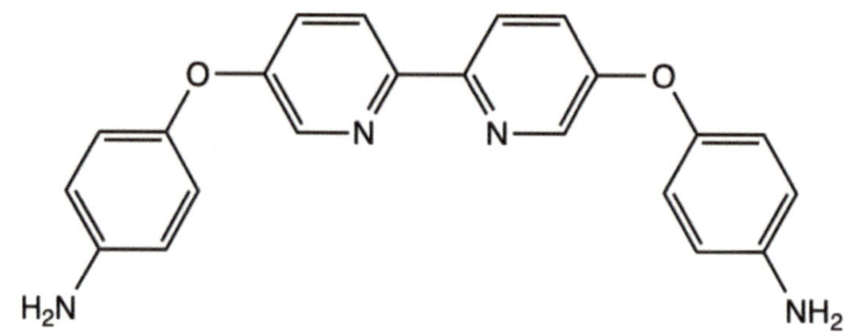

5,5'-bis[(4-amino)phenoxy]-2,2'-bipyridine (BPBPA,4)

Figure 2.
Dipyridyl-containing diamine monomer [3].

3.2 Introduction of sulfonyl group in PI backbone

Tong et al. [4] studied the relationship between molecular structure and properties using a range of modified PIs. In this study, the εr was increased by introducing sulfonyl groups, the loss factor was reduced by introducing flexible bonds, and the T_g was increased by retaining the aromatic structure. The resulting sulfonyl-containing PI with different flexible connections gave high ε_r (4.50–5.98), low loss coefficients (0.00298–0.00426), high breakdown strength (mostly at 500 MV m^{-1} or more) and high heat resistance (T_g: 244–304°C) (**Figure 3**).

For the anhydride 3,3′,4,4′-diphenylsulfonetetracarboxylic dianhydride (DSDA-mDS, each repeat unit contains two -SO2-, which has the highest dipole density), the ε_r was not as high as expected but it can be seen that the two -SO2- units improved the stiffness of the overall chain, hindering the rotation of the dipole. Therefore, in addition to the dipole moment and dipole density, the "effective" dipole is another important factor affecting the value of the ε_r. Compared to ortho-symmetric OPDA-mDS, para-symmetric PI (OPDA-pDS) was more effective for PI with sulfonyl group (OPDA-pDS) in the diamine moiety (para-para bond). The symmetric structure and low free rotation energy barrier facilitate the alignment of the excimer: The ε_r increased to 5.98; the dielectric properties were stable at 150°C; the discharge energy density and charge and discharge efficiency increased to 7.04 J cm^{-3} and 91.3% at 500 MV m^{-1} respectively [4].

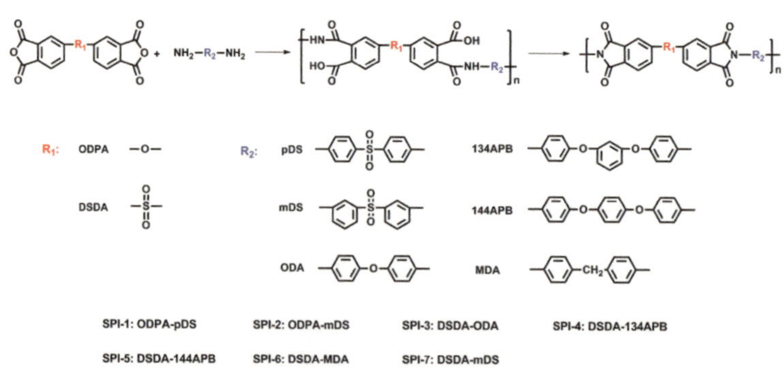

Figure 3.
Synthesis of sulfo-containing PI [4].

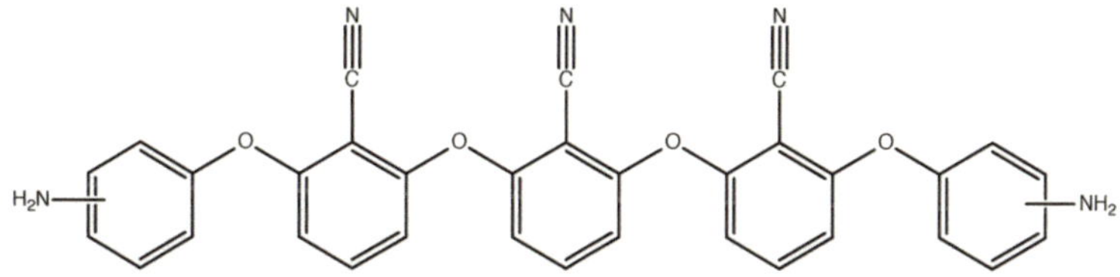

Figure 4.
Diamine monomer containing three nitrile groups [13].

3.3 Introducing nitrile groups in the PI backbone

Due to the high polarity of nitrile groups, special classes of PIs containing nitrile units for piezoelectric and other dielectric applications have previously been studied [11]. Kakimoto et al. [12] reported that attaching a polar -CN side group to a PI could increase its ε_r. Treufeld et al. [11] found that adding a CN dipole to the PI main chain had two main effects: Firstly, because the -CN group is directly connected to the main chain in a 90° configuration, the motion is hindered and generates significant friction with randomly stacked adjacent chains, so dipole motion (such as wobble) is introduced into the PI sample; secondly, by adding the -CN group, the PI becomes more polar, easily contaminated by impurity ions, thereby improving ion mobility. Furthermore, it has been shown that the presence of three nitrile groups on the diamine unit is more effective in improving the ε_r than one nitrile group. Wang et al. [13] studied and synthesized a series of PIs from a diamine synthesized with three nitrile groups (as shown in **Figure 4**) and four commercial dianhydride starting materials. All PIs showed a high T_g, thermal stability and excellent mechanical properties; the PI had a ε_r of 4.7 resulting from the introduction of three highly polar nitrile groups.

3.4 PI grafted with phthalocyanine oligomer at amino terminal

Unlike ordinary composite materials, graft polymers, with good properties, were synthesized by Chen et al. [7]. The copper phthalocyanine oligomer (o-CuPc,

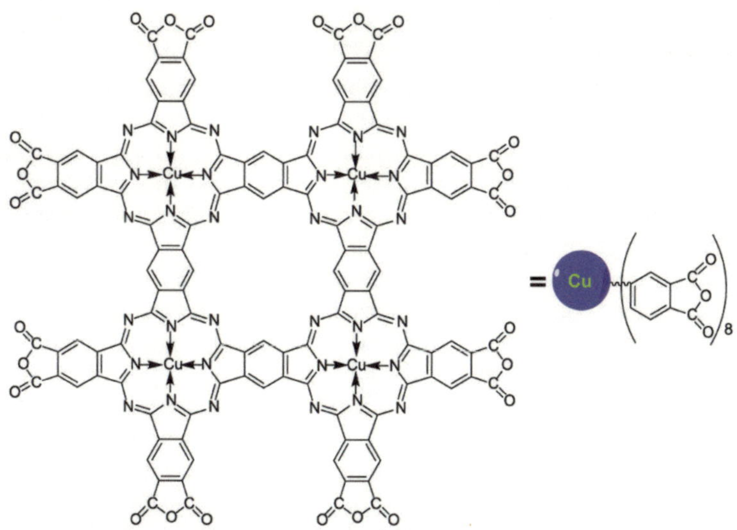

Figure 5.
Synthetic copper phthalocyanine oligomer [7].

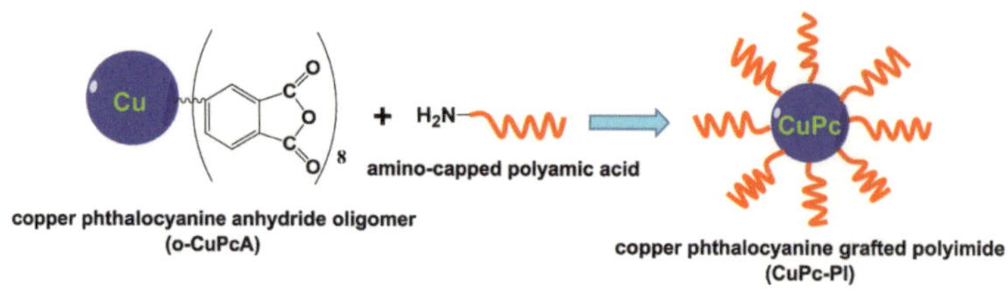

Figure 6.
Schematic diagram of graft reaction [7].

shown in **Figure 5**) is a semiconductor material with unique electrical properties ($\varepsilon_r > 103$) and good thermal stability, used widely in organic optoelectronics, the dye industry, catalysis, electrochromism, and electroluminescence display and other fields. The design and synthesis resulted in a high-ε_r all-organic polymer material, that is, a CuPc-PI homogeneous block copolymer was prepared (see **Figure 6**). The CuPc-PI also showed low dielectric loss, high breakdown strength, high U_e, high thermal stability, and good mechanical properties; its overall performance was higher than the direct use of o-CuPc/PI composites obtained using CuPc as the conductive filler [7].

4. Polyimide-based composite dielectric materials

4.1 Simple compound

Modification of the molecular structure of the polymer can improve its dielectric properties although the effect can be small. Using a simple compounding method with a high ε_r filler (e.g., ceramic filler, conductive filler), a polymer with a high breakdown field strength can be obtained. This procedure has gained acceptance due to its simple preparation method.

The conductive filler polymer-based composite material can attain a high ε_r for relatively small additions of filler, and the large increase in ε_r can be explained by the percolation theory. Adding filler at the percolation threshold will greatly increase the electrical conductivity and ε_r of the composite material, thereby improving the transition layer between the filler and the matrix. Carbon materials such as carbon nanofibers (CNFs), carbon black, carbon nanotubes (CNTs), graphene, and graphite flakes are most commonly used in recent research. Among these conductive fillers, CNTs are a good choice due to their high electrical and thermal conductivity and high aspect ratios. Wu et al. [14] functionalized multi-walled carbon nanotubes (MWCNTs) with carboxyl groups prior to dispersing into PI nanofibers using electrospinning technology. Hot pressing was then performed to produce high-performance PI/MWCNT composites with a high ε_r, good mechanical flexibility, and excellent thermal stability. When the concentration of MWCNT was close to the percolation threshold of 12–14 vol%, the material showed a high ε_r, low breakdown strength, and maximum U_e. When the MWCNT content was 12 vol%, the maximum U_e was 1.957 J cm^{-3}, which was 4.8 times that of pure PI (0.404 J cm^{-3}), and the dielectric loss was less than 0.1. As a two-dimensional nanomaterial, graphene has great potential in the future because it can improve the mechanical, thermal, and electrical properties of polymers. Among these materials, graphene oxide (GO) has also been reported in some articles to improve the mechanical properties and thermal stability of polymer-based composites. Chen et al. [15] prepared pure PI, PI/GO, and PI/reduced GO (rGO) films by in situ

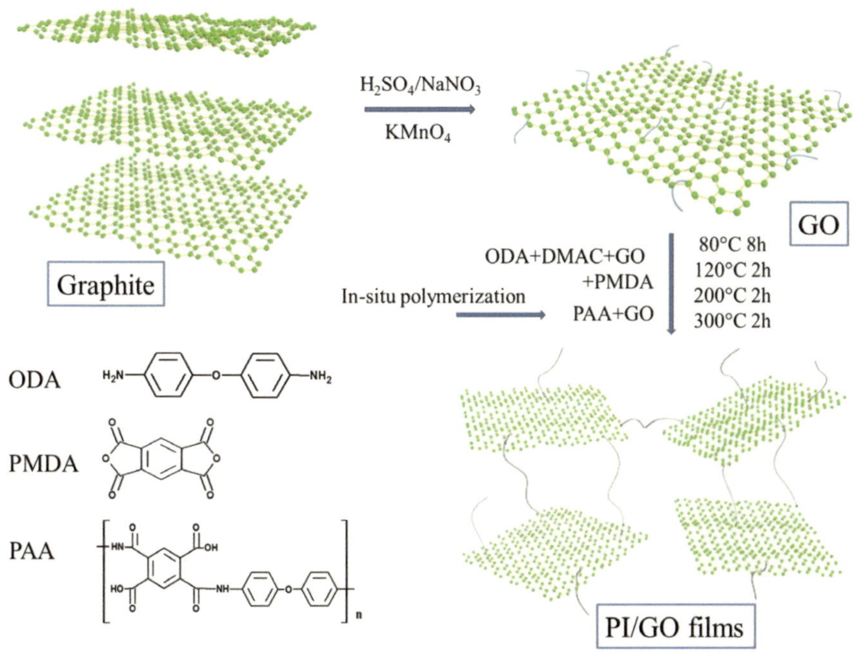

Figure 7.
Schematic illustration of the film preparation procedure for GO and PI/GO composites [15].

polymerization, as shown in **Figure 7**. Among them, PI/GO and PI/rGO films both demonstrated improved thermal stability compared to pure PI films. Furthermore, at 100 Hz, when the mass fractions of GO and rGO were 2 wt%, ε_r were also improved (4.9 and 5.8, respectively).

Adding conductive particles as a filler to the polymer matrix can improve the ε_r of the polymer composite. When the added amount is close to the percolation threshold, the ε_r can be significantly increased. However, as the amount of addition increases, a conductive network is formed, and their dielectric loss will increase sharply. In general, nanostructured BT fillers and BT-based nanocrystals are the more promising materials due to their excellent dielectric and ferroelectric properties [16].

Fan et al. [17] studied the relationship between the ε_r and the temperature for thermosetting PI matrix nanocomposite films containing BT nanoparticles at 10^3 Hz. Two temperature changes were reported, namely heating from 50 to 150°C and cooling from 150 to 50°C to investigate the effects of the transition of the BT crystal phase and the free volume change in PI on the ε_r for BT/PI nanocomposite membranes. Theoretical models were also used to predict the ε_r of composite materials to study the role of the diameter and shape of the nanoparticles. Rajib et al. [18] prepared BT/PI nanocomposites and increased their energy density at high temperatures using different volume fractions to analyze their effect on the dielectric properties. All samples were tested at high temperatures to evaluate their energy storage capacity. The highest U_e was found when the volume fraction of BT was 20% reaching 9.63 J cm^{-3} at 20°C and 6.79 J cm^{-3} at 120°C. As a dielectric material, it is expected to maintain a high energy density value at a temperature of 120°C. A pure PI film prepared by Sun et al. [19] showed high breakdown strength (451 kV mm^{-1}) and high energy density (5.2 J cm^{-3}). The introduction of BT nanoparticles increased the ε_r of the nanocomposite to 6.8, while the dielectric loss was still relatively low (0.012 at 10^4 Hz). However, a small amount of (3 vol%) BT nanoparticles also caused a significant decrease in the breakdown field strength (275 kV mm^{-1}), which greatly reduced the energy density (1.7 J cm^{-3}) of the BT/PI nanocomposite.

Therefore, for BT/PI nanocomposites, future research may concern improvements in the thermal conductivity of nanocomposites and the formation of interpenetrating networks throughout the polymer matrix. Improvements in this area will make nanocomposites less susceptible to breakdown [19]. Wang et al. [20] successfully prepared BT nanowire/PI (BT-NW/PI) and BT nanoparticle/PI (BT-NP/PI) composites with low volume fractions. Due to strong interfacial polarization, the ε_r of BT-NW filled composites was greater than that of BT-NP/PI. The ε_r of the composite containing 5 vol% BT-NW was 6.6 at 100 Hz, which was 94% higher than pure PI (ε_r = 3.4 at 100 Hz) and 22% higher than that of composite containing 10 vol% BT-NP (ε_r = 5.4 at 100 Hz). In addition, BT-NW also significantly improved the U_e of the composite. When the content of BT-NW was 2 vol%, the U_e obtained at 2200 kV cm^{-1} was 1.06 J cm^{-3}, which was 37% greater than pure PI. Therefore, it could be shown that the introduced linear ceramic filler had a positive effect on the dielectric properties and U_e of the composite material [20]. Hu et al. [1] prepared and studied the dielectric properties of a BT nanofiber/PI (BT-NF/PI) composite membrane over the temperature range 20–200°C. The introduction of BT-NF at 9 vol% increased the ε_r for BT-NF/PI to 8.3 while the dielectric loss increased only slightly; these effects could be attributed to dipolar polarization and interfacial displacement of the nanocomposites. The breakdown strength of BT/PI composites containing 1 vol% BT-NF reached 550 kV mm^{-1}, and the discharge energy density reached 5.82 J cm^{-3}. Additionally, the introduction of BT-NF reduced the leakage current and improved the heat conduction. At 1 vol% BT-NF, the PI nanocomposites also exhibited high energy utilization efficiency and good thermal stability. At 150 and 100°C, when the efficiency was greater than 90%, the discharge energy density values were >2.1 J cm^{-3} and ≈4 J cm^{-3}, respectively [1]. The authors used electrospinning to prepare BT-NF while the PI composite membranes were prepared by in situ dispersion polymerization. The dielectric properties of BT-NF/PI composite films in the frequency range of 10^2–10^6 Hz at a temperature of 20–150°C were studied. The results showed that the ε_r of the PI nanocomposite film with 30 vol% BT-NF at 100 Hz increased to ≈27 while the dielectric loss was only 0.015.

Furthermore, the calcination temperature of BT has a significant influence on the ε_r of the PI/BT-nanocomposite film as shown in **Figure 8**. The ε_r of the PI composite film calcined at 1000°C was higher than the PI composite films calcined at 600 and 800°C; when the BT-NF content was 30 vol%, the ε_r of the BT-NF/PI composite film increased to 26.6 [16]. Beier et al. [21] added $Ba_{0.7}Sr_{0.3}TiO_3$ (BST)

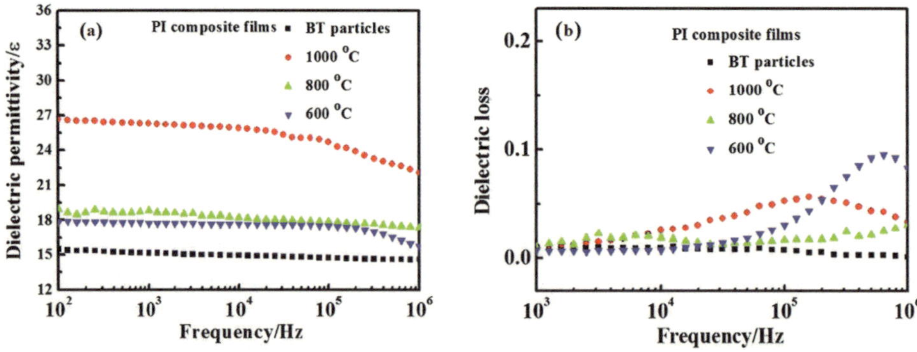

Figure 8.
Frequency dependence of dielectric property of 30 vol% BT nanoparticles (a) dielectric permittivity and (b) dielectric loss measured at room temperature [16].

nanocrystals to the PMDA-1,3-bis(4-aminophenoxy)benzene (BAPB) PI system to generate nanocomposites. Compared with the ε_r (2.8) of pure PMDA-BAPB PI, the ε_r of composites containing 18 vol% BST increased to 6.2; below 1 MHz, the dielectric loss of composite materials with different contents of BST was less than 0.04. At an addition level of 10 vol% BST, the breakdown strength of PMDA-BAPB/BST nanocomposites increased, to reach a maximum value 296 V μm^{-1}, while the energy density of the composite was twice that of pure PMDA-BAPB PI. The observed relative increases in ε_r and breakdown strength together with the reduction in dielectric loss for the nanocomposite with 10 vol% BST are desirable characteristics for practical applications [21]. Wang et al. [22] prepared PI-based composites with good dielectric properties using $CaCu_3Ti_4O_{12}$ (CCTO) and Zr-modified $CaCu_3Ti_{3.95}Zr_{0.05}O_{12}$ (CCTZO) particles as fillers. The results showed that at a filling content of 40 vol%, the ε_r of the CCTZO/PI composite film could reach a value of 70 at 10 Hz, and this was higher than that of the CCTO/PI composite film under the same conditions; at 150°C, the ε_r of the CCTZO/PI composite material reached ≈260 [22].

By changing the design of the inorganic filler, interface problems between the filler and the polymer can be improved, such as poor flake/fiber morphology. The nanosheets can increase the breakdown strength of composites because they provide a uniform insulating center and a curved path for the electrons. Boron nitride nanosheets (BNNSs) have a layered structure like graphene and are wide band gap (6 eV) insulators. Unlike traditional dielectric materials (high-ε_r ceramics and conductive fillers), polymer/BNNS nanocomposites may provide higher breakdown field strengths. Wan et al. [23] prepared three-phase composites of BNN, BT-fibers, and PI (BNNS@BT-fiber/PI) using in situ polymerization. The combination of BNNS and BT fibers can facilitate the dispersion of BNNS nanosheets in BT fibers, thereby improving energy storage performance. When the content of BNNS@BT-fiber was 20% by weight, the ε_r of the composite material was 47.57 at room temperature and 43.03 at 200°C at 100 kHz, demonstrating a reasonable thermal stability. At a BNNS@BT-fiber content of 1 wt%, the maximum U_e of the composite at 3438 kV cm^{-1} was 7.1 J cm^{-3}, that is, about three times that of pure PI [23].

In order to achieve better dispersion and alignment of the filler in the PI matrix, Gu et al. [24] prepared micron boron nitride (mBN)/PI composites by in situ polymerization and electrostatic spinning technology. At 30 wt% mBN, the mBN/PI composite material exhibited a high ε_r (3.77) and low dielectric loss (0.007); the material also showed good thermal stability (λ = 0.696 W m^{-1} K^{-1}), a high temperature index (279°C), and T_g was 240°C [24]. Cheng et al. [25] considered that molybdenum disulfide (MoS_2) had an appreciable band gap and excellent heat resistance, and prepared MoS_2/PI nanocomposite films. Compared with the pure PI film, the ε_r of the composite film was significantly increased, while the dielectric loss remained relatively low. At a filler content of 1 vol%, the breakdown field strength reached 395 MV m^{-1}, while U_e increased to ≈3.35 J cm^{-3}. Furthermore, at 395 MV m^{-1}, the charge and discharge efficiency could still be maintained above 80% [25]. Alumina (Al_2O_3) filler has good insulation performance, high thermal conductivity, and is relatively inexpensive. Therefore, it can be added to the polymer matrix as a filler to improve thermal performance. Choi et al. [26] used 6FDA, 4,4′-methylenedianiline, and bis(3-aminopropyl)-terminated polydimethylsiloxane to prepare PI films with different siloxane content. Since PI-3, PI-4 and PI-5 films were independent and flexible, PI/Al_2O_3 composite films were prepared at different concentrations of Al_2O_3 using these three PIs. The results showed that the thermal conductivity of the composite film increased with increasing Al_2O_3 content. The composite film containing 75% by weight of Al_2O_3 was flexible. The composite film containing 80 wt% Al_2O_3 showed improved thermal conductivity (>1.3 W m^{-1} K^{-1}).

Compared with traditional polysiloxane/Al_2O_3 composite materials, PI/Al_2O_3 composite films demonstrated improved thermal properties [26].

4.2 Surface modification

Because of its simple method, the compounding of fillers and polymers to produce composite materials has become accepted. However, preparation methods, external conditions, and other complications can give rise to many structural defects and electric field concentrations between the two phases of the filler and the polymer matrix. Therefore, surface treatment of the filler using a coupling agent, or decorative insulating, or conductive particles has become a key area of research [27, 28].

Halloysite ($Al_2Si_2O_5(OH)_4{\cdot}2H_2O$) is an aluminosilicate clay, which has a unique tubular structure. It has a high ε_r (6–8), but extremely low dielectric loss (10^{-3}). Because there are moderate hydroxyl groups on the surface that can be chemically modified, and suitable surface modification can be performed, halloysite nanotubes (HNTs) may be an ideal filler for the preparation of dielectric polymer-based composites with high ε_r and low dielectric loss characteristics. Zhu et al. [29] used KH550 (3-aminopropyltriethoxysilane) and polyaniline (PANI) to modify the surface of HNT, and prepared HNT/PI, KH550 modified HNT/PI and PANI-HNT/PI nanocomposite membranes. Among these, at 100 Hz, the PANI-HNT/PI films attained a maximum ε_r of 17.3, while the dielectric loss was only 0.2. Notably, the prepared composite has high breakdown strength (>110.4 kV mm^{-1}), and a maximum discharge energy density of 0.93 J cm^{-3}; these properties could still be maintained at temperatures ≤300°C [29]. Wang et al. [30] prepared a nanocomposite with high thermal conductivity by introducing amide-functionalized MWCNT [MWCNT@p-phenylenediamine (PPD)] into a PEI matrix, as shown in **Figure 9**. Compared with unmodified MWCNT, MWCNT@PPD could participate in the in situ polymerization of PEI to form covalent bonds in the matrix, thereby improving the dispersibility of the filler. This method solved the disadvantages of the traditional CNT acid treatment that can destroy their conjugate structure and greatly affect the aspect ratio. The results showed that the thermal conductivity of nanocomposites containing 4.0 wt% MWCNT@ PPD ≤0.43 W m^{-1} K^{-1} [30].

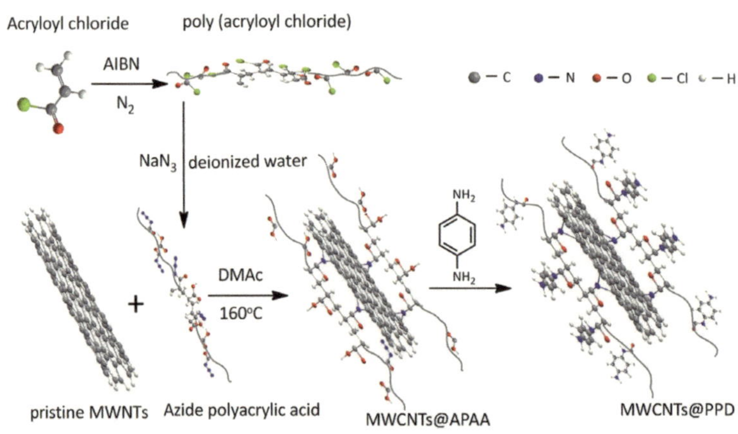

Figure 9.
Schematic for the preparation of multi-walled carbon nanotubes@azide polyacrylic acid (MWCNT@PPD) [30].

Yang et al. [27] investigated the dielectric properties of PI incorporating CCTO/Ag nanoparticles (CCTO@Ag). The use of Ag coating to modify the surface of CCTO nanoparticles increased the conductivity of the intermediate layer, thereby enhancing the space charge polarization and Maxwell-Wagner-Sillars effect, improving the electric field distortion. The results showed that when the content of CCTO@Ag was 3 vol%, the ε_r of PI/CCTO@Ag composites was significantly increased to 103, which was about 30 times the ε_r of pure PI. At the same time, the dielectric loss was very low at 0.018 [27]. Wang et al. [31] used 2-phosphonobutane-1,2,4-tricarboxylic acid (PBTCA) and TH-615 acrylic-acrylate-amide copolymers to modify BT nanoparticles, and then prepared a BT/PI composite film by in situ polymerization. The results showed that this surface modification method could improve the dispersion uniformity of filler particles in the matrix and improve the interfacial compatibility between the two phases. At 10^3 Hz, a BT/PI film modified with 8% PBTCA had a ε_r of 23.5, a dielectric loss of 0.00942, a breakdown strength of 80 MV m^{-1}, and a U_e of 0.67 J cm^{-3}. At the same frequency, the composite modified with 6% TH-615 had a ε_r of 20.3, a dielectric loss of 0.00571, a breakdown strength of 73 MV m^{-1}, and a U_e of 0.68 J cm^{-3} [31]. Due to its unique chemical structure, GO shows great potential in the field of capacitors. The graphene oxide sheet has many hydroxyl groups and epoxy groups on the surface while carboxyl groups are mainly located on the edges. However, most studies involving graphene-based composites have used only marginal carboxyl groups, so the polymer chains were attached to the edges only. Fang et al. [32] made full use of oxygen functional groups to prepare PPD-carboxyl-functionalized graphene oxide (CFGO)/PI composites. The polymer chain was fixed on the base surface, and the graphene oxide sheet was effectively separated. Thermogravimetric analysis (TGA) tests showed that PPD-CFGO/PI composites had good thermal stability below 500°C. When the content of PPD-CFGO was 4 wt%, the ε_r increased to 36.9, which was 12.5 times higher than that of pure PI polymer ($\approx$3.0), the dielectric loss was only 0.0075, and the breakdown strength remained at a high level [32].

However, the covalent functionalization method adopted by Fang et al. [32] reduced the conductivity of graphene by destroying the π-π conjugate structure of graphene. In order to overcome this shortcoming, Feng et al. decorated the surface of rGO with a solid π-π stack by insulating reduced polyaniline (R-PANI) to introduce a space effect and effectively prevent the irreversible agglomeration of rGO. At 1 kHz, the highest ε_r (25.84) was observed in nanocomposite films containing 20 wt% rGO@R-PANI, and the dielectric loss was 0.11. The ε_r and dielectric loss of rGO/PI nanocomposite films were 8.23 and 56.4, respectively. Furthermore, the 5 wt% weight loss temperature for 20 wt% rGO@ R-PANI/PI nanocomposite film was 480°C, indicating that the nanocomposite film has great potential in the field of high-temperature dielectric materials [5]. Yue et al. [5] introduced reduced barium titanate (rBT), sintered in a reducing atmosphere ($95N_2/5H_2$), to PI without using any modifier or surfactant ingredients in the matrix. Surface defects of rBT and interface interactions between two phases caused by the reducing atmosphere lead to an increase in ε_r and U_e. Compared with pure PI, the rBT/PI composite with 30 wt% rBT exhibited the following characteristics: The ε_r at 1000 kHz was ≤31.6 (pure PI = 4.1), the material maintained a low dielectric loss (0.031), the U_e of 9.7 J cm^{-3} at 2628 kV cm^{-1} represented an increase of >400% (for pure PI Ue = 1.9 J cm^{-3} at 3251 kV cm^{-1}) [5].

4.3 Core-shell structure

Recently, much work has focused on introducing an intermediate layer or an insulating shell on the surface of the filler to prevent them from directly connecting to each other. Fillers in composite materials can increase electrical conductivity and cause excessive polarization interfaces. Researchers are also attempting to introduce intermediate layers or oxide shells between fillers to reduce dielectric loss. Studies have also shown that the core-shell structure can achieve a high ε_r, low dielectric loss, and high energy density [28, 33].

Liu et al. [34] synthesized a sandwich-shaped core-shell SiO_2@GO hybrid to prepare a novel SiO_2@GO/PI flexible composite film using in situ polymerization. The dense SiO_2 layer grafted onto the GO surface can effectively suppress leakage current. The results showed that at 40 Hz, the ε_r of the composite material containing 20 wt% SiO_2@GO was as high as 73, which was 21 times that of pure PI (3.0), and the dielectric loss was only 0.39. In order to improve interfacial compatibility, two coupling agents, 3-aminopropyl triethoxysilane and 3-glycidoxypropyltrimethoxysilane (GPTS), were used to modify the surface of SiO_2@GO: At 40 Hz, the ε_r of the GPTS-SiO_2@GO/PI composite increased to 79 and the loss decreased to 0.25. This significant improvement in the dielectric properties was due to the improved dispersibility of the filler following GPTS modification. Wang et al. [28] prepared a core-shell structure of BT@SiO_2 nanofibers by electrospinning, and successfully prepared a nanocomposite membrane composed of core-shell BT@SiO_2 nanofibers and PI. Because SiO_2 has very low dielectric loss (0.00002) and moderate ε_r, using a thin layer of SiO_2 to isolate PI from BT nanofibers can alleviate the local field concentration. The latter is caused by the large difference in ε_r between the concentrations of the two phases, thereby enhancing the breakdown strength of the PI nanocomposite film. Compared with pure PI, the composite film filled with 3 vol% BT@SiO_2 nanofibers had a maximum U_e of 2.31 J cm^{-3} at 346 kV mm^{-1} (pure PI U_e = 1.42 J cm^{-3} at 308 kV mm^{-1}). TGA also showed that below 500°C, BT@SiO_2/PI nanocomposite films had good thermal stability [28]. Wang et al. [35] prepared a core-shell AgNW/PI composite film with high ε_r and low loss (see **Figure 10**). The insulating shell could protect the silver cores from being directly connected to each other, so that when the ε_r of the composite film reached its maximum value (126), the dielectric loss remained at a low level [35].

Weng et al. [33] synthesized a novel core-shell of Ag@Al_2O_3 nanoparticles as conductive fillers and doped them into PI to prepare Ag@Al_2O_3/PI composite films. The composite film containing 10% by weight of Ag@Al_2O_3 had a ε_r of 21, which was seven times higher than that of pure PI (3.1). This increase in ε_r may be due to the high electrical conductivity of the Ag@Al_2O_3 filler, which caused interfacial polarization inside the composite in the applied electric field. Hence, when the mass fraction of Ag@Al_2O_3 was increased to 30%, the maximum value of ε_r was 124 [33].

4.4 Multilayer structure

Most polymer nanocomposites are expected to achieve high energy density by combining the high breakdown strength of the polymer matrix with the high ε_r of the filler. In fact, when the filler is introduced into the polymer matrix, the breakdown strength often decreases, especially when the volume fraction of the composite filler is high, which does not improve the energy density of the nanocomposite. Therefore, there is a need to expand nanocomposites into multilayer structures to compensate for the reduced breakdown strength [36].

Chen et al. [36] designed a three-layer PI composite membrane by combining $KTa_{0.5}Nb_{0.5}O_3$ (KTN) nanoparticles with PI. Pure PI (with high breakdown field strength) was used as the middle layer with KTN/PI nanocomposite as the two outer layers to improve the energy storage performance of the entire composite film. The results showed that the maximum discharge energy density of the triple-layer composite film (t-KPI) was 3.0 J cm^{-3} at 300 kV mm^{-1}, which was much larger

Figure 10.
(a) TEM image of core-shell structured AgNW. (b) TEM image of an individual AgNW. (c) SEM image of core-shell structured AgNW. (d) SEM image of AgNW/PI hybrid film [35].

than the maximum discharge energy density of the equivalent single-layer composite film (1.5 J cm^{-3}, at 210 kV mm^{-1}); at a high electric field of 300 kV mm^{-1}, the t-KPI composite film could still maintain 88% charge and discharge efficiency [36]. Amin Azizi et al. [37] prepared large-scale high-quality hexagonal boron nitride (h-BN) films using vapor deposition technology (CVD) and transferred them to PEI films to synthesize h-BN/PEI/h-BN composite film. As shown in **Figure 11**, this composite film exhibits excellent charge-discharge efficiency and dielectric stability at high temperatures. At 100°C, the discharge energy density of h-BN-coated PEI reached 2.93 J cm^{-3}, and its charge-discharge efficiency was >90%. As the operating temperature increased, its advantages become more obvious. At 200°C, the energy density of h-BN-coated PEI film was 1.19 J cm^{-3}. Rapid cyclic discharge experiments were performed at 150°C and 200 MV m^{-1} to test the stability of h-BN/PEI/h-BN composite films under electric fields and high temperature. The results demonstrated that the h-BN/PEI/h-BN film coated with 19 layers of h-BN did not show any reduction in discharge energy density and charge-discharge efficiency over 55,000 charge-discharge cycles [37].

Chen et al. [38] prepared an amino-modified CNT/PI (NH_2-MWCNT/PI) flexible composite film with a three-layer structure in which a high-dielectric NH_2-MWCNT was inserted between pure PI layers (serving as the bottom and top layers) of the complex. Since the conductive paths of the insulating layer could be effectively isolated, the three-layer composite film showed high ε_r and low dielectric loss. It is worth noting that at 1 kHz, when the NH_2-MWCNT content of the intermediate layer was 10 wt%, the multilayer composite film (P-10-P) gave the highest ε_r of 31.3, while the dielectric loss was 0.0016. In addition, the maximum energy density of the composite membrane containing 5 wt% NH_2-MWCNT in the intermediate layer (P-5-P) was as high as 1.95 J cm^{-3}, which is more than 50% higher than that of pure PI (1.41 J cm^{-3}). The maximum energy density of the composite film P-10-P also remained at 1.31 J cm^{-3} [38]. Among the various films,

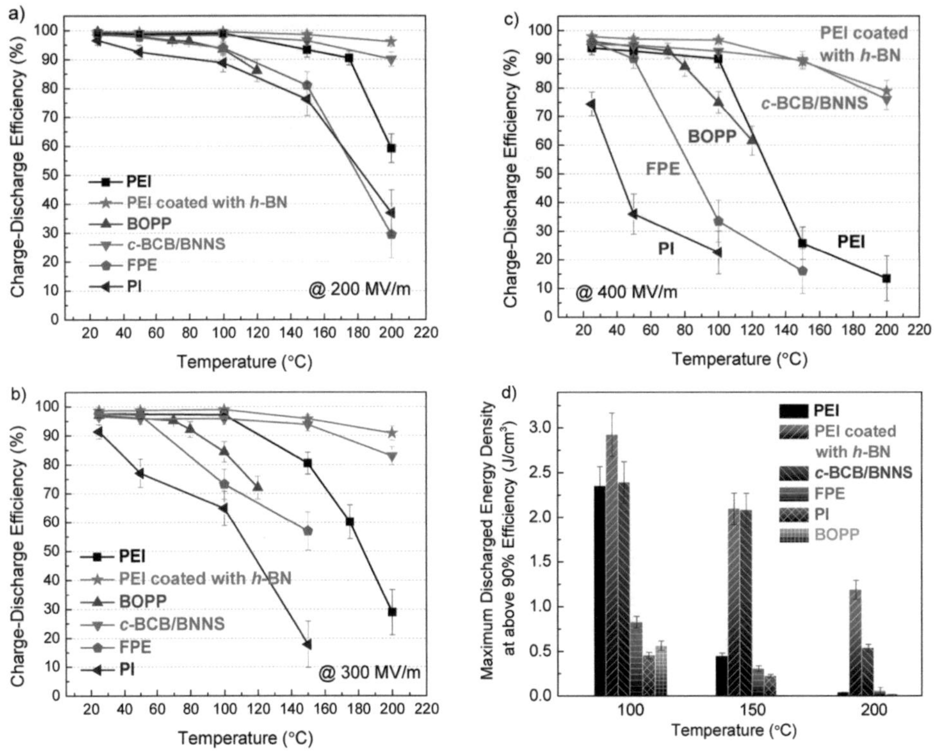

Figure 11.
Charge-discharge efficiency of the dielectrics as a function of temperatures measured at an applied field of (a) 200, (b) 300, and (c) 400 MVm^{-1}. (d) Discharged energy density achieved at above 90% charge-discharge efficiency at varied temperatures [37].

h-BN/PI composite film filled with 5 vol% h-BN as the outer layer could improve the heat dissipation ability of the three-layer composite material, thereby maintaining the dielectric strength and suppressing leakage current at high temperatures. Hence, this sandwich structure composite material had excellent energy storage properties and high temperature stability. At 25 and 150°C, the maximum field strengths of the composite film with a Zr and Ca modified BT (BZT-BCT) content of 1 vol% in the intermediate layer were 360 and 350 kV mm^{-1} respectively, while the storage densities were 2.3 and 1.83 J cm^{-3}, respectively [39]. Zhou et al. [40] proposed a method for preparing high-performance polymer dielectrics at high temperatures (designed roll-to-roll plasma enhanced CVD), which was easily adapted to large-scale production of various surface-functionalized polymer films. In this experiment, they uniformly deposited wide-band gap SiO_2 on the dielectric polymer film at ambient temperature and atmospheric pressure, and their productivity was comparable to that of melt extrusion. The results showed that the introduced SiO_2 layer increased the potential barrier at the electrode/dielectric interface, resulting in a significant decrease in conductivity. Therefore, compared with the pure polymer (see **Figure 12**), the SiO_2-coated film exhibited good high-temperature capacitance performance and had a higher energy storage efficiency (η) value. For example, at 150°C, when $\eta > 90\%$, the maximum U_e values of PEI-SiO_2, PEN-SiO_2, PI-SiO_2, PC-SiO_2, and FPE-SiO_2 composite films were 2.12, 1.75, 1.24, 1.79, and 2.06 J cm^{-3}, which were respectively 236, 672, 510, 1279, and 644% greater than the corresponding pure films. At 100°C, when $\eta > 90\%$, U_e for PEI-SiO_2 was 3.0 J cm^{-3} [40].

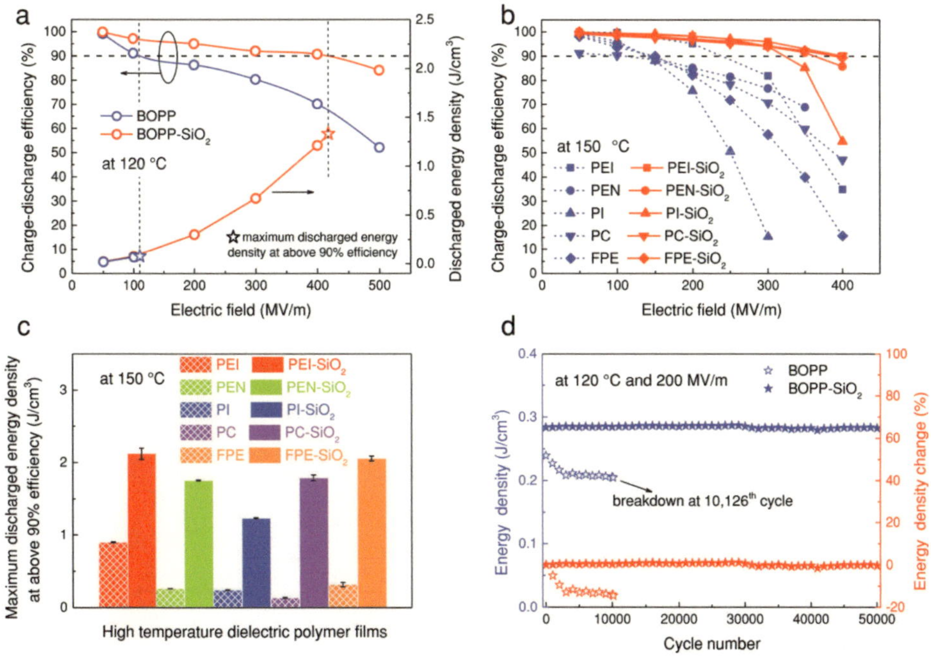

Figure 12.
(a) Charge-discharge efficiency and discharged energy density of BOPP and BOPP-SiO_2 films with 180 nm coating layer on each side of the polymer measured at 120°C. (b) Charge-discharge efficiency of the various dielectric films before and after coating measured at 150°C. (c) Maximum discharged energy density of the various dielectric films before and after coating achieved at above 90% charge-discharge efficiency measured at 150°C. (d) Discharged energy density obtained from cyclic fast discharge tests of pristine BOPP and BOPP-SiO_2 films [40].

5. Conclusion

In conclusion, the high-temperature polyimide dielectric materials used in energy storage application have been summarized, including pure PI, structure modification of PI, PI-based nanocomposites, etc. Many methods for micro molecule dimension and macrostructure design have been analyzed. The reviewed research studies encompassed commercial products progress, material design, and specification, the fundamental theory such as dielectric properties, energy density, and thermal properties. However, the current research for available high-temperature dielectric materials still falls short of industrial application, especially operating under extreme environment conditions, due to the relatively low dielectric permittivity and higher dielectric loss, which severely limit the energy storage density. Moreover, the thermal conductivity is also a limiting factor for high-temperature polymer dielectric materials. Therefore, more fundamental research on developing high-performance intrinsic polymer and high-temperature dielectric phenomena should be focused for future application.

Acknowledgements

This work was financially supported by National Nature Science Foundation of China (No. 51977114).

Conflict of interest

The authors declare no conflict of interest.

Author details

Jun-Wei Zha[1]*, Xue-Jie Liu[1], Yaya Tian[1], Zhi-Min Dang[2] and George Chen[3]*

1 School of Chemistry and Biological Engineering, University of Science and Technology Beijing, Beijing, P.R. China

2 Department of Electrical Engineering, Tsinghua University, Beijing, P.R. China

3 School of Electronics and Computer Science, University of Southampton, Southampton, United Kingdom

*Address all correspondence to: zhajw@ustb.edu.cn and gc@ecs.soton.ac.uk

References

[1] Hu P, Sun W, Fan M, Qian J, Jiang J, Dan Z, et al. Large energy density at high-temperature and excellent thermal stability in polyimide nanocomposite contained with small loading of $BaTiO_3$ nanofibers. Applied Surface Science. 2018;**458**:743-750. DOI: 10.1016/j.apsusc.2018.07.128

[2] Burlingame Q, Wu S, Lin M, Zhang Q. Conduction mechanisms and structure-property relationships in high energy density aromatic polythiourea dielectric films. Advanced Energy Materials. 2013;**3**(8):1051-1055. DOI: 10.1002/aenm.201201110

[3] Peng X, Xu W, Chen L, Ding Y, Xiong T, Chen S, et al. Development of high dielectric polyimides containing bipyridine units for polymer film capacitor. Reactive and Functional Polymers. 2016;**106**:93-98. DOI: 10.1016/j.reactfunctpolym.2016.07.017

[4] Tong H, Fu J, Ahmad A, Fan T, Hou Y, Xu J. Sulfonyl-containing polyimide dielectrics with advanced heat resistance and dielectric properties for high-temperature capacitor applications. Macromolecular Materials and Engineering. 2019;**304**(4):1800709. DOI: 10.1002/mame.201800709

[5] Yue S, Wan B, Liu Y, Zhang Q. Significantly enhanced dielectric constant and energy storage properties in polyimide/reduced $BaTiO_3$ composite films with excellent thermal stability. RSC Advances. 2019;**9**(14):7706-7717. DOI: 10.1039/c8ra10434d

[6] Zhang L, Zhou Y, Mo Y, Zhou Z, Sha Y, Lu Z, et al. Dielectric property and charge evolution behavior in thermally aged polyimide films. Polymer Degradation and Stability. 2018;**156**:292-300. DOI: 10.1016/j.polymdegradstab.2018.06.009

[7] Ma R, Baldwin AF, Wang C, Offenbach I, Cakmak M, Ramprasad R, et al. Rationally designed polyimides for high-energy density capacitor applications. ACS Applied Materials & Interfaces. 2014;**6**(13):10445-10451. DOI: 10.1021/am502002v

[8] Chen L, Ding Y, Yang T, Wan C, Hou H. Synthesis and properties of a high dielectric constant copolymer of a copper phthalocyanine oligomer grafted to amino-capped polyimide. Journal of Materials Chemistry C. 2017;**5**(33):8371-8375. DOI: 10.1039/c7tc03169f

[9] Ho JS, Greenbaum SG. Polymer capacitor dielectrics for high temperature applications. ACS Applied Materials & Interfaces. 2018;**10**(35):29189-29218. DOI: 10.1021/acsami.8b07705

[10] Li B, Jiang S, Yu S, Chen Y, Tang X, Wu X, et al. Co-polyimide aerogel using aromatic monomers and aliphatic monomers as mixing diamines. Journal of Sol-Gel Science and Technology. 2018;**88**(2):386-394. DOI: 10.1007/s10971-018-4800-1

[11] Treufeld I, Wang DH, Kurish BA, Tan LS, Zhu L. Enhancing electrical energy storage using polar polyimides with nitrile groups directly attached to the main chain. Journal of Materials Chemistry A. 2014;**2**(48):20683-20696. DOI: 10.1039/c4ta03260h

[12] Jahani F, Torabi S, Chiechi RC, Koster LJA, Hummelen JC. Fullerene derivatives with increased dielectric constants. Chemical Communications. 2014;**50**(73):10645-10647. DOI: 10.1039/c4cc04366a

[13] Wang D, Kurish BA, Treufeld I, Yang L, Zhu L, Tan L. High-temperature dielectric polyimide films for energy storage applications. MRS Online Proceedings Library (OPL). 2013;**1541**: f05-11. DOI: 10.1557/opl.2013.722

[14] Xu W, Ding Y, Jiang S, Zhu J, Ye W, Shen Y, et al. Mechanical flexible PI/MWCNTs nanocomposites with high dielectric permittivity by electrospinning. European Polymer Journal. 2014;**59**:129-135. DOI: 10.1016/j.eurpolymj.2014.07.028

[15] Chen M, Yin J, Jin R, Yao L, Su B, Lei Q. Dielectric and mechanical properties and thermal stability of polyimide-graphene oxide composite films. Thin Solid Films. 2015;**584**: 232-237. DOI: 10.1016/j.tsf.2015.01.005

[16] Wu Y, Zha J, Yao Z, Sun F, Li RKY, Dang Z. Thermally stable polyimide nanocomposite films from electrospun $BaTiO_3$ fibers for high-density energy storage capacitors. RSC Advances. 2015; **5**(56):44749-44755. DOI: 10.1039/c5ra06684k

[17] Fan BH, Zha JW, Wang DR, Zhao J, Dang ZM. Experimental study and theoretical prediction of dielectric permittivity in $BaTiO_3$/polyimide nanocomposite films. Applied Physics Letters. 2012;**100**(9):092903. DOI: 10.1063/1.3691198

[18] Rajib M, Martinez R, Shuvo M, Karim H, Delfin D, Afrin S, et al. Enhanced energy storage of dielectric nanocomposites at elevated temperatures. International Journal of Applied Ceramic Technology. 2016;**13**(1):125-132. DOI: 10.1111/ijac.12410

[19] Sun W, Lu X, Jiang J, Zhang X, Hu P, Li M, et al. Dielectric and energy storage performances of polyimide/$BaTiO_3$ nanocomposites at elevated temperatures. Journal of Applied Physics. 2017;**121**(24):244101. DOI: 10.1063/1.4989973

[20] Wang M, Li W, Feng Y, Hou Y, Zhang T, Fei W, et al. Effect of $BaTiO_3$ nanowires on dielectric properties and energy storage density of polyimide composite films. Ceramics International. 2015;**41**(10):13582-13588. DOI: 10.1016/j.ceramint.2015.07.153

[21] Beier CW, Sanders JM, Brutchey RL. Improved breakdown strength and energy density in thin-film polyimide nanocomposites with small barium strontium titanate nanocrystal fillers. The Journal of Physical Chemistry C. 2013;**117**(14):6958-6965. DOI: 10.1021/jp312519r

[22] Wang X, Chi Q, Lei Q, Gao L, Yu C. Investigation on dielectric properties of the polyimide-based composite films with high permittivity. IEEE Transactions on Dielectrics and Electrical Insulation. 2014;**21**(4): 1471-1477. DOI: 10.1109/TDEI.2014.004302

[23] Wan B, Yue S, Li H, Liu Y, Zhang Q. Significantly enhanced dielectric and energy storage properties of plate-like BN@$BaTiO_3$ composite nanofibers filled polyimide films. Materials Research Bulletin. 2019;**120**:110573. DOI: 10.1016/j.materresbull.2019.110573

[24] Gu J, Lv Z, Wu Y, Guo Y, Tian L, Qiu H, et al. Dielectric thermally conductive boron nitride/polyimide composites with outstanding thermal stabilities via in-situ polymerization-electrospinning-hot press method. Composites Part A: Applied Science and Manufacturing. 2017;**94**:209-216. DOI: 10.1016/j.compositesa.2016.12.014

[25] Cheng D, Wang H, Liu B, Wang S, Li Y, Xia Y, et al. Dielectric properties and energy-storage performance of two-dimensional molybdenum disulfide nanosheets/polyimide composite films. Journal of Applied Polymer Science. 2019;**136**:47991. DOI: 10.1002/app.47991

[26] Choi J, Nam K, Jin S. Preparation and properties of poly (imide-siloxane) copolymer composite films with micro-Al_2O_3 particles. Applied Sciences. 2019;**9**(3):548. DOI: 10.3390/app9030548

[27] Yang Y, Sun H, Yin D, Lu Z, Wei J, Xiong R, et al. High performance of polyimide/$CaCu_3Ti_4O_{12}$@Ag hybrid films with enhanced dielectric permittivity and low dielectric loss. Journal of Materials Chemistry A. 2015; **3**(9):4916-4921. DOI: 10.1039/c4ta05673f

[28] Wang J, Long Y, Sun Y, Zhang X, Yang H, Lin B. Enhanced energy density and thermostability in polyimide nanocomposites containing core-shell structured $BaTiO_3$@SiO_2 nanofibers. Applied Surface Science. 2017;**426**:437-445. DOI: 10.1016/j.apsusc.2017.07.149

[29] Zhu T, Qian C, Zheng W, Bei R, Liu S, Chi Z, et al. Modified halloysite nanotube filled polyimide composites for film capacitors: High dielectric constant, low dielectric loss and excellent heat resistance. RSC Advances. 2018;**8**(19):10522-10531. DOI: 10.1039/c8ra01373j

[30] Wang X, Zhang C, Du Z. Synthesis of non-destructive amido group functionalized multi-walled carbon nanotubes and their application in antistatic and thermal conductive polyetherimide matrix nanocomposites. Polymers for Advanced Technologies. 2017;**28**(7):791-796. DOI: 10.1002/pat.3979

[31] Wang Y, Wu X, Feng C, Zeng Q. Improved dielectric properties of surface modified $BaTiO_3$/polyimide composite films. Microelectronic Engineering. 2016;**154**:17-21. DOI: 10.1016/j.mee.2016.01.024

[32] Fang X, Liu X, Cui Z, Qian J, Pan J, Lia X, et al. Preparation and properties of thermostable well-functionalized graphene oxide/polyimide composite films with high dielectric constant, low dielectric loss and high strength via in situ polymerization. Journal of Materials Chemistry A. 2015;**3**(18):10005-10012. DOI: 10.1039/c5ta00943j

[33] Weng L, Yan L, Li H, Xia Q, Liu L. Synthesis of Ag@Al_2O_3 core-shell structure nanoparticles and their enhancement effect on dielectric properties for Ag@Al_2O_3/polyimide nanocomposites. Journal of Wuhan University of Technology. 2015;**30**(1):47-50. DOI: 10.1007/s11595-015-1098-5

[34] Feng H, Fang X, Liu X, Pei Q, Cui Z, Deng S, et al. Reduced polyaniline decorated reduced graphene oxide/polyimide nanocomposite films with enhanced dielectric properties and thermostability. Composites Part A: Applied Science and Manufacturing. 2018;**109**:578-584. DOI: 10.1016/j.compositesa.2018.03.035

[35] Wang L, Piao X, Zou H, Wang Y, Li H. High dielectric, dynamic mechanical and thermal properties of polyimide composite film filled with carbon-coated silver nanowires. Applied Physics A. 2015;**118**(1):243-248. DOI: 10.1007/s00339-014-8667-x

[36] Chen G, Lin J, Wang X, Yang W, Li D, Ding W, et al. Three-layer structure design for enhancing the energy efficiency and breakdown strength of $KTa_{0.5}Nb_{0.5}O_3$/polyimide nanocomposite films with high thermal stability. Journal of Materials Science: Materials in Electronics. 2017;**28**(18):13861-13868. DOI: 10.1007/s10854-017-7233-7

[37] Azizi A, Gadinski MR, Li Q, Mohammed AA, Wang J, Wang Y, et al. High-performance polymers sandwiched with chemical vapor deposited hexagonal boron nitrides as scalable high-temperature dielectric materials. Advanced Materials. 2017; **29**(35):1701864. DOI: 10.1002/adma.201701864

[38] Chen Y, Lin B, Zhang X, Wang J, Lai C, Sun Y, et al. Enhanced dielectric properties of amino-modified-CNT/polyimide composite films with a sandwich structure. Journal of Materials

Chemistry A. 2014;**2**(34):14118-14126. DOI: 10.1039/c4ta01818d

[39] Chi Q, Gao Z, Zhang T, Zhang C, Zhang Y, Chen Q, et al. Excellent energy storage properties with high-temperature stability in sandwich-structured polyimide-based composite films. ACS Sustainable Chemistry & Engineering. 2018;**7**(1):748-757. DOI: 10.1021/acssuschemeng.8b04370

[40] Zhou Y, Li Q, Dang B, Yang Y, Shao T, Li H, et al. A scalable, high-throughput, and environmentally benign approach to polymer dielectrics exhibiting significantly improved capacitive performance at high temperatures. Advanced Materials. 2018;**30**(49):1805672. DOI: 10.1002/adma.201805672

Section 6

Breakdown, Aging and Lifetime of Polyimide

Chapter 10

Space Charge Accumulation Phenomena in PI under Various Practicable Environment

Hiroaki Miyake and Yasuhiro Tanaka

Abstract

Polyimide is widely used insulation materials, such as power equipment, motor windings, multi layer insulated, and so on. As the operation environment is high temperature, high humidity, radiation, the dielectric insulation characteristic is decreased compared with pristine one. Especially, the space charge characteristics are obtained big different. Furthermore, the breakdown phenomenon is frequently produced. In this chapter, we discuss the dielectric phenomena through the viewpoints of charge accumulation under the following environment. High temperature, High humidity, DC application, PWM application, Radio-active rays (electron, proton).

Keywords: space charge accumulation, high temperature, high humidity, DC and PWM application, radiation environment

1. Introduction

Normally, you already know the polyimide (PI) and the super engineering plastics derivative from PI are one of best high resistive dielectric materials for any application, for example print board, motor windings, surface materials of spacecraft, and so on. However, sometimes, the PI have breakdown phenomena under high voltage application with high temperature and humidity condition. It is considered that the origin of breakdown is the electric filed enhancement in the bulk or high growth of conduction current during the bulk by the homo charges injection form the electrode, or the hetero charges accumulation due to hydrolysis by surround high humidity environment. Furthermore, concerning the radioactive rays irradiated PI, the huge space charge polarization is produced. And then, sometimes, PI has a breakdown phenomenon due to irradiation.

To understand the dielectric characteristic for the serve condition, it is important that the space charge accumulation phenomena are observed. Therefore, in this chapter, we introduce the space charge accumulation in the bulks under the high electric filed, high temperature, high humidity and radiation environment.

2. Principle of the pulsed electroacoustic (PEA) method

The PEA method is one of the widely used techniques for measuring the charge distribution in dielectrics. The principle of the PEA method is explained at the followings [1].

Figure 1 shows the schematic diagram of the PEA method. A charged sample with thickness a is used, and its both sides are sandwiched by electrodes. The pulse voltage generator is connected to both electrodes. When the pulse electric field e_p (t) is applied with the electrode induced charges σ (0) and σ (a) accumulated on both electrode surfaces due to the accumulated charge ρ in the sample, a pulsed pressure wave p(t) is generated from σ(0), σ(a) and ρ(z) in the sample. Those pulsed pressure wave p(t) is shown by the convolution equation between each accumulated charges (σ(0), σ (a), ρ(d)) and e_p(t). The convolution equation is shown in the following formula.

$$
\begin{aligned}
p(t) &= p_1(t) + p_2(t) + p_3(t) \\
&= \frac{Z_{Al}}{Z_{sa}+Z_{Al}} \cdot \sigma(0) \cdot e_p(t) + \frac{2Z_{Al}}{Z_{sa}+Z_{Al}} \cdot \frac{1}{2} \cdot u_{sa} \cdot \int_{-\infty}^{+\infty} \rho(\tau) \cdot e_p(t-\tau)\, d\tau + \\
&\quad \frac{2Z_{Al}}{Z_{sa}+Z_{Al}} \cdot \frac{Z_{sa}}{Z_{sa}+Z_{BS}} \cdot \sigma(a) \cdot e_p\left(t - \frac{a}{v_{sa}}\right)
\end{aligned}
\qquad (1)
$$

where, Z_{Al}, Z_{sa}, Z_{BS} is the acoustic impedance of Al, sample and backing material, respectively. u_{sa} is the sound velocity of sample.

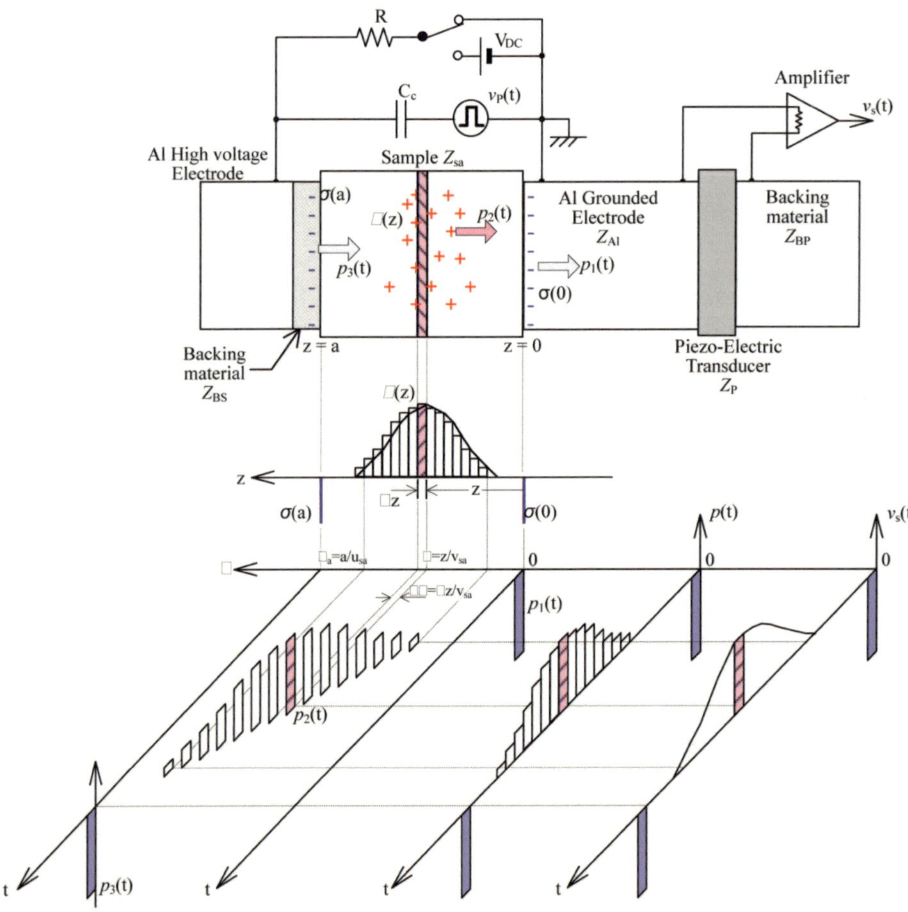

Figure 1.
The schematic diagram of the principle of PEA method.

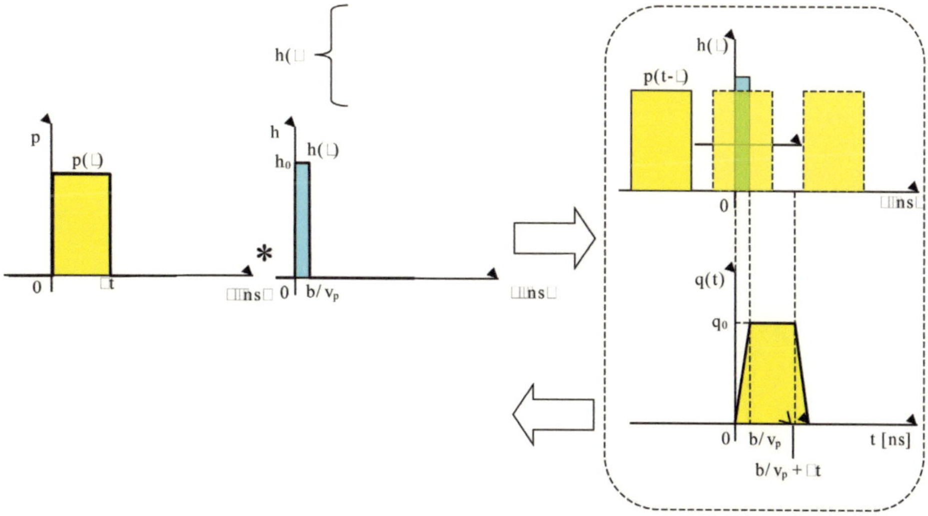

Figure 2.
The schematic diagram of convolution model on the time function between the pulsed pressure wave and the piezoelectric device.

Those pulsed pressure $p(t)$, described the above on the Eq. (1), the intensity of the pressure wave is proportional to the accumulated charges in the sample. The propagation delay of the pressure wave is related to each charge accumulated position.

A piezoelectric transducer is used to detect the pressure wave by transforming into an electric charge signal. The electric charge signal $q(t)$ on the piezo electric device is expressed by the convolution model between pressure wave function $p(\tau)$ and piezoelectric device function $h(\tau)$ as shown in formula (2)

$$q(t) = \frac{v_p}{b} \int_{-\infty}^{\infty} h(\tau) p(t-\tau) d\tau \tag{2}$$

where, v_p and b is sound velocity and thickness of piezoelectric device, respectively. The convolution model is also shown in **Figure 2**. Normally, we observed charge distribution signals obtained by the above convolution model. However, as the real measurement system has a system function, which is the low pass filter between the amplification circuit and and the capacitance of the piezoelectric device, the observed wave form signal is distorted like as the differentiated waveform by the system function.

The details of the principle of PEA are described elsewhere [1].

3. Space charge measurement system

Using the pulsed electroacoustic (PEA) method which is described above, we developed several space charge measurement systems for high temperature. **Figures 3** and **4** shows the developed system [2, 3]. On the **Figure 3**, it is the developed PEA system for high temperature (up to 80 °C). This system is enabled to measure the space charge accumulation in dielectric materials. In this system, temperature (30–80 °C) of silicone oil is control using a band heater and a controller.

Furthermore, as the IGBT using SiC is operated over 150 °C, the space charge measurements under more high temperature are also required. Therefore, we developed newly PEA system for high temperature (up to 150 °C), which is shown in **Figure 4**. From the figure, to measure the space charge distribution at high

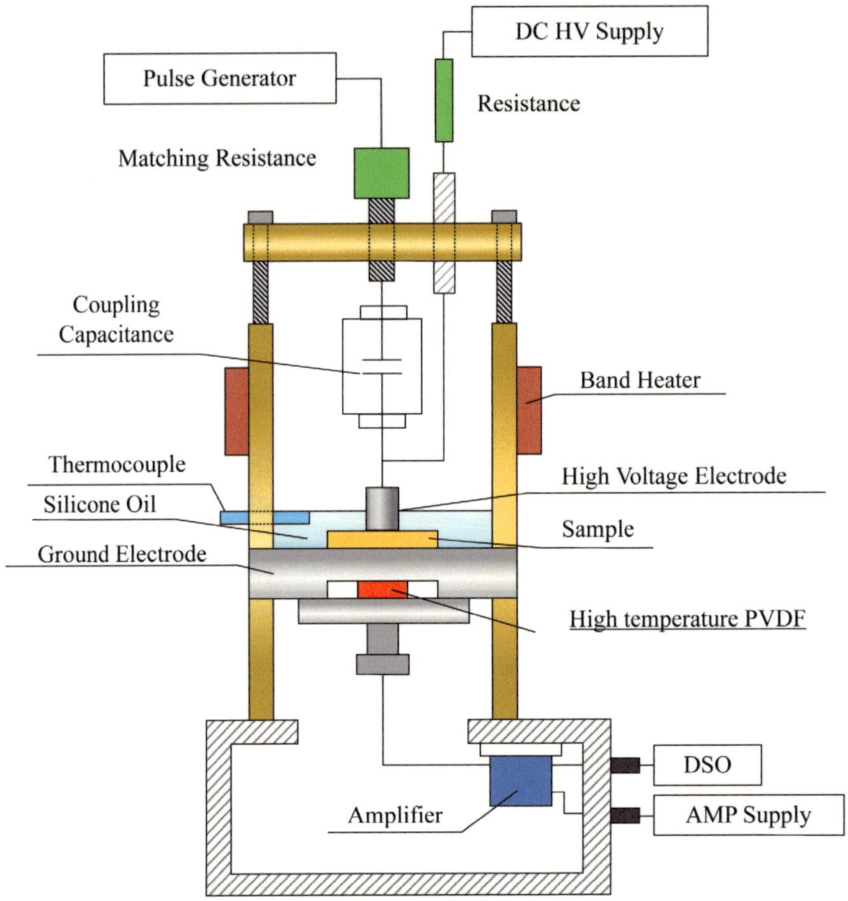

Figure 3.
Schematic diagram of the space charge measurement system for high temperature up to 80 °C.

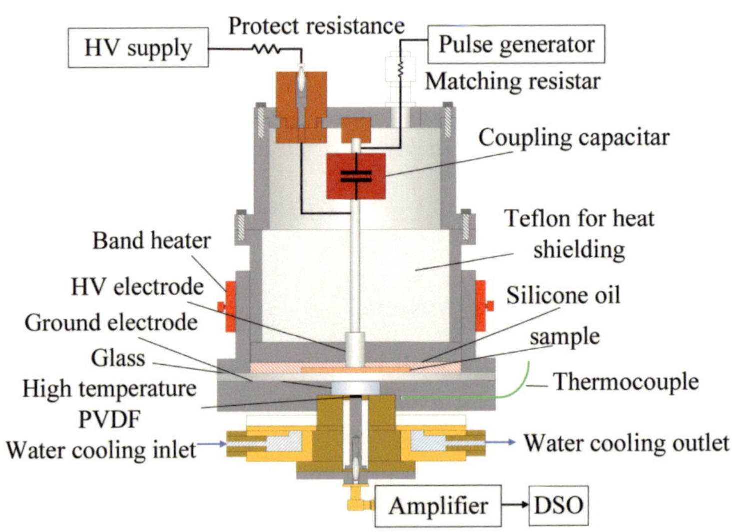

Figure 4.
Schematic diagram of the space charge measurement system for high temperature up to 150 °C.

temperature, the P(VDF/TrFE) is used as the piezo-electric sensor. By combining the sensor and a water-cooling system, the space charge measurement at maximum 150 °C becomes possible. A band heater is fitted to the high voltage electrode unit to increase temperature of the sample. In ordinary PEA measurement, to improve an acoustic impedance matching between the sample and the metallic electrode, a semiconductive (SC) layer, which is used for power transmission cable, is used for a high voltage electrode. However, the base material of this SC layer is usually polyethylene, and it is not available at high temperature. Therefore, in such high temperature measurement, we used a SC layer made of PEEK. In all measurements, the positive voltage was applied to the sample through the upper SC electrode.

4. Heated and humidified sample charge accumulation behavior

In this chapter, we discuss the relationship between space charge distribution and breakdown phenomena of polyimide under heating and humidified condition.

Recently, various and many electronic devices have been used the flexible Print-Circuit Board (PCB) to reduce the size of them. At the moment, polyimide film is usually used for PCB [4]. Polyimide film has an appropriate flexibility and it shows a good insulating performance even at high temperature. On another front, since the devices are improved for downsizing and lightweight every year, it is necessary to reduce the thickness of PCB. When the thickness of PCB becomes thinner, the applied electric field becomes higher. It means that the possibility of electrical breakdown in PCB becomes higher. It is necessary that inverter can control at high frequency and drive at high voltage for high efficiency of inverter control. But inverter serge becomes high voltage in these conditions. The partial discharge happens and insulating covering material becomes depleted. And these insulating materials using electric devices need to become high insulating capacity at high temperature and high humidity. Therefore, it is necessary to understand the mechanism of breakdown in polyimide film to prevent the breakdown. Thus, we tried to measure the space charge distribution in polyimide under heating and humidified condition for understanding the phenomena.

4.1 Samples and measurement procedure

In this chapter, a commercially available polyimide film, Kapton® 500H (nominal thickness 125 μm) supplied from Dupont-Toray Co.Ltd. is used.

For simulating the real usage condition, Heat with humidifying treatment and heat treatment are applied before measuring space charge as pre heat and humidifying treatment. The samples are set in a thermostatic chamber IH 400 manufactured by Yamato Scientific Co., Ltd., (80 °C and 80 or 60%) for one hour.

Furthermore, we prepared the heated sample whose were set in silicone oil boiled at 100 °C in beaker to dry off moisture of sample.

4.2 Results and discussion

Figure 5 shows space charge [2] behaviors in non-treated (1) and heated and humidified (2) sample under DC stress of 60(a), 100(b), 120(c) kV/mm at 80 °C. Measurement interval is 5 s. Amount of accumulated space charges are displayed using color chart. As shown in **Figure 5(a)**, positive and negative homo charges are observed near anode and cathode from the start of measurement in non treated sample. The amounts of homo charges are increasing until about 10 mins later,

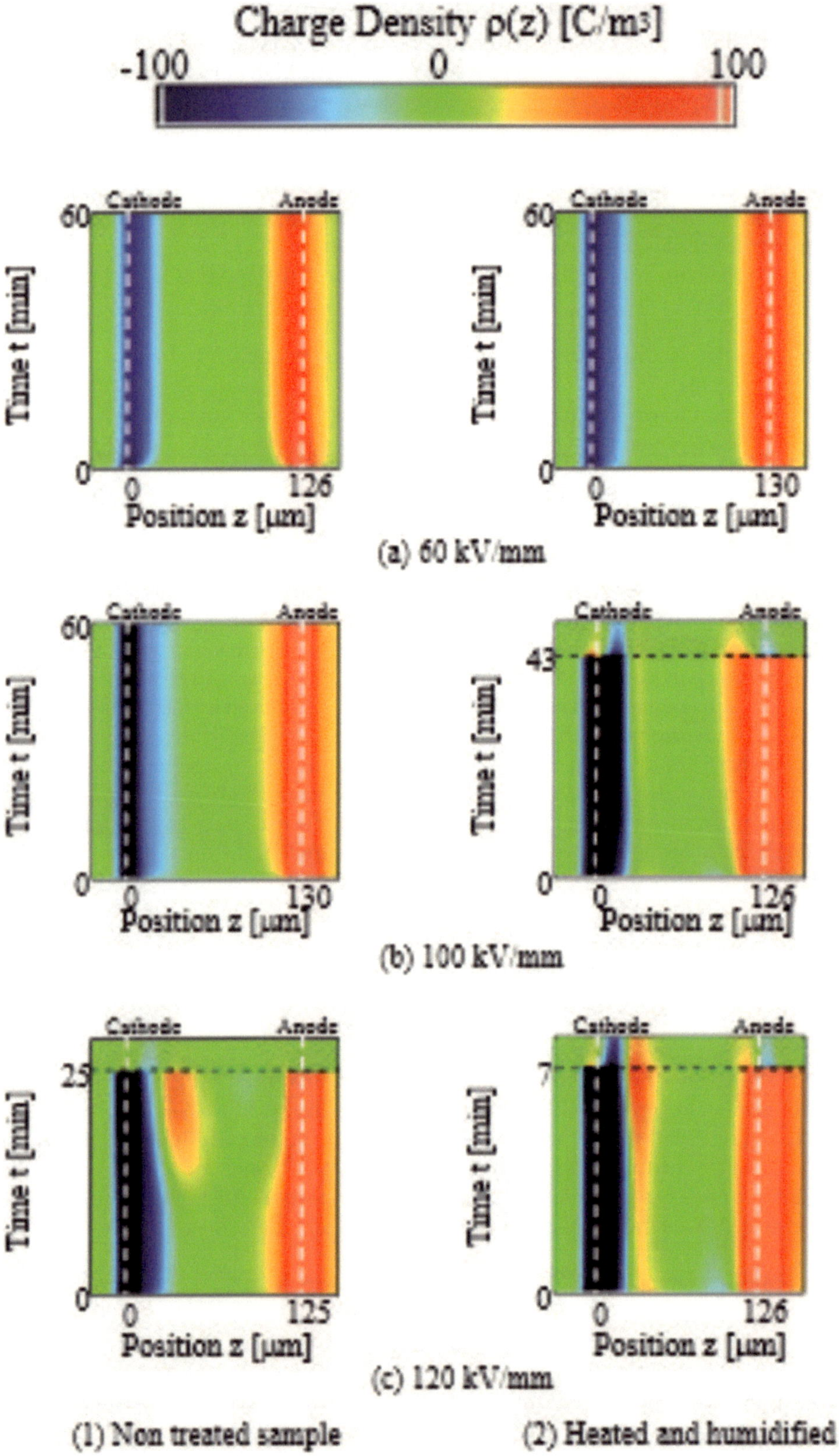

Figure 5.
Space charge behaviors in non treated (1) and heated with humidified (2) sample under DC stress of (a)60, (b)100, (c)120 kV/mm at 80 °C.

but notable changes are not observed after 10 mins. On the other hand, positive and negative homo charges are observed near anode and cathode from the start of measurement in heated and humidified sample. But the amounts of homo charges are increasing until about 5 mins later. In 60 kV/mm, there is no difference point except time to increasing of homo charge and space charge behavior of end of measurement is similar to each other. The time to saturating space charge accumulation in heated and humidified sample is shorter than one in non treated sample. It is appeared that moisture of heated and humidified sample is larger than one of non treated sample. So, moisture of sample is large, it is easily occurs to ingress space charge and migration speed of space charge is more fast.

As shown in **Figure 5(b)**, positive and negative homo charges are observed near anode and cathode from the start of measurement in non treated sample along with space charge behavior of 60 kV/mm(a). But migration speed of space charge is faster and migration position of space charge is larger than one of 60 kV/mm in non treated sample. On the other hands, positive and negative homo charges are observed near anode and cathode from the start of measurement in heated and humidified sample. And positive hetero charge becomes to be observed near cathode. The amount of positive hetero charge is increasing until breakdown at 43 mins later. In our previous research(1), it is observed that positive and negative homo charges are observed near anode and cathode and positive hetero charge becomes to be observed near cathode and breakdown occurs. It is similar process of space charge behavior to heated and humidified sample. And moisture of sample is large, it is easily occurs ingress of space charge and migration speed of space charge is more fast. So, in heated and humidified sample, space charge behavior process until breakdown occurs faster than non treated sample.

As shown in **Figure 5(c)**, homo charges are observed near anode and cathode and positive hetero charge becomes to be observed near cathode and breakdown occurs at 25 mins later in non treated sample. On the other hands, breakdown in heated and humidified sample occurs at 5 mins in same space charge behavior process at non treated sample. However, negative hetero charges are observed near anode from the start of measurement.

In these results, space charge behavior process of non treated sample and heated and humidified sample until breakdown is same. It is said that the breakdown process is promoted by increasing moisture of sample and rising applied electric field. Therefore time to breakdown is shorter than ingress of space charges and migration speed of space charges are faster.

From the above results, we describe process of chare accumulation behavior to breakdown under DC stress. **Figure 6** shows typical process of space charge accumulation behavior in Kapton® under high DC stress (130 kV/mm).

As shown in the figure, a few positive and negative hetero charges are observed near anode and cathode from the start of measurement (charge behavior annotation number 1 in the figure). These hetero charges are same hetero charge in heated sample. Therefore, these space charge accumulations are looked when moisture of sample is low and sample is applied at high electric field. After hetero charges are observed, positive and negative homo charges are observed near anode and cathode and hetero charges decrease (charge behavior annotation number 2 in figure). These homo charges are injected from anode and cathode. When moisture of sample increase, amount of accumulation of homo charges is large. In fact, injection of charge easily happens from electrodes when moisture of sample is large. And charges can easily move in the bulk. And positive and negative hetero charges may disappear by accumulating homo charges. These homo charges decrease and large positive hetero charge accumulated at next period (charge behavior annotation number 3 in figure). And accumulation of positive hetero

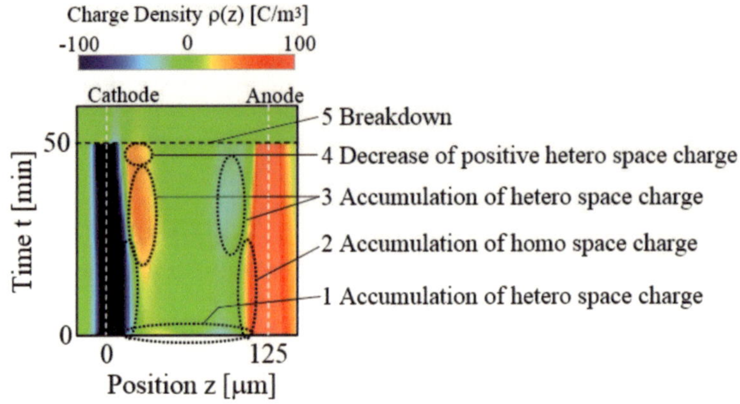

Figure 6.
Typical process of space charge accumulation behavior in Kapton® under high DC stress (130 kV/mm).

charge is larger than that of negative hetero charge. However large hetero charge is observed in **Figure 6**, a few hetero charges can be observed at some sample. Amount of hetero charge is influence at amount of homo charges mentioned at the charge behavior annotation number 2 in figure. Because hetero charge is observed before breakdown happens, this hetero charge promote breakdown. But accentuation of electric field by accumulating these space charges is a little. And so, it is not say that accumulation of space charge is cause which breakdown happens by. But this phenomenon is distinct pre-breakdown phenomenon. After this hetero charge was observed, positive hetero charge decreases a little in this measurement result (charge behavior annotation number 4 in figure). This behavior is not always observed in all samples. In fact, breakdown occurs when accumulation of space charge was saturated or decreased. Therefore, the processes of accumulating of the hetero charge and decreasing ones are equilibrium condition. And electric field around electrode is large by accumulating hetero charge. And given the accumulation of homo charge before hetero charge is observed, homo charge injected from electrode and hetero charge generated from bulk is saturated. In this research, source of hetero charge is not clear up. It is considered that hetero charge is generated by ion breaking the molecular chain or ionization of impure substance of the sample.

5. Charge accumulation behavior in the PI (polyimide) bulk under high temperature (100–150 °C)

Currently, polyimide is demanded for using an insulating material for power electric devices using SiC(silicon carbide) is required [2]. As it is developed enough to be used at high temperature under high electric stress, the insulating materials supporting the device must also exhibit extremely excellent insulation properties at high temperatures under high electric stresses.

In recent years, evaluation methods using storage characteristics of space charge has been as an effective technique to evaluate the character is tics under high electric stress. Under the stress, the current-electric field characteristic often deviates from the Ohm's law and the reason of it is considered that the space charge accumulates in the material [1, 3, 5]. Therefore, if the space charge accumulation state becomes clear, the insulation performance of the material under a high electric field can be evaluated.

Recently our laboratory, has P(VDF/TrFE) (vinylidene fluoride/ ethylene trifluoride copolymer), has been used as a piezoelectric sensor that can be measured even under high temperature, and we can obtaine the space charge distribution in the insulating materials even at relatively high temperatures of up to 150 °C using the developed apparatus. In this report, we introduce some typical measurement results of the space charge accumulation behaviors in the polyimide (PI) obtained using the developed system.

5.1 Samples and measurement procedure

The Kapton®500H (Dupont-Toray Co.Ltd., nominal thickness is 125 μm) was used as measurement sample. Further, heating and humidifying treatment was previously applied to some samples. In this treatment, the sample were kept in a thermostatic chamber IH 400 set at 80 °C with humid of 80% for 1 hour.

Time dependence of the space charge distribution in the sample was measured using the apparatus (up to 150 °C), which is described at the chapter 3, for 100–150 °C. In the measurement, at fast, a dc voltage corresponding to 25 kV/mm was applied to the sample for 30 min at the designated temperature, and the distribution was measured with interval of 5 s by applying a pulse voltage of 200 V with a pulse width of 5 ns. The repetition frequency of the pulse voltage application was 400 Hz and the averaged signal was obtained from 700 signals. Then, the circuit was shorted for 5 min, and the measurement was also carried out using above procedure. After the measurement, next measurement was started under elevated electric stress. The increment of the voltage was corresponding to 25 kV/mm, then the applied stresses were 25, 50, 75, 100, 125, and 150 kV/mm.

5.2 Space charge distribution at 100°C

Figures 7(**A**) and (**B**) show the measurement results of space charge distributions at 100 °C in non-treated and treated PI films respectively. In these figures, (a), (b) and (c) show a time dependence of space charge distribution, a charge density distribution and an electric field distribution profiles, respectively. In figure (a), the charge density is described using a color scale put above each result. In figures(b) and (c) show the profiles at 30 min under each applied electric stress.

From the results of PI shown in **Figure 7(A-a)**, accumulation of positive and negative homo charges was observed in the vicinity of both anode and cathode electrodes, respectively, under the applied stress of 100 kV/mm or less in the untreated PI sample. When the stress became 125 kV/mm or more, accumulation of positive hetero charge was observed on the cathode side. Finally, a dielectric breakdown was observed at about 10 min later after 150 kV/mm was applied to this sample. In PI sample, the breakdown occurred after such positive hetero charge accumulation was also observed in other reports [1, 3]. It can be considered that the positive hetero charge accumulation may be a kind of "sign" for the breakdown in PI sample under dc high stress. On the other hand, in treated PI, as shown in **Figure 7(B-a)**, an accumulation of positive homo charge was observed on the anode side from under 25 kV/mm, and it spread towards the cathode side with increase of the applied stress. Then, finally in this sample, a breakdown was observed at about a few minutes later after stress of 125 kV/mm was applied. Judging from the electric field distribution in non-treated PI shown in **Figure 7(A-c)**, while the electric field near electrode seem to be enhanced slightly by the hetero charge accumulation, it can hardly expect to be a main reason for the breakdown. On the other hand, as shown in **Figure 7(B-c)**, a relatively large electric field enhancement was observed

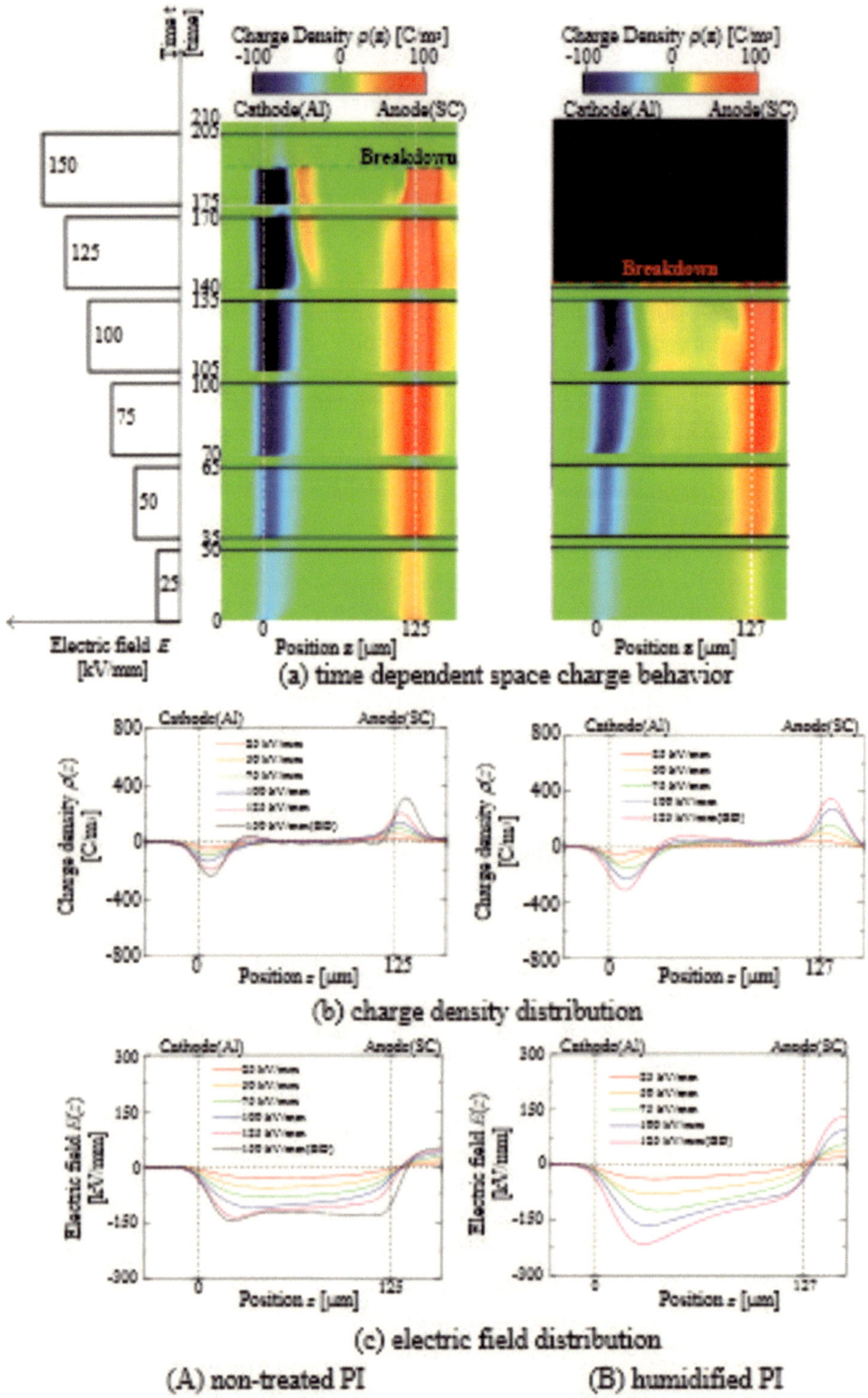

Figure 7.
Space charge distribution in PI at 100 °C.

in the treated PI near the cathode. Therefore, the enhancement might affect the breakdown under lower applied stress in the treated PI than that observed in the non-treated PI.

5.3 Space charge distribution at 150 °C

Figures 8 shows the measurement results of space charge distribution at 150 °C in PI. From the measurement results of PI in **Figure 8(A-a)**, accumulation was not observed under the stress of 75 kV/mm or less. While the positive homo charge was

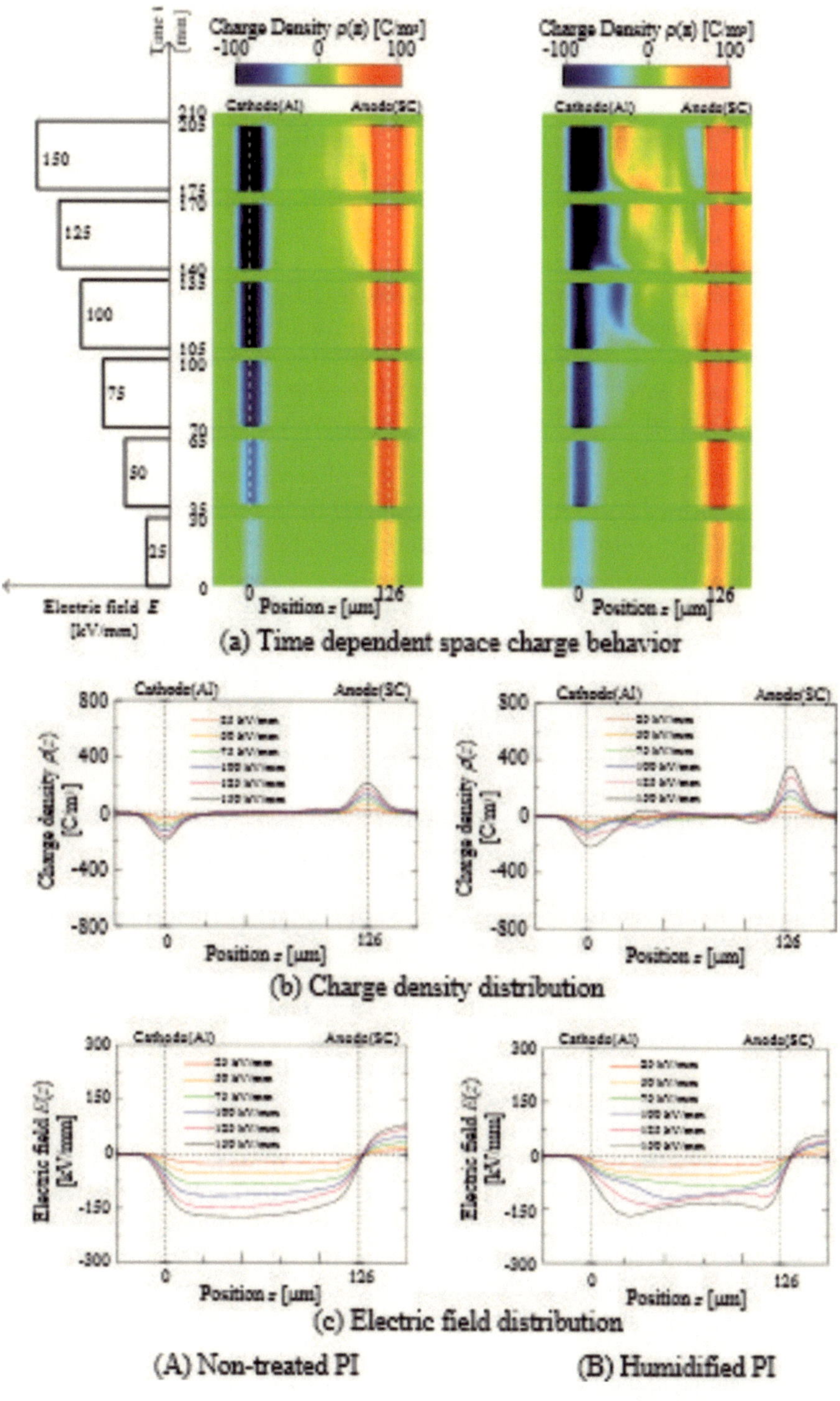

Figure 8.
Space charge distribution in PI at 150 °C.

observed under the stress of 100 kV/mm or more, the amount of it was not so large and breakdown was not observed by the end of measurement. In this case, the electric field was not distorted by the accumulation of charge as shown in **Figure 8(A-c)**. On the other hand, in the measurement results of the treated PI sample, positive charge accumulation near the anode was observed under 50 kV/mm or more as shown in **Figure 8(B-a)**. When the applied stress increased, the positive charge distribution spread towards the cathode, then the negative homo charge accumulation was also observed near the cathode under 100 and 125 kV/mm. Furthermore, in this condition, a negative charge accumulation was also observed near the anode. Finally, the negative and the positive charges were located near the cathode and the anode, respectively, at the end of the measurement. The electric field distribution in this sample was, of course, distorted by the accumulated charges, and the electric field near the cathode and the anode were enhanced to 150 kV/mm as shown in **Figure 8(B-c)**, but the increments of them are not so large.

This process observed in the treated PI at 150 °C is similar to that in non-treated PI at 100 °C, except for the breakdown. The space charge accumulation seems to be affected by the moisture content in the sample, as mentioned above. Therefore, the measurement at 150 °C was also affected by it. In other words, the moisture in both of the non-treated and treated PI sample might be vaporized by the heating. Similar results were obtained in annealed PI sample in silicone oil [1].

6. Charge accumulation behavior in the PAI bulk under simulated high voltage PWM application and high temperature (80 °C)

Recently, the needs of fast measurement are demanded, day by day [5–7]. It is because that many high-speed voltage applications are widely used all over the world. The motor system is one of the examples. The PWM signal is driven by over several tens of kHz. Therefore, under the above situation, the space charge measurement is also required under such high frequency rectangular voltage. Therefore, we can solve such requirement by cooperatively controlling the pulse voltage and the application high voltage.

In this chapter, measurement results in PAI using cooperative operation system between the pulse voltage and the application voltage are introduced.

6.1 Samples, PWM voltage application and space charge measurement timing during PWM application

All measurements were carried out at 80 °C using the high temperature PEA system which is already described in **Figure 3** at chapter 3. In this section, we used commercially available polyamide-imide (PAI) films as measurement samples, which are actually used as the covering insulating materials for the general motor windings. Since the PAI are usually distributed in varnish form, we made a film shape sample from the varnish. The thicknesses of the films were about 100 μm.

Figure 9 (a), **(b)** and **(c)** show a schematic model of the applied half-wave of AC, unipolar square rectified wave and unipolar square rectified wave with surge shape voltages to the samples, respectively. In this figure, the timings of the application of pulse voltage, which is applied to the sample to obtain the measurement signal, are also described using green lines. The peak voltage of it was corresponding to the average electric field of 50 kV/mm in each measurement. Also, the peak surge voltage of it was corresponding to the electric field of 60 kV/mm in each measurement. Frequency of the half-wave of AC, unipolar square rectified wave, unipolar surge square wave rectified wave voltage was 50 Hz. Duty ratio of the square wave voltage was basically

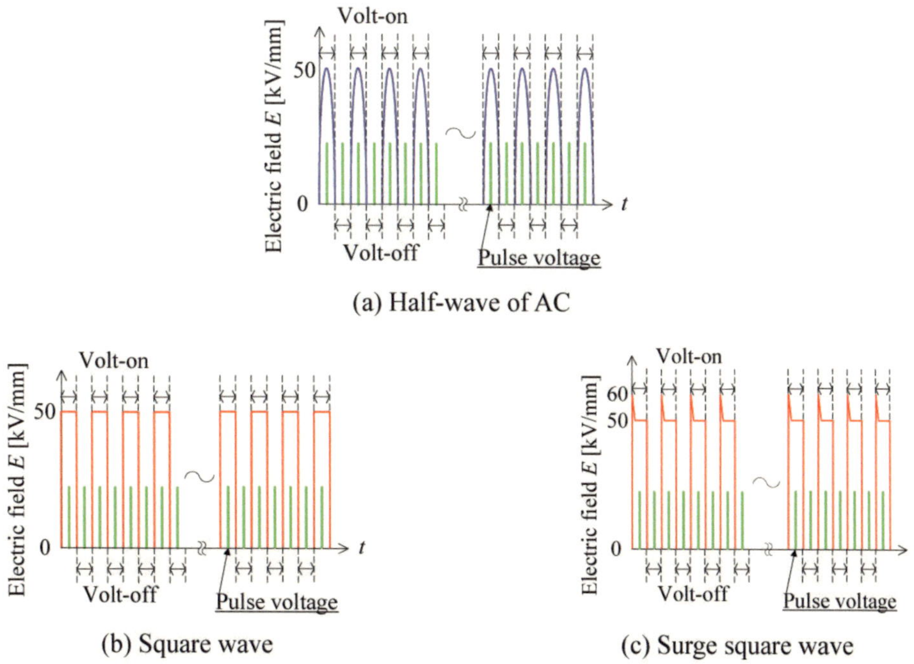

Figure 9.
Time sequence of various wave voltages and pulse voltage applications for PEA measurement. ((a) Half-wave of AC, (b) square wave and (c) square wave with surge shape).

50%. To generate waveforms mentioned above, low voltage waveforms are creates digitally using a computer at first, then the waves are amplified using a high voltage amplifier. Therefore, the timing of the pulse voltage application is also controlled using the computer. The pulse voltage to observe the space charge distribution was applied to the sample under the voltage application ("volt-on" duration) and the short circuit ("volt-off" duration) conditions alternatively. To obtain a clear space charge distribution, it is usually used an averaging technique to reduce noise. In this measurement, we tried to create one space charge distribution under voltage application by averaging the sequential 200 signals obtained at "volt-on" durations. In the case to obtain one space charge distribution under short circuit condition, 200 sequential "volt-off" signals were also averaged. Since the frequency of the square wave voltage was 50 Hz, one averaged signal was composed of 200 signals during 4 s. It means that the observation interval of the space charge distribution was 4 s under "volt-on" and "volt-off" conditions, alternatively. To simulate a practical test condition, some measurements were also using carried out samples, which were kept at a high temperature (80 °C) with a high humidity (80%) condition for 60 min before the measurement.

6.2 Space charge accumulation results under various voltage wave form

Figure 10 (A), **(B)**, **(C)** and **(D)** show the measurement results of time dependent space charge distributions in PAI by applying DC, half-wave of AC, unipolar square rectified wave and square rectified wave with surge shape voltages, respectively. In this figure, (a), (b), (c), (d) and (e) show a time dependent charge density distribution using a color chart at "volt-on", a time dependent charge density distribution using a color chart at "volt-off", profiles of a charge density distribution at "volt-on", a charge density distribution at "volt-off", and electric field distribution at "volt-on", respectively. In these figures, red, black and blue lines ware obtained at just after (4 s), 2 min and 5 min later after the beginning of the voltage application.

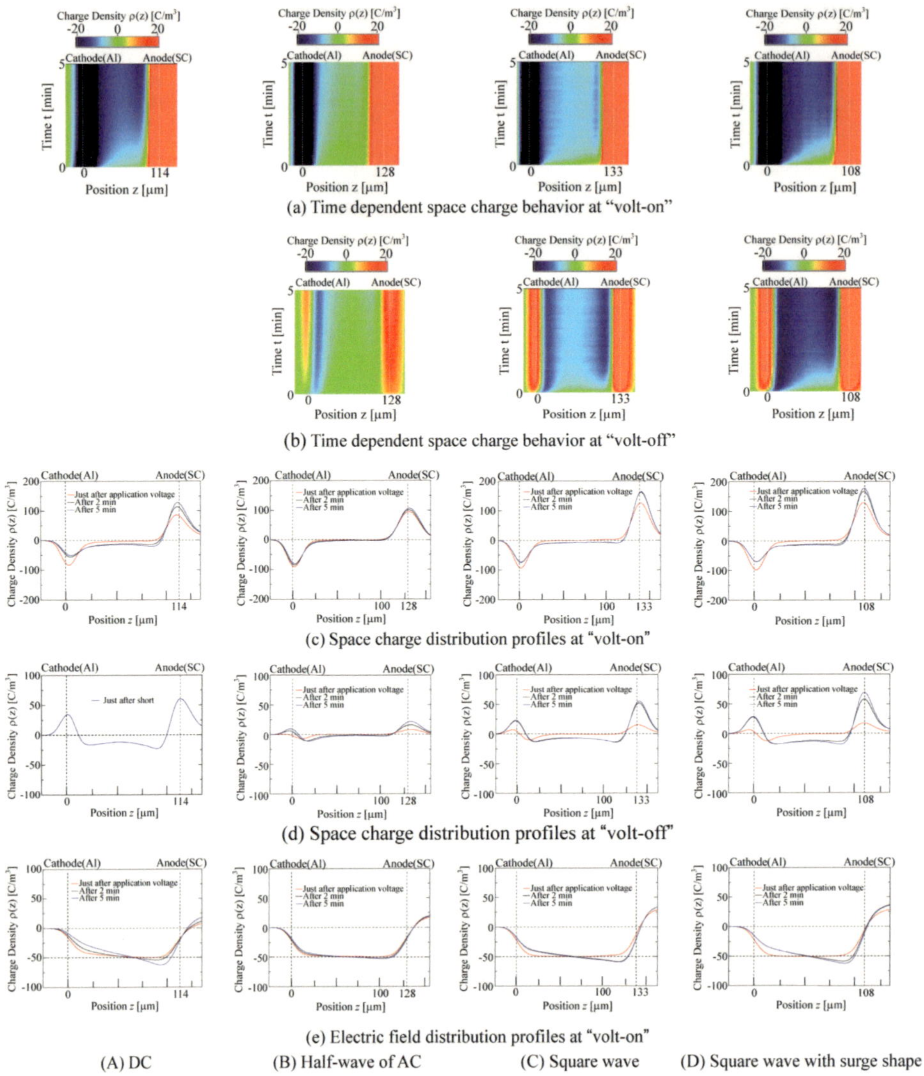

Figure 10.
Space charge distributions in PAI films under various voltage application (50 kV/mm).

As shown in **Figure 10(A-d)**, **(C-d)** and **(D-d)**, in the cases by applying the DC, the square wave and surge voltages, it is found that homogeneous negative charge accumulations across the bulk of the sample were observed from 2 min later after the voltage applications, and it was obviously recognized from the result shown in **Figures 10(A-a)**, **(C-a)** and **(D-a)**. As shown in **Figure 10(A-a)**, the accumulation of the negative charge seemed to be stable after 5 min later. By the negative charge accumulation, the electric field was distorted near the anode as shown in **Figure 10(A-e)**. On the other hand, in the case of result obtained by applying the half-wave of AC voltage, while a negative charge accumulation near the cathode was observed as shown in **Figure 10(B-b, d)**, the amount of it was small and it was only located near the cathode. By the negative charge accumulation, the electric field was distorted near the anode as shown in **Figure 10 (B-e)**, but it was very tiny.

Figure 11 (a) and **(b)** show the maximum electric field and amount of charge in the sample calculated from the distributions obtained by applying DC,

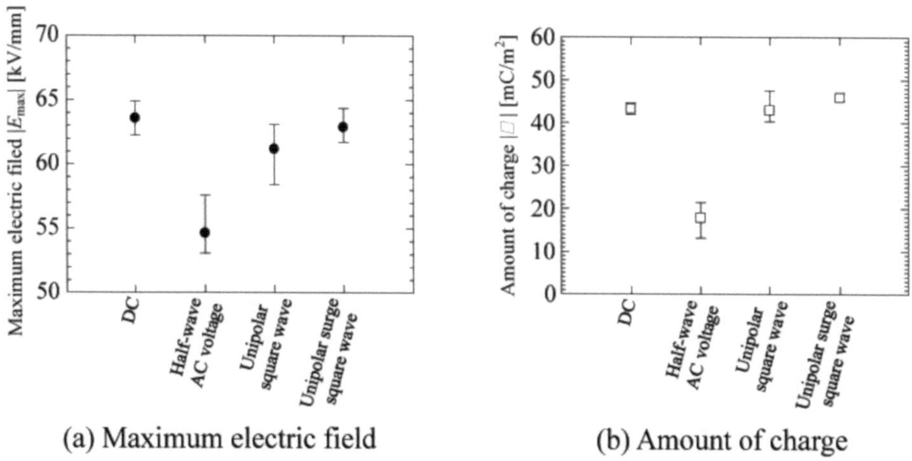

Figure 11.
Maximum electric field and amount of accumulated charge.

half-wave of AC, square wave, square wave with surge shape voltages. As shown in **Figure 11 (a)**, it was found that the maximum electric field value is larger in the order of DC, square wave with surge shape, square wave and half-wave of AC voltages. Furthermore, it was found that the charge accumulation amounts were large in the order of square wave with surge shape, DC, square wave and half-wave of AC. However, there was almost no difference among them expect for that obtained by applying half-wave of AC. From the above, it is clear that the space charge is not accumulated, nor consequently the electric field is not distorted by applying the half-wave of AC to the sample.

7. Polyimide for space application

Polyimide (PI) is also used for the materials of multilayer insulators (MLIs) of spacecraft. Many researchers studied space charge characteristics in PI irradiated by electron [8, 9]. And recently, the space charge behavior in PI irradiated by proton is also focused. It is because that many satellite had operation anomaly due to charging by radiation and discharging [10]. In this chapter, few examples of charge accumulation behavior in PI irradiated by radio-active rays are explained.

7.1 Negative charge accumulation in PI irradiated by an electron

The space charge [8] distributions in Kapton®500H (Dupont-Toray Co.Ltd., nominal thickness is 125 μm, a surface with Al evaluation layer) are irradiated by an electron are discussed. The electron beam irradiation condition is 40 and 60 keV, with 40 nA/cm^2 current density in 10^{-5} Pa order vacuum chamber. The in-site space charge measurements were carried out using PIPWP method [1, 8, 11]. Charge distributions and integrated charge amount in the bulk are shown in **Figures 12** and **13**, respectively. The measurement interval is 30 s.

In the **Figure 12**, the charge profile waveform is the data of maximum negative charge accumulation in the bulk during 20 min irradiation. From the figure, we can see the negative charge accumulation in the bulk. Such negative charges distributed from the irradiation surface to depth of 34, 60 μm (with the broken perpendicular line). Those accumulated positions have a good correlation with the calculated penetration depth using Karz and Penfold's experimental equation.

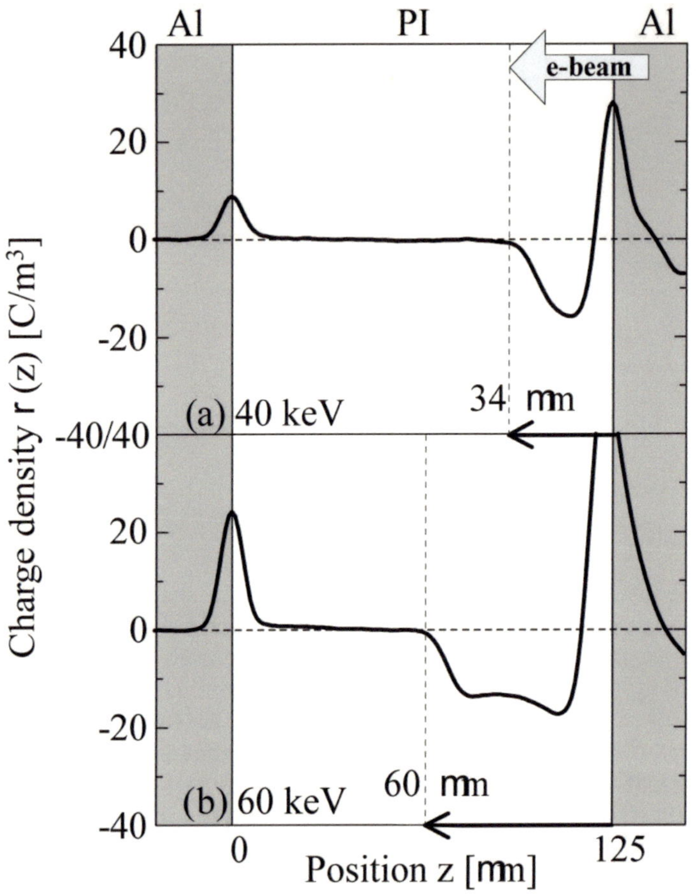

Figure 12.
Negative charge accumulation in PI under electron irradiation.

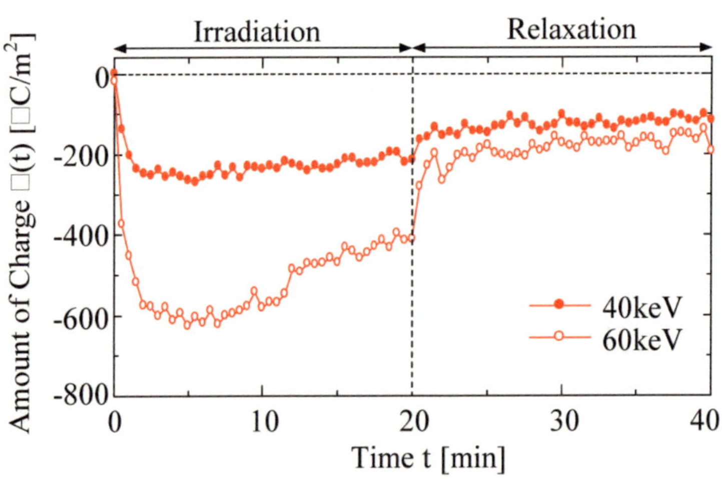

Figure 13.
Total negative charge amount in the bulk of PI under And after electron irradiation.

Form **Figure 13**, those negative charges were saturated during the irradiation time progress. And in the case of 60 keV irradiation, those negative charges were decreased in spite of irradiation progressing. The total accumulated charge amount in the bulk is decided due to the balance between the amount of charges supplied by the irradiation and the amount of charges released due to change of the bulk conductivity. Therefore, in this case, it is considered that the amount of accumulated charges was reduced because the amount of change in conductivity inside the bulk became largely dominant.

7.2 Positive charge accumulation in PI irradiated by a proton

In this section, the space charge distribution in PI irradiated by a proton is explained [12]. **Figure 14** shows space charge distribution in PI irradiated by a proton with several irradiation energy. PA and PB shows the polyimide materials, and the differential between PA and PB is the fabrication company. The irradiation condition is 1.0–2.0 MeV for energy with 0.3–30 nA/cm^2 under 1×10^{-5} Pa, and those irradiation times is 30 min. The space charge measurement was carried out with each 30 s for the irradiation and 10 min relaxation following from the above irradiation. The irradiation was carried out using 3 MeV tandem type ion accelerator facility of Takasaki Advanced Radiation Research Institute of National Institutes for Quantum and Radiological Science and Technology (QST), 3.75 MV Van de Graff of High Fluence Irradiation Facility at the University of Tokyo and the space environment examination facility at Japan Aerospace Exploration Agency (JAXA).

From **Figure 14**, those space chare distributions show the only maximum charge accumulation during irradiation with the 30 nA/cm^2 current density. In the figure, the irradiation direction is described on the right side of the figure. From the figure, it is found that positive charges are accumulated in the bulks under all irradiation condition. The charge accumulated position is moved become from the bulk near the irradiation side to middle of the bulk with irradiation energy progress. The

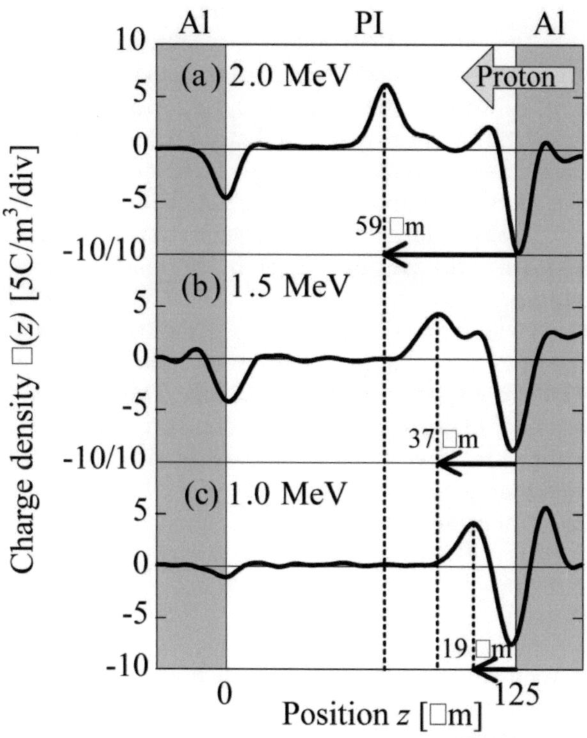

Figure 14.
Positive charge accumulation in PI under proton irradiation.

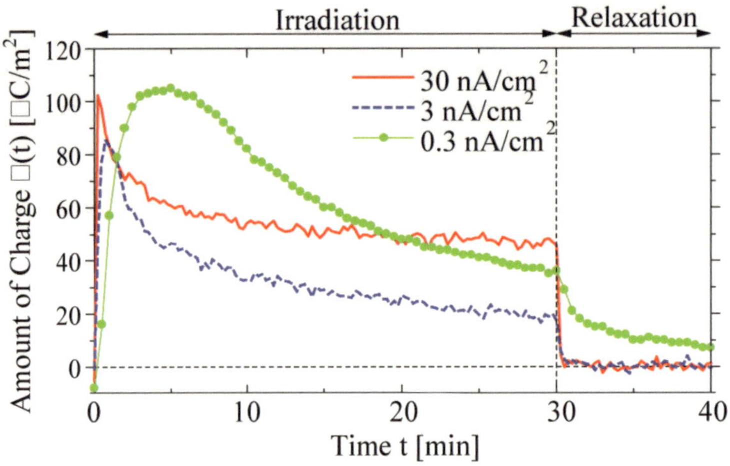

Figure 15.
Total positive charge amount in the bulk of PI under And after proton irradiation.

ideal penetration depths calculated by the SRIM code are shown with broken line in the bulk on the figure. From these results, we found that the calculated penetration depth has a good correlation with charge accumulated position. Therefore, it is considered that the irradiated proton is the origin of those positive accumulated charges in the bulk.

Figure 15 shows the time dependence of the amount of accumulated positive charges in those bulks irradiated by 2.0 MeV proton. Those are obtained by integration calculation of the charge distributions shown in **Figure 14**. From the figure, concerning accumulation behavior, the increases of the amounts of charges saturate within 3 minutes from the start of irradiation. The accumulated positive charges saturated for a brief time with the current density increasing. After the saturations, the amounts of charges decreased with irradiation time progress.

Furthermore, while we also observed that those accumulated positive charges remained after irradiation with low current density 0.3 nA/cm^2, no charge was observed immediate after irradiation with high current density irradiation.

From the above charge accumulation penonena, we considered that the origin of those phenomena was produced due to the generation of radiation induced conductivity (RIC). Form the previous research, we could find that conducitibity κ of the PI irradiated a proton with an energy of 2 and 1.5 MeV is 10^3 and 10^2 higher than non-treated sample, respectively. It is though that the strength of activation is depended on the irradiation energy. Furthermore, the charge accumulation phenomena may be strongly affected by proton dose. The origin of the RIC is considered activation of material, sission of molecular chain, ionization, and vacancy [13]. Furthemore, after the irraidaion, we could confirm the condirctivity was 10^4 times larger than prisitn sample's. It is one of evidence about molecule chain modification such as mentioned above.

8. Conclusions

It this chapter, we introduced the dielectric phenomena of PI and PAI materials using space charge measurement results using the PEA method. And you could understand the PEA method is very useful for understanding the charge accumulation phenomena. However, the PI is applied for various condition, it is necessary that the measurement system should be modified for the condition and environment applicable. And then, we could discuss the dielectric phenomena. From the described above, although we could understand the PI has a good insulation performance, since the characteristics may deteriorate depending on the environment, an analysis simulating the usage environment is required.

Acknowledgements

This research was partly supported by a KAKENHI from The Ministry of Education, Culture, Sports, Science and Technology, Japan. We are also grateful to the Tokyo University and National Institutes for Quantum and Radiological Science and Technology (QST) for providing us the proton irradiation facility according with their facility share program.

Author details

Hiroaki Miyake* and Yasuhiro Tanaka
Tokyo City University, Japan

*Address all correspondence to: hmiyake@tcu.ac.jp

References

[1] T. Takada, "Acoustic and Optical Methods for Measuring Electric Charge Distributions in Dielectrics", IEEE Trans. DEI, Vol.6, No.5, 1999, pp. 519-547

[2] Tomoyuki Ishii, Takamasa Takiwaki, Hiroaki Miyake, Yashuhiro Tanaka, Tatsuo Takada, Toshikatsu Tanaka, "Space Charge Formation in Polyimide Film under High DC Voltage at High Temperature", IEEJ Trans. on Fundamentals and Material., Vol. 133, No. 3, pp.91-97 (2013)

[3] Y. Tanaka, *M. mima*, Y. Narita, H. Miyake and Y. Tanaka, "Space Charge Accumulation Characteristics in Super Engineering Plastics under DC Stress at High Temperature," 2019 IEEE Conference on Electrical Insulation and Dielectric Phenomena (CEIDP), Richland, WA, USA, 2019, pp. 793-796, doi: 10.1109/CEIDP47102.2019.9010550.

[4] E. Sugimoto, "Application of Polyimide films to the Electrical and Electronic Industries in Japan", IEEE Electrical Insulation Magazine, Vol.5, pp.15-23 (1989)

[5] Y.Kishi, T. Hashimoto, H. Miyake, Y. Tanaka and T. Takada, "Breakdown and Space Charge Formation in Polyimide Film under DC High Stress at Various Temperature", J. phys. Conference Sevies, Vol.183,012005, 2009

[6] *M. mima*, H. Miyake and Y. Tanaka, "Measurement of space charge distribution in polyamide-imide film under square wave voltage practical environment for inverter-fed motor," 2018 IEEE 2nd International Conference on Dielectrics (ICD), Budapest, 2018, pp. 1-4, doi: 10.1109/ICD.2018.8514747.

[7] Maimi Mima, Tokihiro Narita, Hiroaki Miyake, Yasuhiro Tanaka, Masahiro Kozako, Masayuki Hikita, Influence of Space Charge Accumulation by Pre-stress on Partial Discharge Inception Voltage, IEEJ Transactions on Fundamentals and Materials, 2020, Volume 140, Issue 5, Pages 276-284, Released May 01, 2020, Online ISSN 1347-5533, Print ISSN 0385-4205, https://doi.org/10.1541/ieejfms.140.276, (in Japanese)

[8] K. Nagasawa, M. Honjoh, H. Miyake, R. Watanabe, Y. Tanaka and T. Takada, "Charge Accumulation in Various Electron-Beam-Irradiated Polymers", IEEJ Trans. Erectrical and Electronic Engineering, Vol. 5 No. 4, pp. 410-415 (2010)

[9] C. Perrin, V. Griseri, C. Inguimbert and C. Laurent, "Analysis of internal charge distribution in electron irradiated polyethylene and polyimide films using a new experimental method", J. Phys. D: Appl. Phys., Vol. 41, No. 20, 205417 (2008)

[10] H. C. Koons, J. E. Mazur, R. S. Selesnick, J. B. Blake, J. F. Fennell, J. L. Roeder and P. C. Anderson, "The Impact of the Space Environment on Space Systems", Proceedings of the 6th Spacecraft Charging Technology Conference, Air Force Research Laboratory, pp.7-11, (1998)

[11] Y. Tanaka, M. Honjo, S. Maruta and T. Takada, "Observation of Charge Accumulation Behavior in Polyimide Film under Electron Beam Irradiation", Proc. 37th Symposium on Electrical and Electronic Insulating Materials and Applications in Systems, A-2, pp. 21-24 (2006) (In Japanese)

[12] R. Uchiyama, T. Hara, T. Homme, H. Miyake and Y. Tanaka, "Degradation phenomena of electric property on polymeric materials films irradiated by proton," 2012 Annual Report Conference on Electrical Insulation

and Dielectric Phenomena, Montreal, QC, 2012, pp. 641-644, doi: 10.1109/CEIDP.2012.6378862.

[13] H. Miyake, V. Griseri, T. Mori, Y. Tanaka, G. Teyssedre and C. Laurent, "Physicochemical analysis for fluorinated polymer films irradiated by proton," 2017 IEEE Conference on Electrical Insulation and Dielectric Phenomenon (CEIDP), Fort Worth, TX, 2017, pp. 106-109, doi: 10.1109/CEIDP.2017.8257545.

Chapter 11

Lifetime of Polyimide under Repetitive Impulse Voltages

Yan Yang and Guangning Wu

Abstract

Polyimide (PI) is a commonly used insulating material to resist surface discharge, for instance, as turn-to-turn insulating material in inverter-fed motors driven by pulse width modulation (PWM) converters. Under the effect of repetitive impulse voltages, PI is expected to withstand surface partial discharge (PD) during service. However, lifetime under repetitive impulse voltages is much shorter than that under AC voltages due to storage effect of charges. Many approaches have been proposed to improve lifetime of polyimide under repetitive impulse voltages, such as using nanocomposites, surface modification, and structure design. In this chapter, we will discuss the lifetime of polyimide under repetitive impulse voltages and corresponding theoretical mechanism, together with modification approaches and their effects on lifetime improvement.

Keywords: lifetime, repetitive impulse voltages, surface discharge, nanocomposites, surface modification

1. Introduction

Polyimide (PI) is used as insulating material for resist surface discharge because of its excellent electrical, thermal, and mechanical characteristics [1, 2]. PI films are usually expected to withstand surface discharge caused by repetitive impulse voltages. For example, surface partial discharge occurs during lifetime of turn-to-turn insulation of inverter-fed traction motors due to the driven pulse width modulation (PWM) converters, which has become one of the most important factors for surface degradation and final premature failures of the insulation of inverter-fed traction motors. Therefore, the PD-dependent lifetime of PI films under repetitive impulse voltages needs to be improved to meet the requirement for rapid development of industry applications [3, 4].

To provide a comprehensive understanding to lifetime of PI under repetitive impulse voltages and to clarify the emerging problems, we review the recent progress in partial discharge characteristics, degradation, and charge storage effect under repetitive impulse voltages. Particular attention is paid on lifetime improvement approaches, such as using nanocomposites and surface modification.

2. Lifetime of polyimide under repetitive impulse voltages

2.1 Lifetime and partial discharge characteristics of polyimide under repetitive impulse voltages

Figure 1 shows the system for testing lifetime and partial discharge characteristics of polyimide under repetitive impulse voltages. Bipolar continuous square impulse voltage was conducted. The amplitude, risetime, duty cycle, and frequency of applied voltage were 1 kV, 40 ns, 50%, and 10 kHz, respectively. The electrodes used in the test were designed according to the standard of ASTM 2275 01, including a rod electrode, a plate electrode, and an insulating board [5]. By this configuration, we can obtain the voltage endurance of solid electrical insulating materials subjected to surface partial discharges. The PD signal was measured by an UHF antenna.

Figure 2 shows the lifetime of PI under repetitive impulse voltages, which decreases sharply with increasing temperature. The lifetime values are much shorter than that under AC voltages [6].

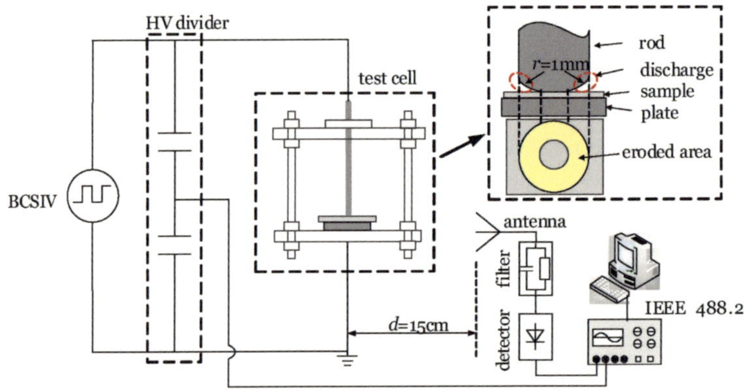

Figure 1.
Schematic diagram of the PD aging test system.

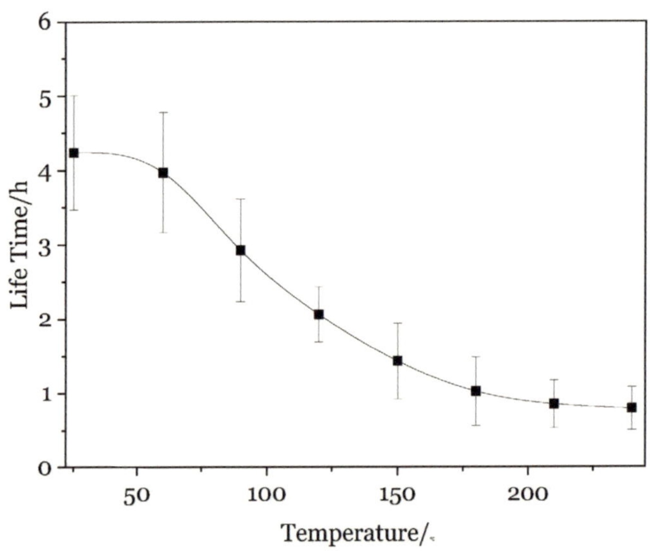

Figure 2.
Lifetime of PI under repetitive impulse voltages [6].

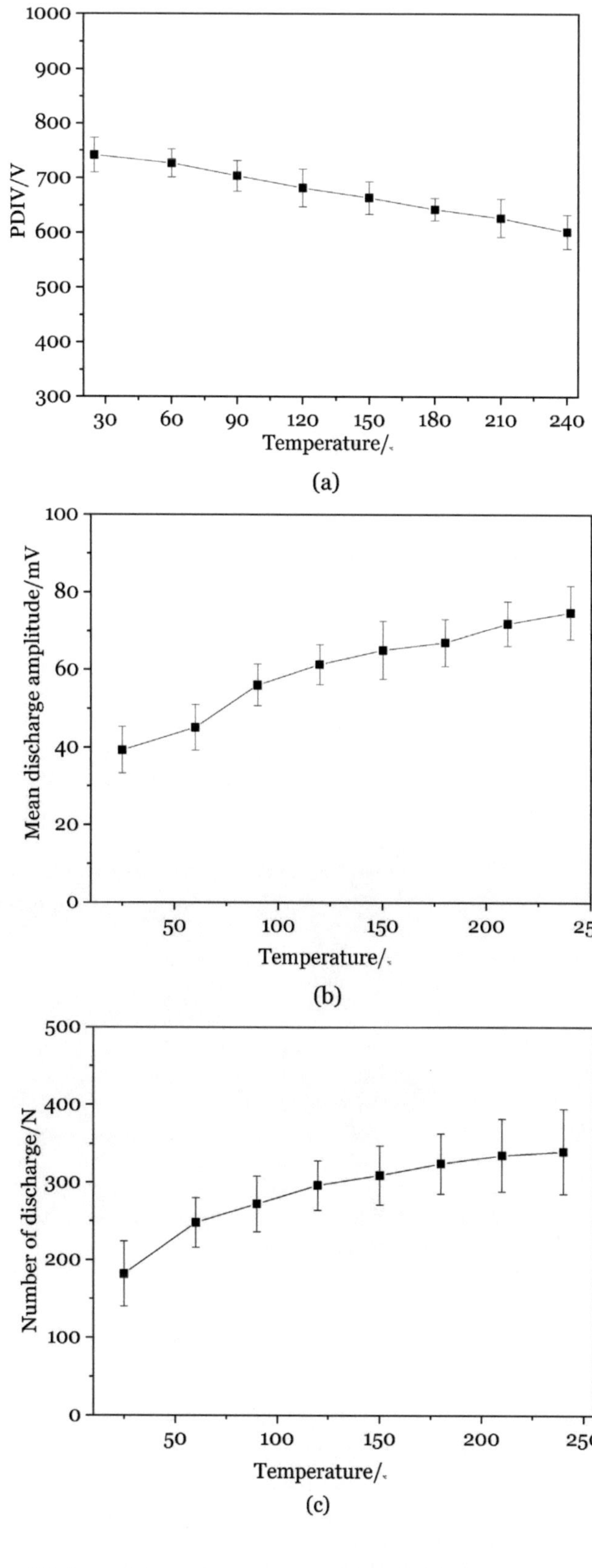

Figure 3.
PD characteristics of PI films with increasing temperature. (a) PDIV, (b) PD amplitude, (c) PD number.

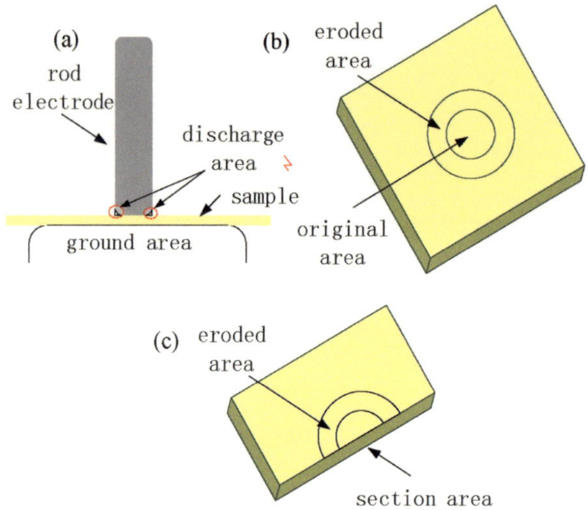

Figure 4.
Sample discharge area and eroded area of sample [6].

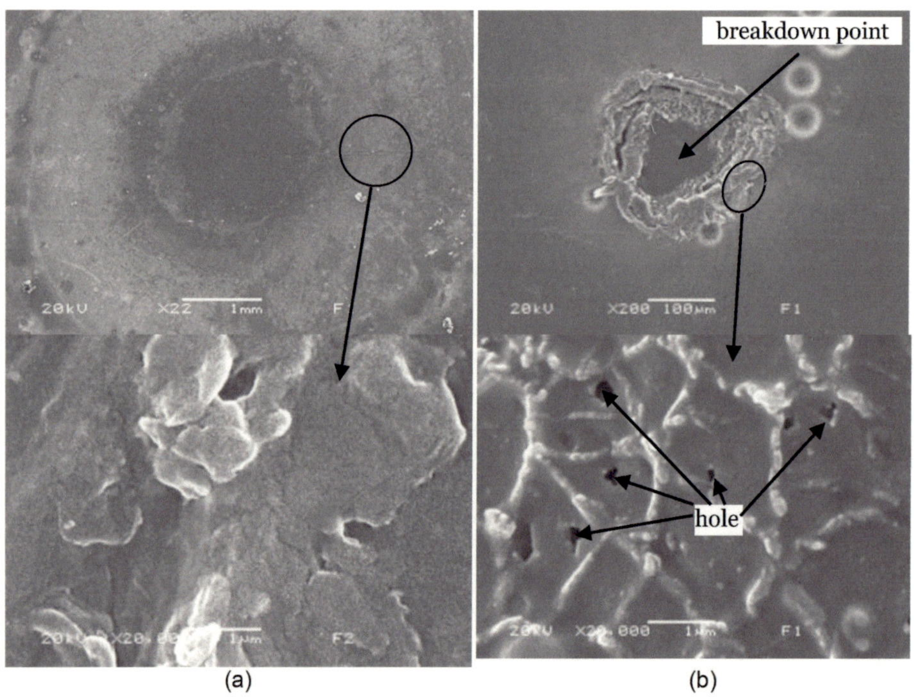

Figure 5.
Microtopography of PI films after 3-h aging and breakdown. (a) 3-h aging and (b) breakdown.

Figure 3 shows the PD characteristics, including the PDIV, PD amplitude, and PD number, of PI films under repetitive impulse voltages as a function of temperature. PDIV decreases with increasing temperature; on the contrary, PD amplitude and PD number increase with increasing temperature. All these parameters are higher than that under AC voltages [6].

2.2 Partial discharge-induced degradation under repetitive impulse voltages

Due to the electrode structure as shown in **Figure 2**, surface discharge would occur in the air gap between the sample and the upper rod electrode during lifetime testing. Certain region on the film surface, as shown in **Figure 4**, would be corroded by surface discharge generating high-energy electrons, ultraviolet rays, and high-activity chemical groups [7, 8]. The microtopography of PI films after 3-h aging and breakdown (lifetime: 6 h and 10 min) under repetitive impulse voltages is shown in **Figure 5**. An obvious circle can be seen on the surface of aged PI film, which is eroded as the consequence of surface discharge, resulting in surface roughness and morphology changing. From the microscopic view of the breakdown point as shown in **Figure 5(b)**, the mesh structure and voids can be observed on the surface microtopography around the breakdown point obviously, indicating degradation of PI occurred during discharge.

2.3 Storage effect of charges under repetitive impulse voltages

Under repetitive impulse voltages, partial discharge-induced degradation would be intensified because of charge storage effect. We discuss the electrical field distribution on discharge region of the sample surface, as shown in **Figure 6**, where E_i is the electric field in the air gap between the sample and the edge of the upper electrode. When external electric field reverses, accumulated charges on the film surface will enhance the electric field in the air gap between the rod and the sample, as shown in **Figure 6**. Assuming the external electric field is negative, the charge-induced electric field in the air gap E_q is proportional to the difference between the charges on the rod electrode Q_{up} and that on the film surface Q_{down}, namely, E_q ($Q_{up} - Q_{down}$). Thus, the effective electric field in the air gap can be obtained, $E_i = E_t + E_q$, where Et is the external electric field. Since the charges on the film surface are unable to dissipate in such short rising time, the Q_{down} can be considered as a constant. When the external electric field reverses and becomes

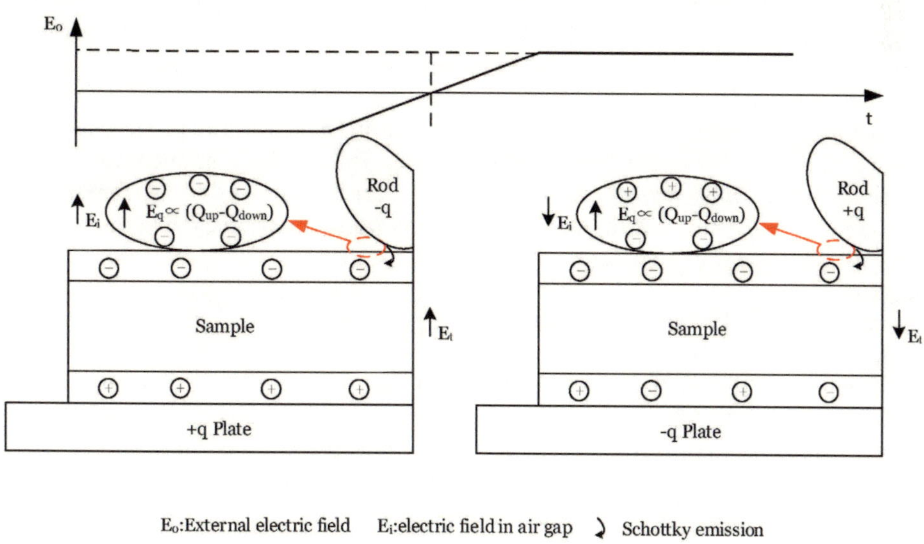

Figure 6.
The electric field change in air gap when external electric field inverses from negative to positive side [9].

positive, the E_q is proportional to the sum of the charges on the rod electrode Q_{up} and that on the film surface Q_{down}, namely, E_q (Q_{up} + Q_{down}). As the charge continues accumulating, the value of E_i can increase to a critical value, and then PD would take place.

3. Approaches for lifetime improvement

3.1 Polyimide nanocomposites

PD characteristics and lifetime repetitive impulse voltages would be improved by using polyimide nanocomposites. **Figure 7(a)** shows the PDIV of PI and PI/Al_2O_3 nanocomposites. PDIV increases with increasing nano-alumina content. The PDIV of PI/Al2O3 (8 wt.%) increases from 693 to 782 V (12.8% increase) compared with that of PI. This means the discharge threshold is much higher for PI/Al_2O_3 nanocomposites. **Figure 7(b)** gives the PD amplitude and number recorded in 1 min, which both decrease with increasing nano-alumina content, revealing that the decrease in PD intensity is attributed to the effect of nano-alumina.

As shown in **Figure 8**, the PD resistance is enhanced significantly as the content of nano-alumina increases. The lifetime values of PI/Al_2O_3 with nano-alumina content of 0, 2, 5, and 8 wt.% are 37.8, 52.1, and 61.9 min, increasing by 92.8, 165.8, and 215.8% respectively, compared with that of PI. This demonstrates that the nano-alumina helps to prolong the lifetime of PI.

In polyimide nanocomposites, the melting point of nano-alumina particles is higher than that of polyimide. Under the effect of surface discharge, nanoparticles float on the film surface, forming a protective layer to prevent the polyimide matrix from further erosion. This layer, therefore, is able to mitigate the PD effects on films and block the erosion path, which can help avoid the same local point being eroded continuously. In contrast, without nanoparticles, the erosion will follow a specific path and lead to breakdown much faster. This is why there are a number of micro-cavity groups in PI/Al_2O_3 nanocomposites but only a few big cavities in the PI films, as shown in **Figure 6 c** and **d**. Thus, the incorporation of nanoparticles improves the physical characteristics of PI/Al_2O_3 nanocomposites, because nanoparticles can act as obstacles in the erosion path, and then the degradation of polymer can be reduced, and prolonged lifetime of PI/Al_2O_3 nanocomposites can be obtained subsequently.

In addition, some acid compounds (e.g., the amic acid and the nitric acid) are generated during PD aging. Meanwhile, new bonds in PI/$A_{12}O_3$ nanocomposites are the weakest, subsequently the ether linkage, imide ring, and then aromatic ring. Those weakest bonds in interfacial regions would be destroyed at the first stage of aging, so the dissociations of polyimide molecules would be reduced. Because of the chemical bonding between nanoparticles and PI molecules, the degradation of polymer can be reduced, and prolonged lifetime of PI/Al_2O_3 nanocomposites can be obtained consequently.

Moreover, higher surface conductivity facilitates surface charge dissipation and leads to low PD intensity. Lower trap density is responsible for the low possibility of charge recombination and the corresponding released energy. All these factors improve the electrical characteristics of PI/Al_2O_3 nanocomposites resulting in reduced degradation of polymer and longer lifetime.

3.2 Surface modification

Lifetime of PI under repetitive impulse voltages would be further improved by surface modification. **Figure 9** shows the lifetime of PI and PI/Al_2O_3

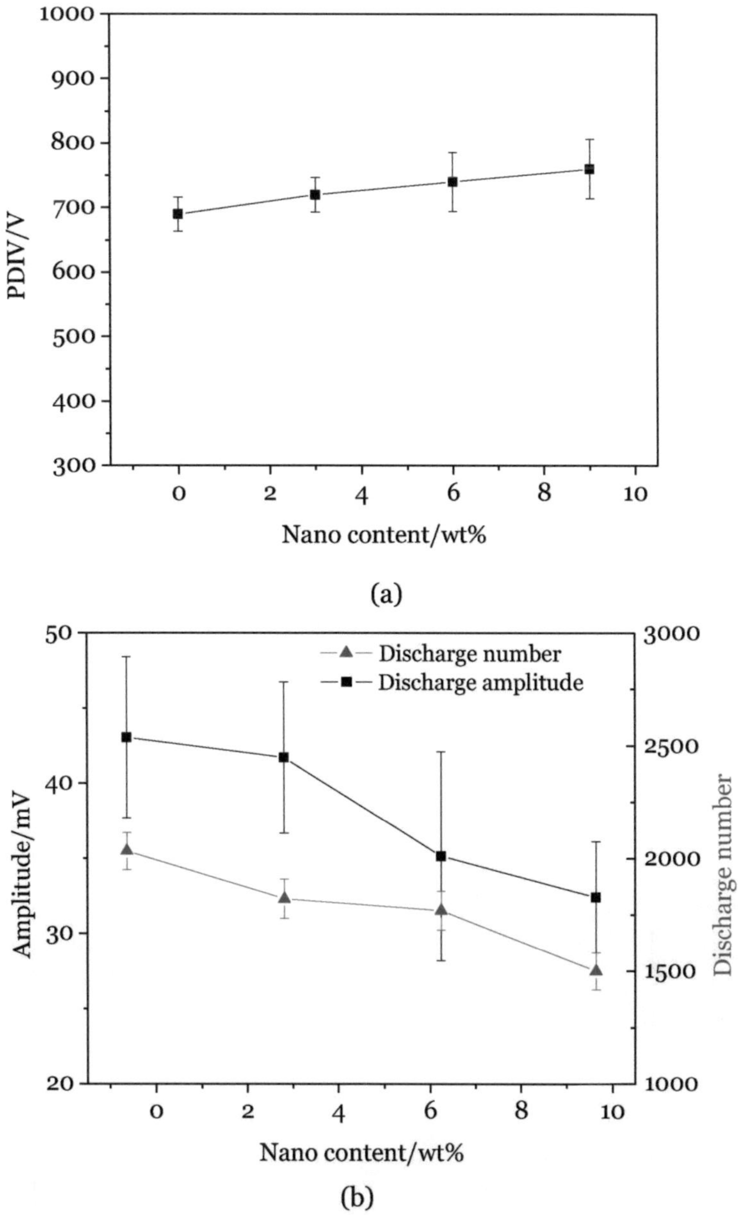

Figure 7.
PD characteristics of PI and PI/Al_2O_3 nanocomposites. (a) PDIV and (b) PD amplitude [9].

nanocomposites under repetitive impulse voltages, after being treated by nonthermal plasma in atmospheric air with different treating times. For both PI and PI/Al_2O_3 nanocomposites, lifetime can be prolonged by plasma treatment. Apparent improvement of lifetime can be seen in the first 20 s of treating time. However, the lifetime decreases with increasing treating time, when the treating time is longer than 20 s. When the treating time exceeds 30 s, the lifetime decreases sharply, by reaching the values even shorter than that of the untreated samples. For plasma-treated samples, with shallower trap energy level, less charge accumulation facilitates charge dissipation and local electric field mitigation, leading to suppressed PD intensity and longer lifetime. Active groups containing oxygen and nitrogen, which are introduced during plasma treatment, are responsible for trap energy

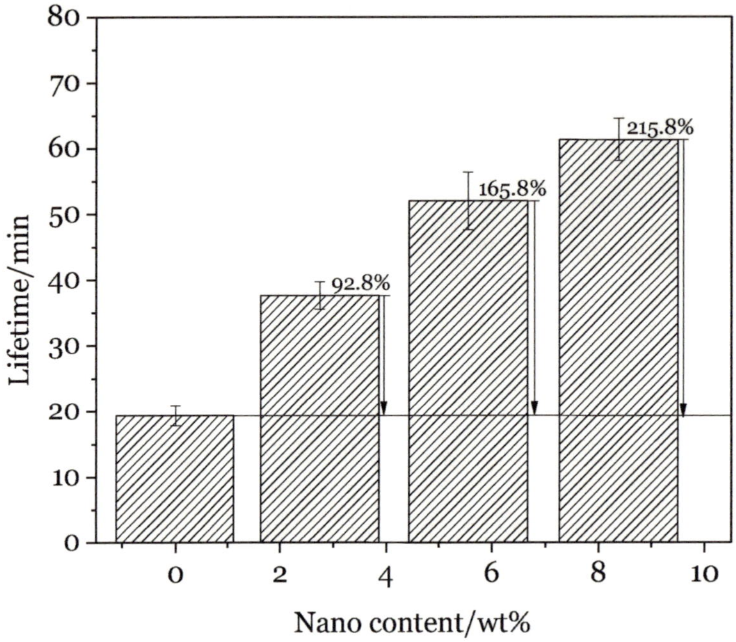

Figure 8.
Lifetime of PI and PI/Al_2O_3 nanocomposites under PD aging [9].

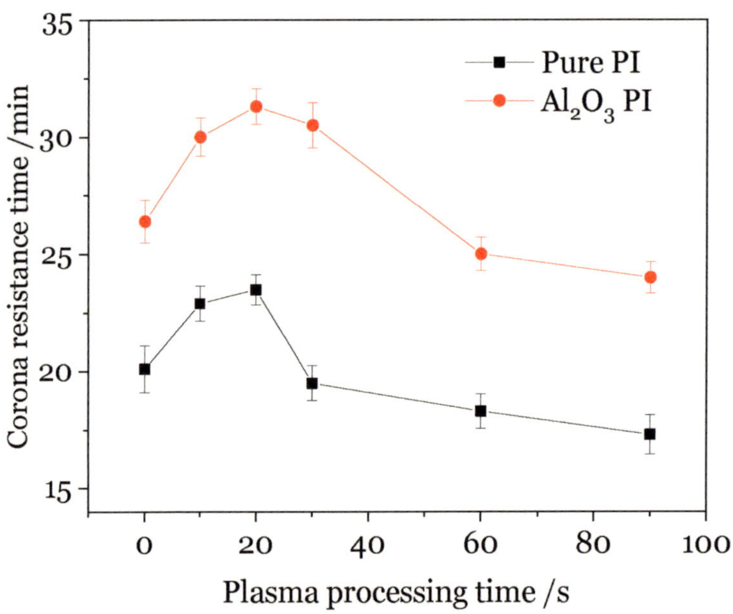

Figure 9.
Lifetime of PI and PI/Al_2O_3 nanocomposites with different plasma treating times under repetitive impulse voltage.

distribution modification of PI films. An appropriate treating time is the key factor to introduce reactive groups on film surface and prolong the lifetime.

3.3 Structure design

Structure design is another effective approach to improve lifetime of PI under repetitive impulse voltages. **Figure 10** shows the lifetime of PI, PI/Al_2O_3

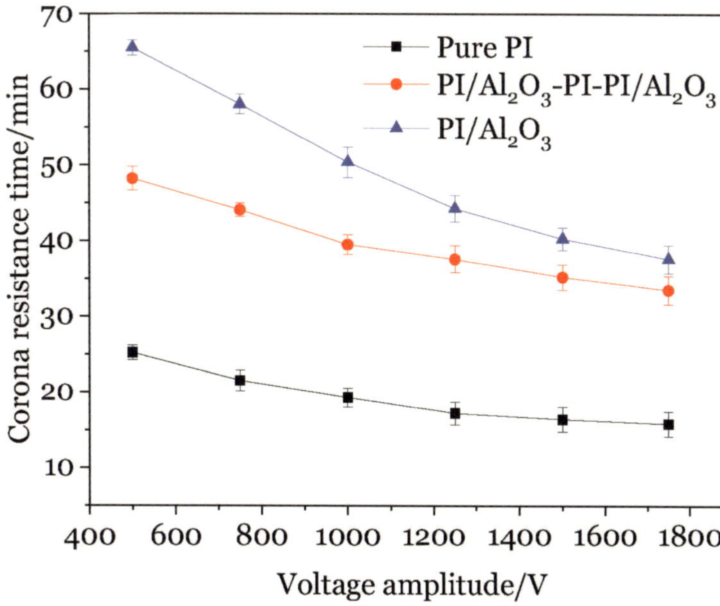

Figure 10.
Lifetime of PI, PI/Al_2O_3 nanocomposites, and PI/Al_2O_3-PI-PI/Al_2O_3 films under repetitive impulse voltages as a function of applied voltage (room temperature) [10].

nanocomposites, and PI/Al_2O_3-PI-PI/Al_2O_3 films under repetitive impulse voltages as a function of applied voltage. For all samples, lifetime decreases with increasing applied voltage. As the thickness of these three types of samples is all 60 mm, the lifetime of PI/Al_2O_3 nanocomposites are the longest. However, the results shown in **Figure 10** were tested at room temperature. Considering the effect of temperature, lifetime of PI/Al_2O_3-PI-PI/Al_2O_3 films is longest when the temperature is higher than 90°C, as shown in **Figure 11**. For applications such as inverter-fed motors, the

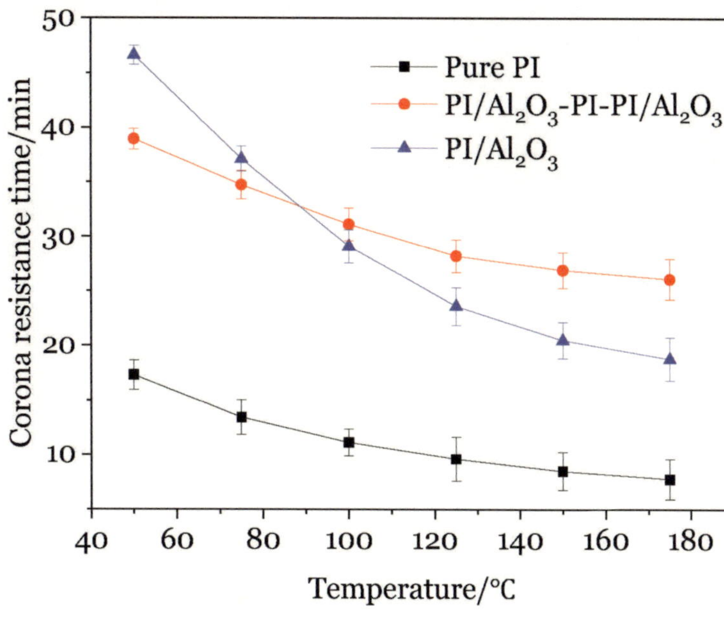

Figure 11.
Lifetime of PI, PI/Al_2O_3 nanocomposites, and PI/Al_2O_3-PI-PI/Al_2O_3 films under repetitive impulse voltages as a function of temperature (1 kV) [10].

operating temperature of PI is usually higher than 90°C; thus layered structure is an important approach to improve lifetime for industry applications.

4. Conclusion

PD-dependent lifetime of PI under repetitive impulse voltages is a key parameter for its industry applications. Due to charge storage effect under repetitive impulse voltages, partial discharge-induced degradation would be intensified, resulting in a shorter lifetime. By using nanocomposites, surface modification, and structure design, the lifetime would be prolonged.

Acknowledgements

This work was financially supported by National Natural Science Foundation of China (NSFC) (under Grant 51837009 and 51907167) and State Key Laboratory of Electrical Insulation and Power Equipment (EIPE18212).

Author details

Yan Yang* and Guangning Wu
School of Electrical Engineering, Southwest Jiaotong University, China

*Address all correspondence to: yangyanyy@swjtu.edu.cn

References

[1] Zhou L, Wu G, Gao B, et al. Study on charge transport mechanism and space charge characteristics of polyimide films. IEEE Transactions on Dielectrics and Electrical Insulation. 2009;**16**(4):1143-1149

[2] Luo Y, Wu G, Liu J, et al. PD characteristics and microscopic analysis of polyimide film used as turn insulation in inverter-fed motor. IEEE Transactions on Dielectrics and Electrical Insulation. 2014;**21**(5):2237-2244

[3] Kaufhold M, Aninger H, Berth M, et al. Electrical stress and failure mechanism of the winding insulation in PWM-inverter-fed low-voltage induction motors. IEEE Transactions on Industrial Electronics. 2000;**47**(2):396-402

[4] Akram S, Wu G, Gao GQ, et al. Cavity and interface effect of PI-film on charge accumulation and PD activity under bipolar pulse voltage. Journal of Electrical Engineering and Technology. 2015;**10**(5):2089-2098

[5] Min D, Cho M, Khan AR, et al. Surface and volume charge transport properties of polyimide revealed by surface potential decay with genetic algorithm. IEEE Transactions on Dielectrics and Electrical Insulation. 2012;**19**(2):600-608

[6] Yang LUO. Study on Damage Characteristics of Polyimide Film Caused by PD and Material Modification. Chengdu, China: Southwest Jiaotong University; 2014

[7] Park SJ, Lee HY. Effect of atmospheric-pressure plasma on adhesion characteristics of polyimide film. Journal of Colloid and Interface Science. 2005;**285**(1):267-272

[8] Yang Y, Yin D, Xiong R, et al. Ftir and dielectric studies of electrical aging in polyimide under AC voltage. IEEE Transactions on Dielectrics and Electrical Insulation. 2012;**19**(2):574-581

[9] Zhong X, Wu G, Yang Y, et al. Effects of nanoparticles on reducing partial discharge induced degradation of polyimide/Al_2O_3 nanocomposites. IEEE Transactions on Dielectrics and Electrical Insulation. 2018;**25**(2):594-602

[10] Jian ZHU. Structure Design of Polyimide Nanocomposite Film and Research on its Electric Heating Performance. Chengdu, China: Southwest Jiaotong University; 2017

Chapter 12

Electrical Endurance of Corona-Resistant Polyimide for Electrical Traction

Tao Han and Andrea Cavallini

Abstract

This paper shows the behavior of several dielectric properties of corona-resistant polyimide tapes as a function of thermal (high temperatures) and ambient stress (high humidity coupled with high temperature). The main goal of the investigation is to understand and explain the evolution of the partial discharge endurance of the tapes as a function of nonelectrical aging. The results indicate that polyimide tapes are very stable up to temperatures of 320°C. However, the most elevated temperatures reduce the partial discharge endurance to 50% of the original value. This is probably due to morphological changes in the (outer) nanostructured layers of the tapes.

Keywords: electrical drives, insulation reliability, partial discharges, corona-resistant insulation, thermal aging

1. Introduction

Polyimide tapes are used to manufacture rotating machines. Due to their relatively large cost, these tapes are mostly in special machines exposed to high temperatures [1, 2]. As an example, machines used for drilling oil wells are insulated using PI enamels or tapes.

The situation is changing, however. Transport electrification is demanding more compact machines, characterized by higher power densities. This implies that electrical machines will be designed having higher frequencies, temperatures, and voltages compared with the current standards. This evolution places PI insulation in the forelight. As a matter of fact, aircraft turbine-mounted generators and train machines are already experiencing high temperatures, and many manufacturers already used PI-based commercial products (e.g., Kapton). Indeed, the use of PI wires has been challenged for aircraft wiring due to its sensitivity to hydrolysis, wet and dry arcing (it is speculated that PI exposed to arcing is prone to catch fire prior than reaching a failure [3, 4]).

PI tapes can be used as slot liners (generally in form of laminates as NKN), phase separators, as well us around conductors to realize the phase-to-ground, phase-to-phase, and turn-to-turn insulation. In the last case, adhesive tapes are usually wrapped around the conductor. Complete adhesion is achieved by microwave heating of the conductor first and then infrared heating of the outer part. Eventually, the conductor is cooled down rapidly by immersion.

To date, most electrical machines are inserted in power electronic drives and are fed by a power electronic converter (inverter). The waveform provided by inverters is not the standard sinusoidal waveform, but a train of voltage impulses whose fundamental frequency is equal to the frequency of the reference sinusoidal voltage waveform. This type of voltage is sketched in **Figure 1**.

The advantages of PWM inverters are so many that nowadays only very few motors are still connected to the AC grid directly. As an example, the frequency of the drive is regulated in a straightforward way by changing the frequency of the reference voltage waveform. Furthermore, the maximum frequency of the drive can be much higher than that of the power grid, being the latter selected based on the limits of large rotating machines and with transmission losses and limits as a target. Going to higher frequencies ensures the possibility of shrinking the magnetic circuit of the motor, reducing the weight and the volume of the machine. Both advantages (controllability, high power densities) are vital for transport electrification to ensure that the drive works smoothly using a light actuator.

The “dark side” of PWM waveforms, however, is the electrical stress they impose on the insulation [5–8]. The voltage at the motor terminals can increase due to wave reflections when the voltage impulses reach the machine. This is because the connecting cables have a characteristic impedance of 20–30 Ω, whereas the machine characteristic impedance can be up to 400 Ω. Furthermore, a phenomenon known as “double pulse,” which happens when two phases commutate simultaneously, can further raise the applied voltage. The theoretical limits of these phenomena are (a) the voltage can be doubled when only reflections occur and (b) the voltage can be four times larger if both reflection and double pulse occur (**Figure 2**).

Furthermore, the impulses must propagate within the winding of the machine, which are complex inductive capacitive networks. Without going into details, it is important to understand that, during the rising and falling flanks of the voltage impulses, the first coil of the machine will withstand most of the inverter voltage, whereas if Nc is the number of coils in the machine, all coils will be subjected to the same voltage V/Nc when subjected to AC voltage waveforms.

This change in the electrical stress levels brought about a radical change in the failure mode of low-voltage rotating machines. Prior to the advent of power electronics, insulation mostly failed due to thermal aging. The eventual failure mode was the opening of crack where a large leakage current, able to melt the dielectric,

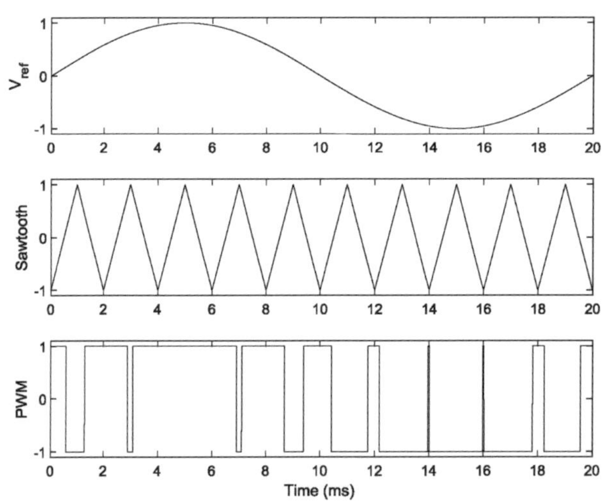

Figure 1.
PWM voltage waveform.

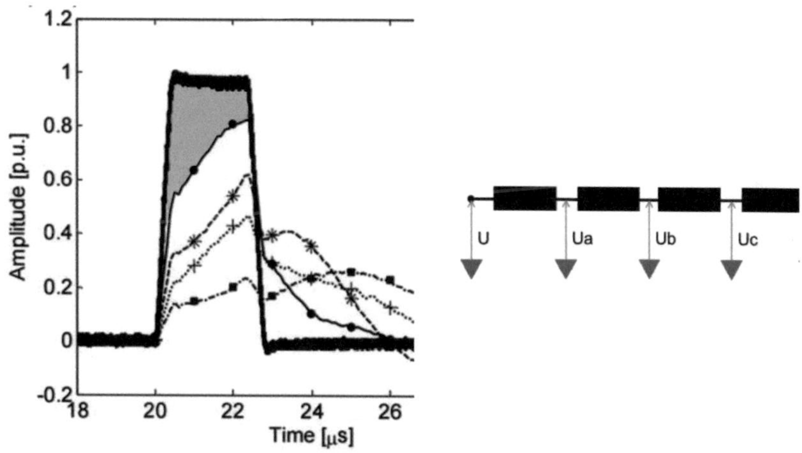

Figure 2.
Potential at a five-coil low-voltage motor terminal (U) and at the interconnection between adjacent coils (Ua, Ub, Uc, Ud). In gray, the voltage drops across the first coil.

could flow. Under power electronics, if the electrical stress exceeds a critical threshold, partial discharges are incepted inside the winding. These discharges can occur with a very large repetition frequency, comparable with the switching frequency of the power converter.

The insulation that is most vulnerable to partial discharges is the turn insulation [9, 10], as often enamels or tapes having limited thickness are used (thus, it is easier to puncture the insulation, and, moreover, the electric fields are the highest in the machine when inverters are used). Dielectrics made of purely organic polymers degrade quickly under PD bombardment as the C—C and C—H bond energies are 3–4 eV, and a large percentage of electrons in the discharge have sufficient energy to cause dissociative electron attachment (DEA) leading to bond breaking and free radical formation [11, 12]. After long enough, this process leads to the puncture of the insulation.

Due to these considerations, dielectrics used for the turn insulation of rotating machines have been nanostructured, meaning that inorganic particles of nanometric size were added to the base polymer [13, 14]. Nanostructured insulation for rotating machines is often indicated as corona-resistant insulation (e.g., CR wires, CR tapes, CR laminates, etc.).These particles tend to form a barrier to the progressive degradation induced by partial discharge activity (as an example, the formation of ceramic-like layers has been observed on the outer surface of CR winding wires). CR PI tapes have been proposed as a solution to improve the reliability of inverter-fed machines [15, 16]. Besides these tapes can be of use also in medium-voltage machines, where the large fields in the phase-to-ground insulation can lead to partial discharges in proximity of the conductors if the insulation detaches from the conductor due to the combined effect of high temperatures and thermal cycling (in that case, an air pocket can form in proximity of the conductor, leading to the inception of partial discharges that, in the long term, can cause a failure of the turn insulation).

The superior performance of CR insulation compared to standard insulation has been proven in literature [13, 16]. However, all the tests were conducted on pristine CR insulation. As this book is being written, activities are under way to test CR winding wires using repetitive voltage impulses and temperatures of 155°C. The limit of this approach is that CR insulation systems are compared in the absence of thermal aging. As a matter of fact, the partial discharge endurance tests are relatively short, and the oven temperature is well below the class temperature (that can be 180°C, 200°C, 220°C, or 240°C). This implies that the dielectric will not undergo a significant amount of oxidation or other forms of chemical degradation. Besides, the impact of hydrolysis will not be accounted for. As a result, it will not be easy to extrapolate the results to practical applications, as the combined electrical-thermal-ambient stress will differ substantially from the tests experienced during the tests.

In order to show how these stresses can alter the endurance of PI insulation to PD activity, we shall discuss in the forthcoming sections experiments carried out on CR Kapton films subjected to thermal and ambient stress and tested at regular times for PD endurance. The results will highlight the most likely phenomenon leading to a significant reduction of the endurance properties.

2. Test structure

2.1 Test sequence

The tests were programmed in sequences, to show the progressive deterioration (if any) of the PI properties. The test logic is shown in **Figure 3** and follows the ideas reported in [17]. Aging is subdivided in aging sub-cycles followed by diagnostic sub-cycles. During the diagnostic sub-cycles, measurements apt to infer the degradation of the PI films as electrical insulators are carried out. The samples that during the tests are modified by the test itself are discarded, and aging continues with the remaining samples. Therefore, PI films were prepared in large numbers to ensure that aging tests could be performed on different samples to highlight the degradation of PI dielectric properties (and not the effect of previous tests).

2.2 Measurements

The tests were performed on corona-resistant PI tapes (Dupont Kapton 200 CR), having nominal thickness of 50 μm. During the diagnostic sub-cycles, the following quantities were measured/reported (**Figure 4**):

1. Visual appearance of the sample.

2. Thickness evaluated by means of a micrometric screw.

3. Dielectric spectroscopy in the range of 0.1–3 MHz. The specimens were extracted from the aging cell and covered with silver paint. Since elevated temperatures could affect the silver paint, samples were discarded after testing.

4. Bulk DC conductivity using a three-electrode system. The electric field was 15 kV/mm.

5. PDIV measurements were carried out using a 50 Hz sinusoidal supply voltage. A rod/plane electrode configuration was designed and manufactured following [18]. The PD sensor was a high-frequency current transformer (HFCT). A Techimp PDBase II detector (bandwidth 50 MHz, sampling rate 200 MSa/s, and vertical resolution 12 bit) was employed to detect PD. The sensitivity was better than 5 pC. During the PDIV tests, the samples were held in a climatic chamber at 20°C, 25% relative humidity. The voltage was increased in steps of

50 V each lasting 30 s. PDIV measurements were repeated eight times, each time changing the position of the rod electrode. The PDIV measurement setup is outlined in **Figure 1**. Samples were discarded after testing.

6. Partial discharge endurance tests were carried out on all the samples. The supply voltage was a 1 kHz unipolar square wave. The peak-to-peak voltage was 7 kV. The rod/plane electrode configuration is the same used in the PDIV tests. Samples were discarded after the tests [18].

2.3 Aging stresses, levels, and timings

A first round of tests was performed placing the sample in the oven at a temperature of 270°C for 864 h. Indeed, as it will be exposed in the next section, 270°C is too a low temperature to produce a remarkable aging in PI samples (**Table 1**). Therefore, a second aging cycle lasting 294 h was performed. Knowing the potential negative impact of hydrolysis on PI films, this cycle was carried out placing the samples in a climatic chamber at 80°C with a relative humidity of 80%. Eventually, a last cycle lasting 432 h was carried out placing the samples in the oven at 320°C.

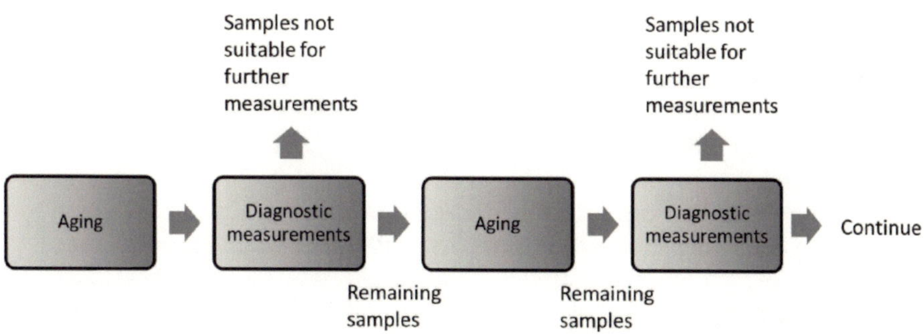

Figure 3.
Test plan for thermal aging at 270°C.

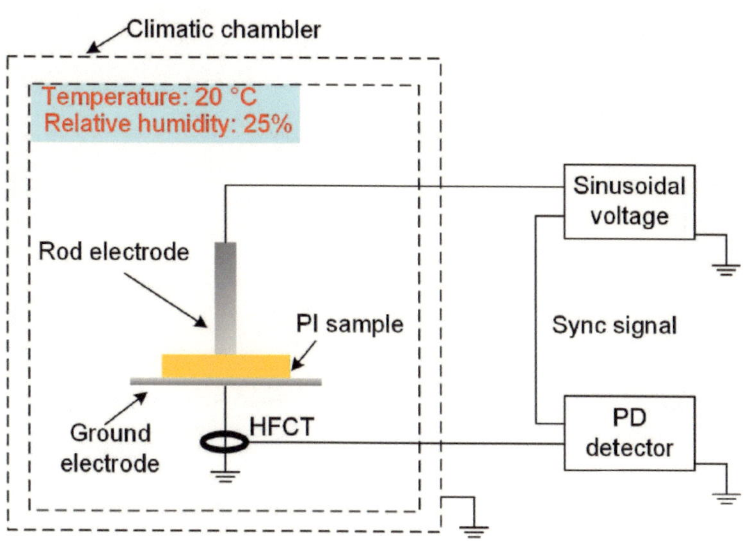

Figure 4.
Test cell for PDIV tests.

Aging stage	Time (h)	Condition
1	864	270°C
2	294	80°C and 80% RH
3	432	320°C

Table 1.
Thermal aging processes.

3. Results

3.1 Dielectric spectroscopy

To check the dielectric characteristics during each aging stage, the dielectric spectroscopy test was employed. **Figures 5** and **6** show the real and imaginary parts of the permittivity (ε' and ε"). It can be found that both ε' and ε" change to some extent in stage 1. More notable increases can be found in aging stage 2, particularly for frequencies below 100 Hz. ε' reaches a value of ~7.0 at low frequency. During aging stage 2, the outer layer (nanocomposites) was exposed in high temperature and humidity, where hydrolysis and water absorption will occur as a result [19, 20]. The polar production after hydrolysis and absorbed water (with ε' of ~80) contributes to the increased ε' and ε". After, in stage 3, the sample was kept in oven at 320°C. The acid and amine groups formed by hydrolysis can be recombined, and the absorbed water will be removed in a short time [21]. As a result, the dielectric response was reverted to be near to that of PI without aging.

To get more details of the changing dielectric properties, ε' and ε" of some selected frequencies are shown in **Figures 7** and **8**. As shown in **Figure 7**, the increasing ε' can be found in the first 576 h in stage 1, while it decreases till the end of this stage. A peak value of ε' at each frequency can be confirmed in stage 2, which indicates that the aging parameters of stage 2 affect the dielectric properties

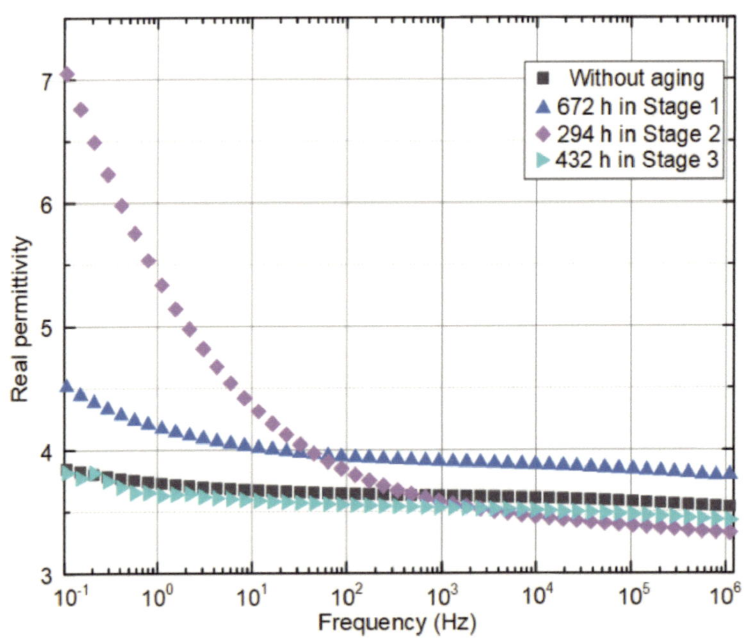

Figure 5.
Real permittivity at the end of each aging stage.

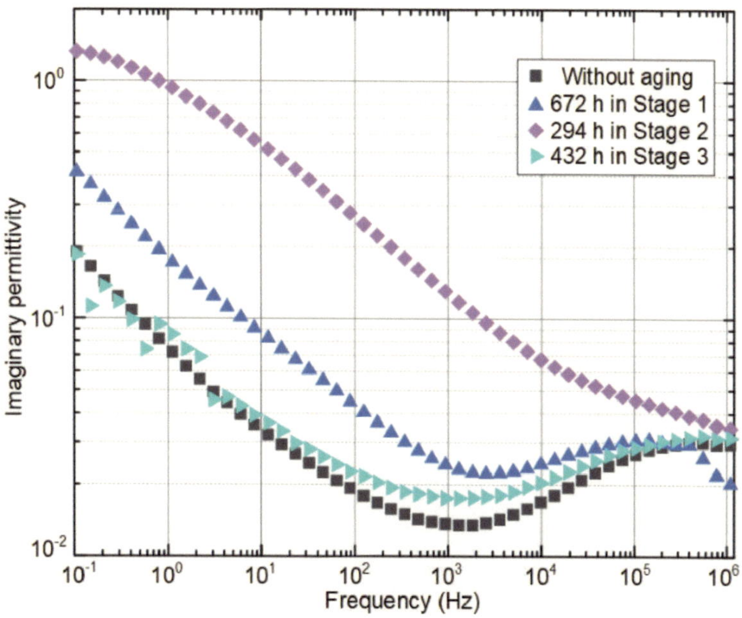

Figure 6.
Imaginary permittivity at the end of each aging stage.

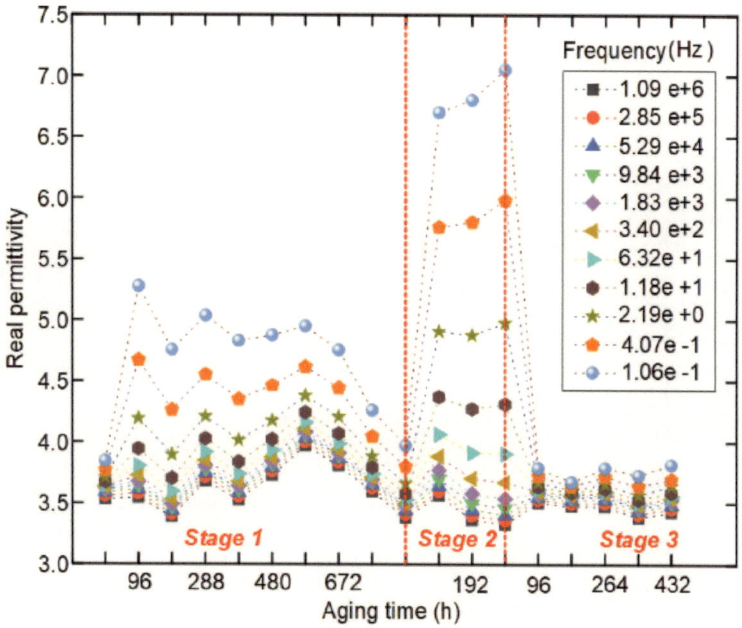

Figure 7.
Trend of ε' at different frequencies and different aging stages.

obviously. In contrast, ε' at each stage is stable in the whole stage 3. As shown in **Figure 8**, a similar tendency can be found in ε''.

3.2 Bulk conductivity

Bulk conductivity of PI in the three aging stages is reported in **Figure 9**. Before aging, PI samples have a conductivity below 10^{-15} S/m. During stage 1

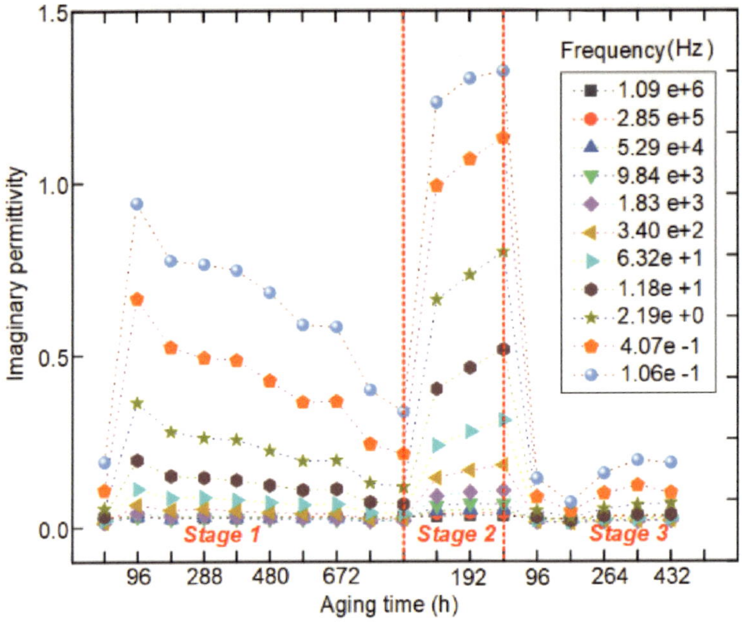

Figure 8.
Trend of ε" at different frequencies and different aging stages.

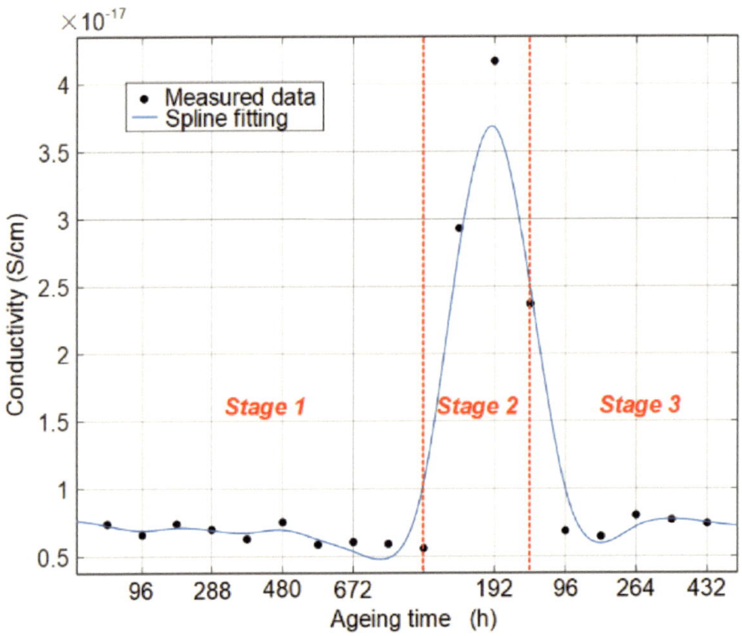

Figure 9.
Change of bulk conductivity with aging stage.

(thermal aging at 270°C) there might be a decrease of conductivity (~25%). After hydrolysis and moisture absorption (stage 2, T = 80°C, RH = 80%), the conductivity increases by an order of magnitude. In stage 3 (thermal aging at 320°C), the moisture is removed, and the chemical structure is restored [21]. The conductivity returns to the initial value, a finding in agreement with the results reported in [19].

In summary, the conductivity of PI is stable with thermal aging up to 320°C. However, high humidity levels in the operating environment might cause an increase of conductivity, particularly if the machine has not been used for some time.

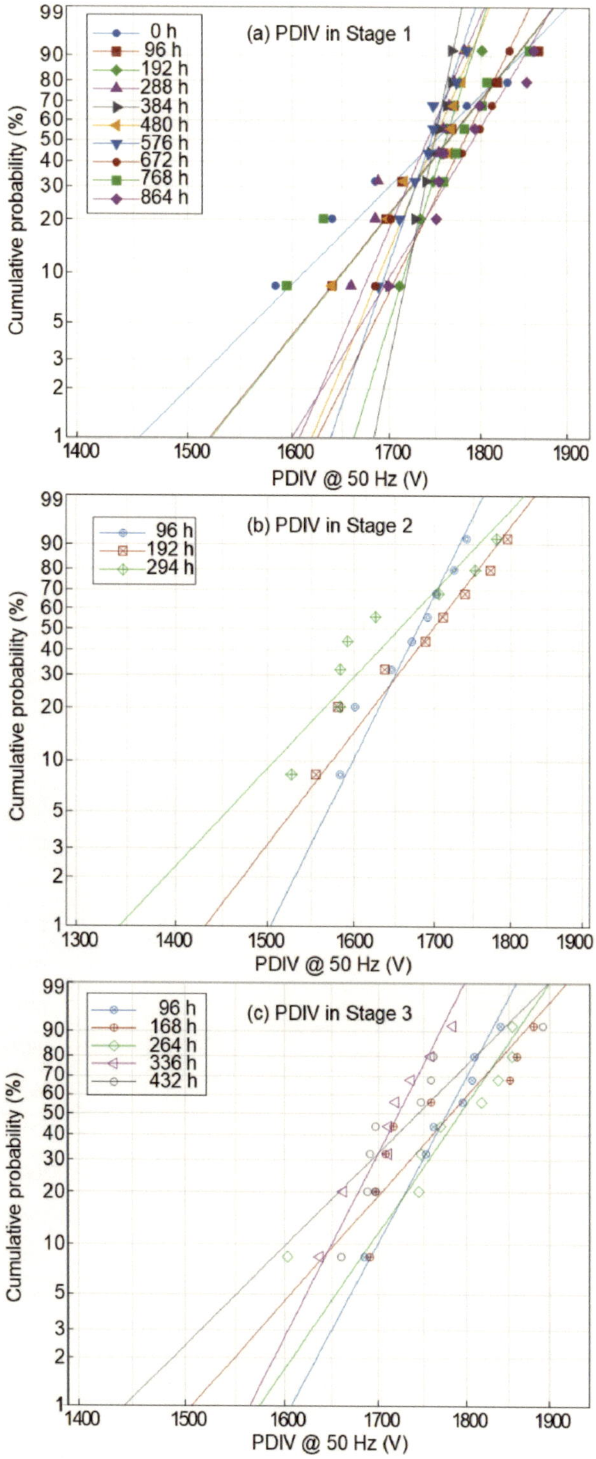

Figure 10.
Weibull plots of PDIV results in all aging stages.

3.3 PDIV results

PDIV tests were carried out with AC of 50 Hz. Eight specimens were tested at each aging time. The peak-to-peak value of applied voltage is recorded for the PDIV. All the data of each aging time are Weibull plotted and shown in **Figure 10**. The B10 parameter is employed here, which means the 10% probability of PD occurrence. Besides, the Weibull scale parameter (α) and shape parameter (β) are also employed and shown in **Table 2**.

From the estimation of Weibull plot, B10 and α can be calculated at each aging time. In stage 1 shown in **Figure 10(a)**, B10 changes from 1590 (at 0 h) to 1715 V (at 192 h), while α varies from 1750 to 1810 V. It indicates that B10 is more scattered after the thermal aging at 270°C. In stage 2 shown in **Figure 10(b)**, both B10 and α show similar behavior, while in stage 3 (**Figure 10(c)**), both B10 and α are scattered, compared with the other two stages.

B10 can be employed to show the reliability of PI in severe working conditions. As shown in **Figure 11**, this "reliability" can be divided into three stages corresponding to the three aging stages. B10 increases sharply in the first 200 h, and then it becomes stable till the end of stage 1. With the beginning of stage 2, the high humidity makes B10 drop linearly with aging time, which indicates the risk of PI application in wet environment. As mentioned above, the high temperature in stage 3 will recombine the PI chains and remove the moisture in the bulk, resulting in the recovery of B10. However, a decrease of B10 occurs after its recovery, from 1680 to 1640 V.

Aging stage	Time under stress (h)	B10 (Vpp)	α (Vpp)	β
(I) T = 270°C	0	1590	1780	19.9
	96	1646	1786	27.6
	192	1715	1770	70.8
	288	1663	1754	42.4
	384	1710	1756	83.6
	480	1664	1764	38.8
	576	1692	1755	61.5
	672	1689	1797	36.3
	768	1615	1790	21.9
	864	1702	1808	37.5
(II) T = 80°C, RH = 80%	96	1583	1696	32.5
	192	1551	1725	21.2
	294	1514	1682	21.3
(III) T = 320°C	96	1684	1794	35.6
	168	1660	1803	27.2
	264	1652	1817	23.6
	336	1642	1737	39.8
	432	1641	1766	30.7

Table 2.
PDIV statistics during aging (N = 8).

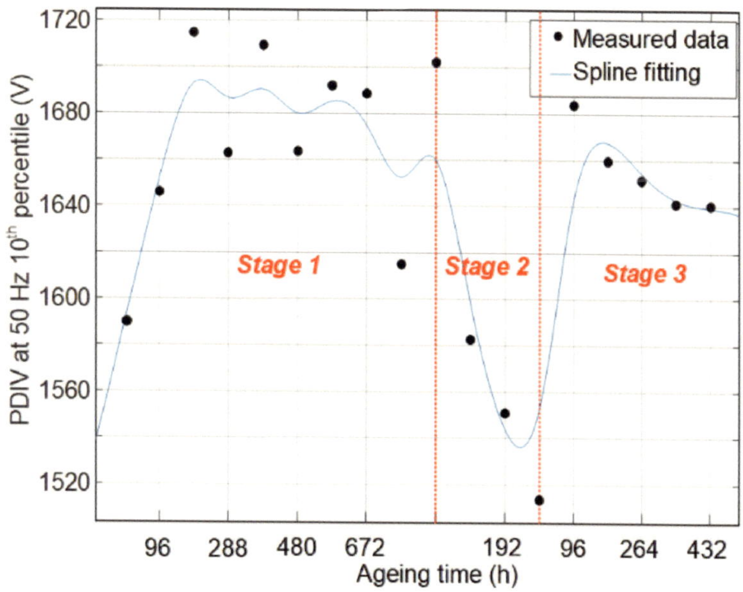

Figure 11.
Trend of PDIV B10 of at 50 Hz during aging.

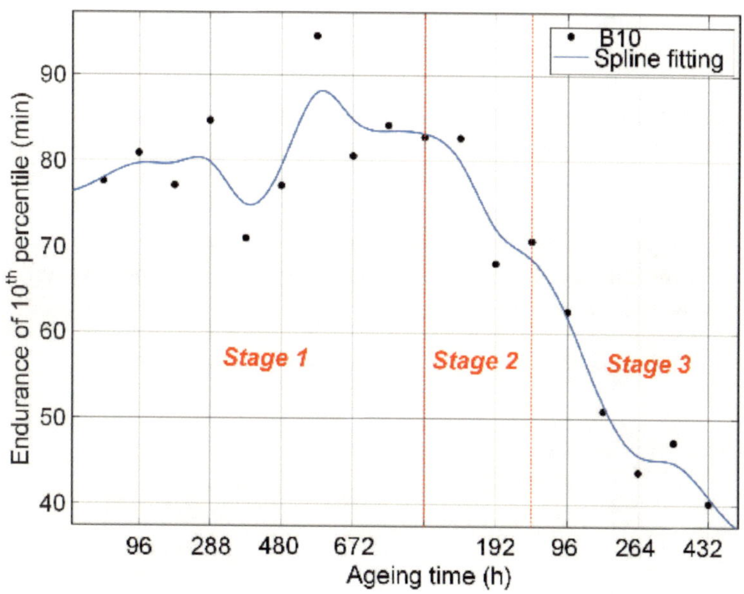

Figure 12.
B10 of PD endurance in all the stages.

By checking **Figures 5, 7** and **11**, one can find that B10 is closely related to the change of ε'. A capacitor model can be used here to explain this relation. In this model, the air gap between rod electrode and sample surface is equivalent to a capacitor, the PI tape to another one. With the increasing ε' of PI, the electric field in air gap will increase simultaneously. For instance, the ε' at 50 Hz increases 8% (from ~3.8 to ~4.0) in stage 2; B10 in **Figure 11** decreases from 1660 to 1520 V (~9%) correspondingly.

3.4 PD endurance under impulse voltages

PD endurance time tests were carried out under a square wave with a peak-to-peak voltage of 7 kV. After the application of this voltage, a corona will be found around the rod electrode, destroying the PI tape and leading to breakdown. This voltage parameter was selected for the purpose of achieving breakdown within less than 2 h.

Eight samples were tested at each aging time. The Weibull plots were employed, and B10 of endurance time was calculated. B10 of all the stages are shown in **Figure 12**. B10 in stage 1 varies from 70 to 95 min; this fluctuation is maybe caused by the limitation of sample quantity. In stage 2, the trend is obvious. B10 decreases from 83 to 68 min. While in stage 3, B10 declines linearly with aging time, from 68 to 40 min. It can be found that thermal aging in stage 1 will not change the endurance of PI. The reason of decreasing endurance in stage 2 is assumed to be the reduction of PDIV. However, endurance in stage 3 also decreases, while the PDIV is recovered to the initial level. This change is more likely related to the material changes inside the sample caused by thermal aging, which will be discussed in the next section.

4. Discussion

The excellent results achieved by Kapton 200 CR exposed to high temperature could have been, somehow, predicted by a thermogravimetric analysis (TGA). The weight loss of Kapton tapes is negligible below 500°C, as shown in **Figure 13**, confirming the high thermal stability of the tapes.

The Fourier transform infrared (FTIR) spectroscopy of pristine and aged samples is reported in **Figure 14**. Without knowing exactly the chemical structure of the tapes, it is difficult to trace back the FTIR peaks to the corresponding chemical species. Yet, the peak sequence does not appear to be modified in an appreciable way through the series of aging cycles, suggesting that the chemical structure is stable. This is another confirmation that the Kapton 200 CR tapes did not change in a remarkable way their structure during the aging.

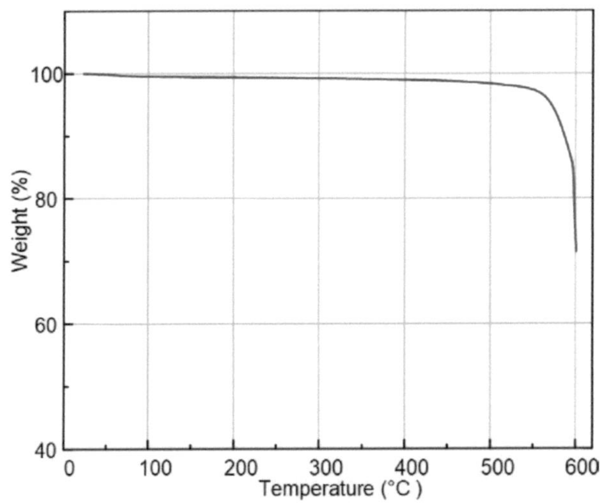

Figure 13.
TGA test of PI samples.

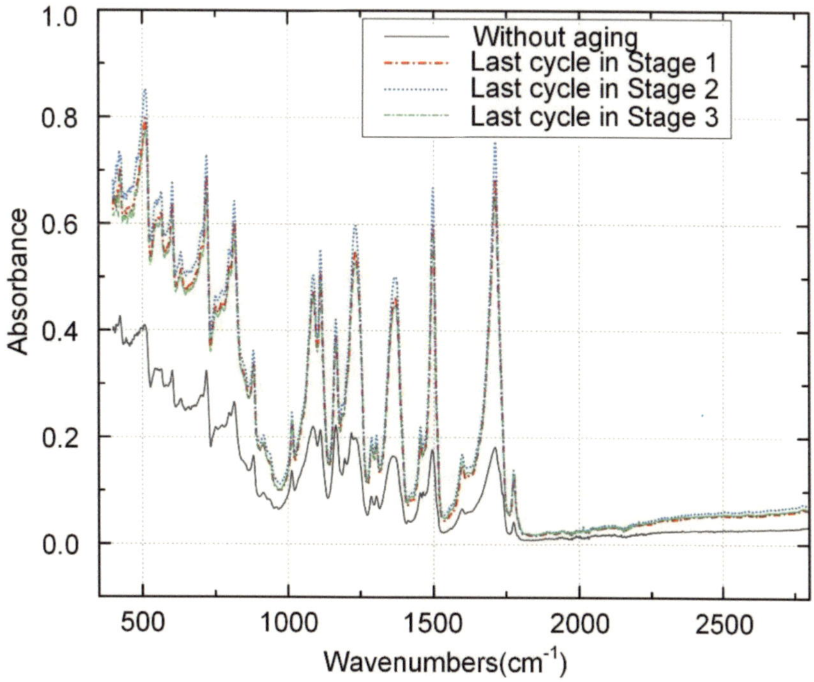

Figure 14.
FTIR spectroscopy of pristine and aged Kapton samples.

Both TGA and FTIR suggest that little changed in the PI tapes during changes. Yet, partial discharge endurance halved, with a sharp reduction in stage 3 (300°C). As the chemistry of the material seems to play a marginal role in this phenomenon, it was decided to inspect the morphology of the samples. In a nanostructured dielectric, inorganic nanoparticles are bonded to the polymeric matrix through a coupling agent able to confer a good stability to the resulting structure. Above a critical temperature, the bonds between the host matrix and the nanoparticles can break, enabling the nanoparticles to rearrange. The scanning electron microscope (SEM) imaging was thus used to investigate whether this kind of phenomena was taking place in the sample tapes. To do that, SEM pictures of the surface and of the cross section of both pristine and aged tapes were taken. The pictures of the cross section were obtained by immersing the tapes in liquid nitrogen and then cracking them.

The SEM pictures of pristine and aged sample surfaces at two magnification levels (5000× and 50,000×) are shown in **Figure 15**. The 5000× image does not reveal any specific sign of aging. The 50,000× picture highlights the presence of nanoparticles (white spots) in contrast with the host matrix. No appreciable sign of deterioration can be observed.

Figure 16 shows the cross section of the samples. At the lowest magnification level (5000×), it can be appreciated that the tapes are probably a three-layer structure. It can be guessed that the central layer is pure PI, conferring elasticity to the tapes (in general, nanostructured materials tend to be less elastic than the host matrix). The outer layer consists of nanostructured PI, to ensure electrical endurance.

At the lowest magnification level (50,000×), signs of reorganization of the nanostructured layers can be observed. Apparently, the nanoparticles appear more

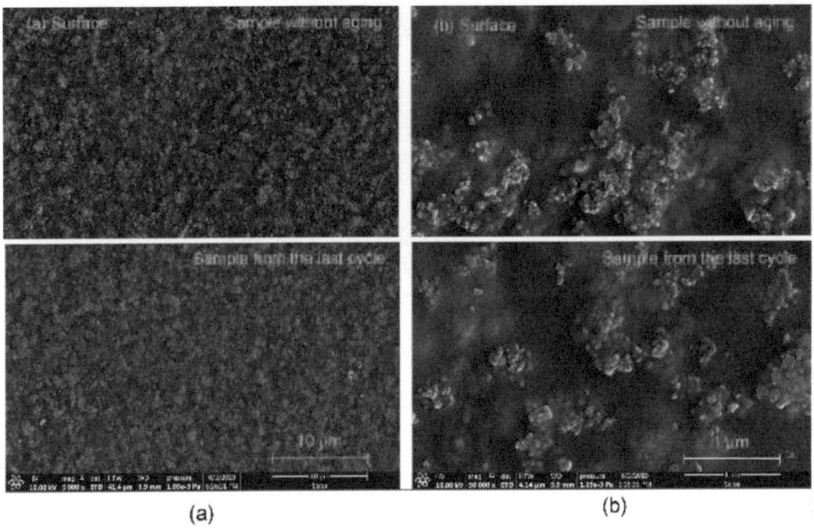

Figure 15.
SEM images of the surface of samples virgin and at the end of aging stage 3. (a) and (b) are pictures obtained with magnification of 5000× and 50,000×, respectively.

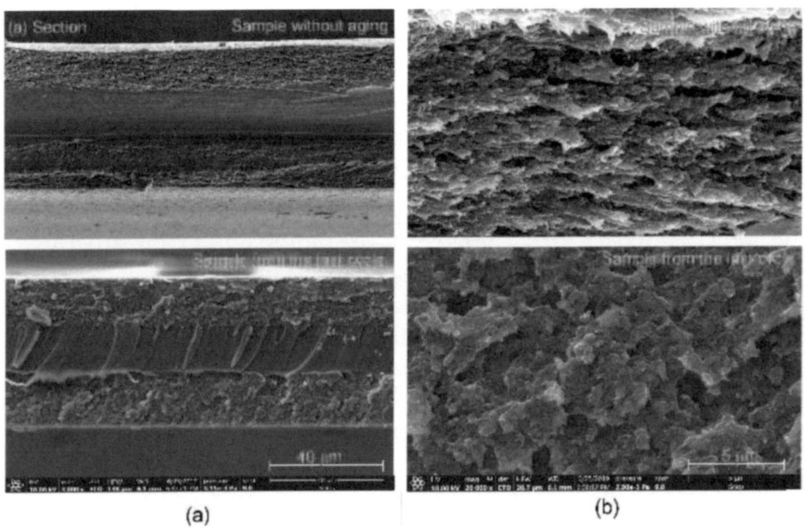

Figure 16.
SEM images of the crack surface of samples virgin and at the end of aging stage 3. (a) and (b) are pictures obtained with magnification of 5000× and 50,000×, respectively.

randomly oriented and with the tendency to form agglomerations. Therefore, regions where the host matrix becomes predominant are created. In these regions, partial discharge erosion can take place more easily leading to a reduction of the partial discharge endurance of the tapes.

These morphological changes could be addressed to the temperatures above 300°C which might have caused the stripping of the nanoparticle/host matrix bonding agent. Besides, the outer layer might have a differential thermal expansion coefficient compared with the central layer. At a microscopical level, the different expansions of the layers might have helped the distortion of the structure (the orientation of the clusters in the picture at 5000× seems to change from horizontal to skewed after aging).

5. Conclusions

Polyimide tapes have excellent properties that make them ideal choice to manufacture the turn insulation of form wound machines (both low-voltage hairpin machines used, e.g., in traction electrification and medium-voltage machines). Indeed, the final insulation class of the insulation system could be lower than that of the polyimide tapes, as it will be difficult (or extremely costly) to find a whole set of materials (enamels, slot liners, phase separators, wedges, sleeving, varnish or resin, conductive armor tape, stress control tapes) able to achieve the same thermal performance of the turn insulating. The results reported here confirm the capability of PI tapes to withstand extreme temperatures for long times. Exposing PI to temperatures above 300°C, moreover, has a positive effect as the acid and amine groups formed by hydrolysis reconnect, restoring the polymeric chains. Indeed, hydrolyzed PI displays a higher relative permittivity than the pristine tapes (4 versus 3.5 at 50 Hz), leading to a reduction of the partial discharge inception voltage (90%). A possible negative effect observed after thermal aging at 270°C is a general increase of the relative permittivity that would also lead to a reduction of PDIV, although more moderate than that induced by hydrolysis.

The electrical endurance tests highlight the excellent capability of PI corona-resistant (CR) materials to withstand partial discharge bombardment. The apparent short times (from 80 min at the beginning of the test to 40 min after the three stages of aging) should be weighed against the large electrical stress used for the tests (7 kV/100 μm = 140 kV/mm within the PI), a stress level hardly experienced in electrical machinery. Thermal aging seems to have a limited impact on the electrical endurance unless the PI is exposed to temperatures above 300°C for long times. By inspecting pristine and aged tapes using the scanning electron microscope, the electrical endurance of the CR PI tapes was explained by a three-layer structure of the tapes: a central layer of PI serves the purpose of conferring flexibility to the tapes, and two nanostructured outer layers ensure resistance to partial discharge bombardment. Aging at 300°C seems to have an impact on the morphology of the nanostructured layers, leading to a substantial (50%) reduction of the electrical endurance of the tapes.

Author details

Tao Han[1] and Andrea Cavallini[2]*

1 School of Electrical and Information Engineering, Tianjin University, Tianjin, China

2 University of Bologna, Bologna, Italy

*Address all correspondence to: andrea.cavallini@unibo.it

References

[1] Liaw DJ, Wang KL, Huang YC, Lee KR, Lai JY, Ha CS. Advanced polyimide materials: syntheses, physical properties and applications. Progress in Polymer Science. 2012;**37**(7):907-974. DOI: 10.1016/j.progpolymsci.2012.02.005

[2] Abadie MJM. High Performance Polymers–Polyimides Based–From Chemistry to Applications. Riheca, CA: IntechOpen; 2012

[3] Piloian M. What are the disadvantages of Kapton insulated wire?. Available from: https://www.interconnect-wiring.com/blog/disadvantages-kapton-insulated-wire/

[4] Paterson A. Kapton A Dangerous Aircraft Wiring Product implicated in aircraft electrical fires. Available from: http://vision.net.au/~apaterson/aviation/kapton_mangold.htm

[5] Persson E. Transient effects in the application of PWM inverters to induction motors. IEEE Transactions on Industry Applications. 1992;**28**:1095-1101

[6] Oliver JA, Stone GC. Implications for the application of adjustable speed drive electronics to the motor stator winding insulation. IEEE Electrical Insulation Magazine. 1995;**11**(4):32-36

[7] Stone GC, Campbell S, Tetreault S. Inverter-fed drives: Which motor stators are at risk? IEEE Industry Applications Magazine. 2000;**6**:17-22

[8] Wheeler JCG. Effects of converter pulses on the electrical insulation in low and medium voltage motors. IEEE Electrical Insulation Magazine. 2005;**21**(2):22-29

[9] Kaufhold M, Borner G, Eberhardt M, Speck J. Failure mechanism of the inter-turn insulation of low voltage electric machines fed by pulse-controlled inverters. IEEE Electrical Insulation Magazine. 1996;**12**(5):9-16

[10] Yin W. Failure mechanism of winding insulation in inverter-fed motors. IEEE Electrical Insulation Magazine. 1997;**13**(6):18-23

[11] Sanche L. Nanoscopic aspects of electronic aging in dielectrics. IEEE Transactions on Dielectrics and Electrical Insulation. 1997;**4**(5):507-543

[12] Serra S, Montanari GC, Mazzanti G. Theory of inception mechanism and growth of defect-induced damage in polyethylene cable insulation. Journal of Applied Physics. 2005;**98**(3):034102

[13] Hayakawa N, Okubo H. Lifetime characteristics of nanocomposite enameled wire under surge voltage application [feature article]. IEEE Electrical Insulation Magazine. 2008;**24**(2):22-27

[14] Fabiani D. Accelerated degradation of AC motor insulation due to voltage waveforms generated by adjustable speed drives. [Ph.D. Thesis], 2002, Gedit Edizioni, Bologna, May 2003

[15] Corona Resistant Kapton CR Takes Electrical Insulation Design and Reliability to New Levels [Online]. Available from: https://www.dupont.com/products/kapton-cr.html

[16] Wu GN, Wu J, Zhou L, Gao B, Zhou K, Guo X, et al. Microscopic view of aging mechanism of polyimide film under pulse voltage in presence of partial discharge. IEEE Transactions on Dielectrics and Electrical Insulation. 2010;**17**(1):125-132. DOI: 10.1109/TDEI.2010.5412010

[17] Rotating Electrical Machines—Part 18-41: Partial Discharge Free Electrical Insulation Systems (Type I) Used in

Rotating Electrical Machines Fed from Voltage Converters—Qualification and Quality Control Tests. IEC TS. 2007;60034:18-41

[18] IEC 60343. Recommended Test Methods for Determining the Relative Resistance of Insulating Materials to Breakdown by Surface Discharges. 1991

[19] Khazaka R, Locatelli ML, Diaham S, Bidan P. Endurance of thin insulation polyimide films for high-temperature power module applications. IEEE Transactions on Components, Packaging and Manufacturing Technology. 2013;**3**(5):811-817

[20] Li L, Bowler N, Hondred PR, Kessler MR. Statistical analysis of electrical breakdown behavior of polyimide following degrading processes. IEEE Transactions on Dielectrics and Electrical Insulation. 2011;**18**(6):1955-1962

[21] Punderson JO, Heacock JF. Polyimide film insulation for aerospace wire and cable: Why long-term performance exceeds some limited laboratory projections. In: Proceedings of the 34th International Wire and Cable Symposium. 1985. pp. 19-21

Section 7

Future Trends in Polyimide Functionalization for High Voltage Applications: Nanocomposite Structuration and Surface Treatments

Chapter 13

Synthesis Process Optimization of Polyimide Nanocomposite Multilayer Films, Their Dielectric Properties, and Modeling

Shakeel Akram, Jérôme Castellon, Serge Agnel and Jean-Pierre Habas

Abstract

Polymer nanocomposite-based dielectric materials are playing a vital role in the area of electrical insulation research and developments. The nanoparticle dispersion and interface region are the crucial parts of these developments. This chapter begins with the description of physical properties and their derived nanoparticles of polyimide (PI) films. Then, the detailed synthesis process of PI/nanocomposite multilayer film and its optimization is discussed in this chapter. Several factors in the synthesis process, which can influence the quality of the film, are discussed. After synthesis, the dielectric properties such as space charge were measured, and the results are compared with single and multilayer PI/nanocomposite films. Simulations and modeling help to shed light on the experimental results and create an understanding of polymer nanocomposite properties. Therefore, the PI/nanocomposite multilayer 3D model based on boundary conditions obtained from SEM/TEM images of synthesized samples was also constructed and simulated in COMSOL multiphysics software. The nanoparticle agglomeration and the impact of nanoparticle dispersion on the electrical properties of the material are described in detail in this model. The results demonstrate that the nanoparticle dispersion is improved by using a thin layer of PI/nanocomposite on PI film. As a result, fewer space charges and low electric fields are observed in multilayer films.

Keywords: polyimide nanocomposite, multilayer insulation, synthesis optimization, dielectric properties, polyimide nanocomposite modeling

1. Introduction

Electrical insulations are the key components for electric motors, which are used in space crafts and electric trains. Polyimide (PI) films as an insulating material are used in such motors, which are commercially available in single- and multicoated forms. Over the past few years, global market has shown a great interest in the application of nanodielectrics, especially in the field of electrical insulating materials. Various research results have claimed that polymer nanocomposite materials can improve dielectric properties for electrical insulation applications [1–4]. The key role played by the nanoparticle dispersion and interface region are essential parts of these improvements. Further conditions to advances are the size and type of nanoparticles properly chosen and distributed into the polymer matrix. Initially, it seemed like magic that everything is possible by using nanodielectrics, which later proved wrong after understanding the exact working principles of polymer-based nanocomposites, though several questions still need to be solved [5]. This has motivated the author to explore in this field further and find those principles by using experimental and simulation work on improving the dielectric properties of polyimide-based nanocomposite.

This chapter begins with a description of the physical properties of polyimide and its derived nanoparticles. Afterward, a detailed synthesis process of polyimide nanocomposite single and multilayer films is outlined leading to the synthesis process optimization. Polyimide nanocomposites are the leading component in the advancement of electric motors and generator's insulating materials. However, nanoparticle dispersion is the primary concern to improve polymer nanocomposite dielectric properties. In this work, polyimide-based nanocomposite single- and multilayer films are synthesized and characterized in detail. The preparation of polyimide nanocomposite is a complex process with many variables involved. Therefore, it is vital to know the right chemistry when dealing with it. Several methods were probed before an optimal synthesis process was found. A detailed synthesis process optimization is described at the end of this chapter to understand all variables that can alter the dielectric properties of the polyimide nanocomposite films.

1.1 Polyimide insulation films and their physical properties

Thermo-oxidative polyimide films by DuPont are in the market since the 1960s. These low dielectric constant thin films are highly corona resistive, are thermally stable, and have higher breakdown strength electrically and mechanically [6]. Such unique characteristics have made polyimide films to use in large industrial applications such as aerospace, automotive, and microelectronic devices. The fire resistance property of PI and low dielectric constant has made it possible to isolate metal lines and reduce electromagnetic interference effect in electronics and signal processing devices [6]. Polyimide is also used in electric motor insulation for high-speed trains. Polyimide is a high-temperature organic class of polymer that is mechanically robust and thermally stable based on stiff aromatic backbones [6]. The main functional groups in polyimide structure are aromatic ring, amide, and ether groups. There are several monomers and methods available to synthesize the polyimide. Therefore, a slight change in the monomer's structure and synthesis process can alter the physical properties of PI films significantly. Polyimides are chemically closed structure polymers that are nonreactive to many chemicals such as solvents and oils. Polyimides are intrinsically resistive to heat and flame retardants. PI is also resistive to acids but avoids to use in alkalis and inorganic acid environment. The remarkable radiation resistant property of PI has made it an ideal material to use in outer space radiation environment and in nuclear reactors, where PI is used alone as well as in composite forms. The changes in the dimension of material per 1°C rise in temperature are called the coefficient of thermal expansion. PI exhibits higher values of thermal expansion coefficient than other polymers. PI undergoes numerous phase changes to 400°C during the imidization process from polyamic acid solution to thin solid films. PI is a thermally stable polymer that has a very high value of T_g and only 5% weight loss above 400°C [6]. Therefore, it is a very suitable material for packaging applications. PI films have higher mechanical strengths. The stress-strain results have sown that the flawless PI films have mechanical strength in between 100 and 200 MPa and the elongation at break in between 10 and 25% [6]. The mechanical vibrations in electric motors and metal conductor's contact in electronic packaging applications can cause severe damage to the mechanical strength of PI films. If the films are brittle and the applied force due to mechanical vibrations crosses the fracture limit, then the internal cracks or defects can break the insulation. The brittleness of PI films can be controlled during the synthesis process by using different monomers and imidization temperature and time [6]. The physical properties of polyimide films at room temperature are shown in **Table 1**.

1.2 Properties of polyimide in electrical engineering

1.2.1 Dielectric constant and dielectric loss

When the dielectric material is subjected to an electric field, it becomes polarized due to the movement of induced and permanent electric dipoles [7]. For ideal insulation, the movement of dipoles should be zero or very low to block the conduction current. The value of dielectric constant (ε') defines the polarization ability of dielectric material. The movement of dipoles in an alternating electric field causes the loss of energy known as a dielectric loss (ε''). The conduction loss and dielectric loss are two significant losses that are responsible for energy loss in a dielectric material. The movement of charges determines the conduction loss, while the movement of dipoles determines the dielectric loss, the movement of dipoles causes the energy dissipation as the polarization switches its direction in an alternating electric field. The polarization lags the alternate electric field to produce heat, and dielectric loss increases at the relaxation frequencies. Therefore, the value of dielectric constant reduces quickly at relaxation frequencies because the polarization is not able to keep pace with the alternating electric field, as illustrated in **Figure 1**. An efficient insulating dielectric material blocks the conduction with a minimum dissipation of energy. The materials with a higher value of dielectric constant usually have a higher dielectric loss. The energy loss in dielectrics can be used to heat the food in a microwave oven. The orientational polarization in water frequency is utilized for this process, which is close to the relaxation or resonance frequency. It means water molecules absorb a lot of energy, which later dissipated to heat the food. The dielectric constant of polyimide films varies from 3.0 to 3.8 according to the structure and composite fillers added into it [7].

The relative permittivity is composed of two parts: the real part denoted as ε' and the imaginary part indicated as ε''. The ratio of these two values is defined as the dissipation factor and given as follows:

$$\tan\delta = \frac{\varepsilon''}{\varepsilon'} \tag{1}$$

Properties	Typical value at 25°C	Test methods
Tensile strength, MPa	231	ASTM D-882-91
Density, g/cc or g/mL	1.42	ASTM D-1505
Glass transition temperature (T_g), °C	360–410	Differential calorimetry
Thermal conductivity, W/m.K	0.12	ASTM F-433
Electrical breakdown strength, kV/mm	280	ASTM D-149
Dielectric constant at 1 kHz	3.4	ASTM D-150
Dissipation factor at 1 kHz	0.0020	ASTM D-150
Volume resistivity, Ω.cm	1.5E17	ASTM D-257

Table 1.
Properties of polyimide.

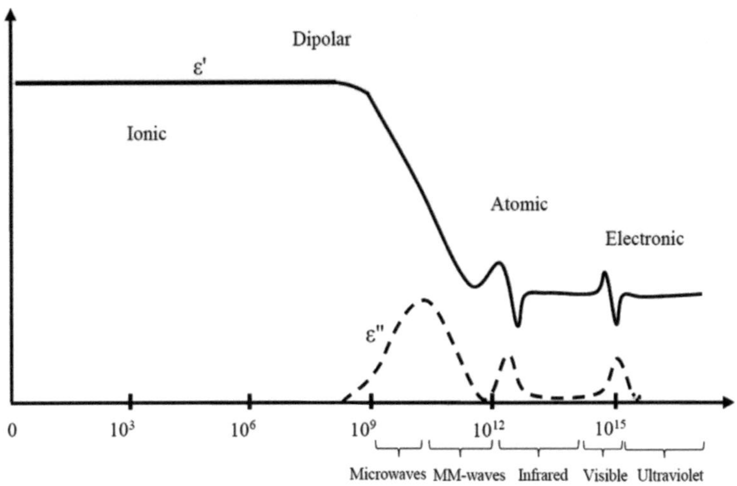

Figure 1.
Dielectric loss vs. frequency.

Typically, PI films have dissipation loss in between 0.001 and 0.02 [8]. The low tan δ value indicates that PI loses less electrical energy. The low dielectric constant and low dielectric loss make PI films suitable to use in electrical signal packaging applications to avoid signal interference.

1.2.2 Conduction current

Conduction current attributes to different polarization and depolarization processes happening inside the material. The complete polarization process can be presented as Eq. (2) [9].

$$i_p = i_i + i_a + i_c \tag{2}$$

where i_i is the instantaneous current due to the displacement polarization, i_a represents the relaxation polarization current, and i_c presents the conduction current due to the conductivity of the specimen. The Simons and Tam theory represents that the depolarization current is a superposition of different relaxation processes depending on the trap levels [9]. There are several polarizations due to dipole relaxation process that may take place in a dielectric material as follows:

1. Electronic polarization
2. Ionic polarization
3. Orientation polarization
4. Interfacial/space charge polarization
5. Hopping polarization

1.2.3 Dielectric breakdown strength

The breakdown strength is the ability of dielectric material to oppose electric field stresses without any insulation breakage or passing a certain amount of leakage current. The value of dielectric breakdown strength can be found as

DOI: http://dx.doi.org/10.5772/intechopen.91206

the applied voltage at which the electric breakdown occurred. The breakdown strength may vary by varying the temperature, moisture, and defects inside the material. The PI films that we synthesized have breakdown strength between 150 and 260 kV/mm [8, 10]. Due to its high electrical breakdown strength and mechanically tough properties, PI as an insulation can be useful to utilize in a high voltage industry applications, where electrical instruments can produce strong electrical field and leakage current to damage the insulation. Dielectric strength can vary due to the following reasons:

1. Nonhomogenous nature of sample thickness

2. Decrease with an increase in temperature

3. Decrease with an increase in frequency

4. Decrease with an increase in humidity

1.2.4 Charge transport phenomena

The PI films can be amorphous or crystalline, depending on their synthesis chemistry. In these regions, the trap energy of electrons varies according to the band structure [11]. The high electric field stresses create more trap levels in the insulation, especially at the top and bottom surface of samples near to the electrodes. These trap levels reduce toward the interior regions. In **Figure 2**, a thin PI film is placed between two electrodes. The layer near to anode can act as hole transport layer, and the layer near to cathode can act as electron transport layer. The intersection region of these layers can provide enough space for the recombination of electrons and holes that are injected by electron

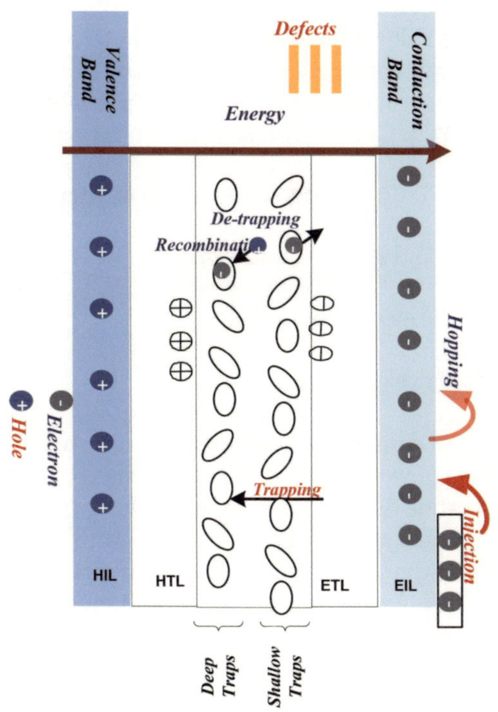

Figure 2.
PI film layer structure, hole injection layer (HIL), hole transport layer (HTL), electron transport layer (ETL), electron injection layer (EIL).

injection layer (EIL) and hole injection layer (HIL), respectively. As illustrated in **Figure 2**, the injection of holes from HIL and the injection of electrons from EIL move toward hole transport layer (HTL) and electron transport layer (ETL), respectively.

The charges move from one electrode to the opposite polarity electrode; during the movement, some charges are trapped, some detrapped, and some recombined in the interfacial region due to charges already exist in these regions. The polarity of these interfacial regions depends on the electronic state of the adjacent electrode. In case of PI nanocomposite multilayer structure, the accumulation of charges can depend on the following [10]:

1. The mobility of electrons and holes
2. The distinctive charge barriers at the interface
3. The charge injection rate
4. The polarity of the existing charges at the interface
5. Permittivity and conductivity contrast at interface

Therefore, the discontinuity of electronic state distribution causes an additive trap at the interface for the transportation of charges.

1.2.5 Corona discharge resistance and material degradation

The ionization of air in the form of an electrical discharge is known as corona discharge. This ionization occurs when the electric field crosses the threshold value of air breakdown. Partial or surface discharge is also a kind of corona discharge because it is half discharged to bridge between the electrodes [12]. Corona discharge appears in the form of a lightning color. In high voltage laboratory, corona discharge can be originated using a simple rod to plane electrodes with an air gap between them. Extensive research work has been studied on the appropriate electric field stress to set up corona discharge [12]. Corona discharge depends on the following parameters:

1. Air or gas ionization limit and free path for gas molecules
2. Electrode geometry and surface conditions
3. Distance between the electrodes
4. Insulation structure design flaws

Corona discharges can damage the insulation significantly. With the growing demand for high voltage power supplies, it is becoming common that the insulating materials face severe corona discharges. Corona discharge produces an eminently energized plasma of charged species and emits UV light. These plasma discharges emit charged particles on the surface of insulation and deteriorate the surface physically and chemically. The increase in temperature and humidity can accelerate the corona discharge intensity.

1.3 Physical properties of nanodielectric-based polyimide

For the last decade, the research has shown that when two materials in which one part is inorganic nanoparticles are combined to form a nanodielectric material, which may have superior properties than single ones. The combined nanodielectric materials are known as polymer nanocomposites, when the base material is polymer matrix and the adding fillers are nanoparticles. Polymer nanocomposites have been used widely in academic research and industry [13]. The properties of nanocomposites change due to the large surface to volume ratio of the nanoparticles. The addition of nanoparticles into the base polymer matrix modifies the physical properties of composites. If the size of the nanoparticles is less than the critical length scale, then the physics of nanocomposite changes significantly. It has become one of the most reliable materials in electrical engineering since the first time the term nanocomposite was introduced in 1984 and since then it has been warmly accepted by the scientific community [13]. "Nanometric dielectrics," later named "Nanodielectrics" in 2004 by M.F. Frechette, is a nowadays popular term in the research community also known as nanocomposites made by the inclusion of nanometer size of nanoparticles in a polymeric matrix for dielectric applications [13]. In the beginning, the glass and ceramics were very common to use as solid dielectric materials. But, over time the power supplies demand increased abruptly, and these insulating materials were not enough to fulfill the demand. Therefore, new dielectric materials such as natural and synthetic polymers successfully overcome conventional dielectric materials. These new polymer-based dielectric materials have superior properties and lighter in weight to use in different applications.

1.3.1 Inorganic filling particles

Fillers are often fine-grained nanometer- or micrometer-sized particles and fibers, which can be made of organic or inorganic materials. Generally, such particles and fibers are named as conventional sized fillers according to their size range. Composites formed by using these fillers are labeled as conventional sized filled composites, for example, nanocomposites and microcomposites. Recently, filler materials have been used to improve electrical, mechanical strength, and thermal properties [5]. Polymers are chosen in electrical engineering according to the required application, and their properties can be modified by adding inorganic fillers in it. The quantity of fillers in the base polymer matrix is still under question from the literature that which amount by weight is most suitable to improve the desired properties. Therefore, we only use very low filler quantity of nanoparticles into the base polyimide. The common used organic polymers and their derived inorganic fillers are shown in **Table 2** [6].

Chemical family	Examples
Inorganics	
Oxides	Al_2O_3, SiO_2, MgO, ZnO, TiO_2, glass
Hydroxides	$Al(OH)_3$, $Mg(OH)_2$
Silicates	Talc, mica, nano clays, asbestos
Salts, compounds	$CaCO_3$, $BaSO_4$, $CaSO_4$, $BaTiO_3$, $SrTiO_3$
Metals	Al, Ag, Sn, Au, Cu
Nitrides, carbides	AIN, BN, Si_3N_4, SiC
Organics	
Carbon	Carbon fibers, carbon black, graphite fibers, carbon nanotubes
Natural polymers	Cellulose fibers, wood flour, flax
Synthetic polymers	Polyimide, polyester, polyethylene, polypropylene

Table 2.
Organic and inorganic fillers and their chemical family.

1.3.2 Particles' size, shape, and types

Various results of electrical, mechanical, and thermal properties have announced that the polymeric nanocomposites can be invaluable when a small amount of nanosize particles is appropriately chosen and dispersed into the base polymer matrix [14]. Composite material properties change dramatically due to the size, shape, and type of nanoscale particles. These nanoparticles have individual mechanical, electrical, and thermal properties [11]. Based on the size and type of particles, nanomaterials properties are shown in **Figure 3** and they are categorized as follows in **Figure 4** [15, 16].

1. Zero-dimensional (0-D)

2. One-dimensional (1-D)

3. Two-dimensional (2-D)

4. Three-dimensional (3-D)

If the number of the dimensions that are in the nano range is considered, exfoliated clay will be regarded as 0-D nanoparticles, because the diameter of these particles are in nanometer range. In order to utilize the ceramic nanoparticles in a better way with polymers, a potential studied has been done recently, especially in the field of the synthesis process and surface science. The most common studied ceramic nanoparticles are silica (silicon dioxide—SiO_2), alumina (Al_2O_3), titania (TiO_2), zirconia (ZnO), or silicon carbide (SiC).

1.3.3 Nanoparticle dispersion and distribution

The homogenous dispersion and distribution of nanoparticles are the major factor to improve nanocomposite dielectric material properties. A vital part of preparing

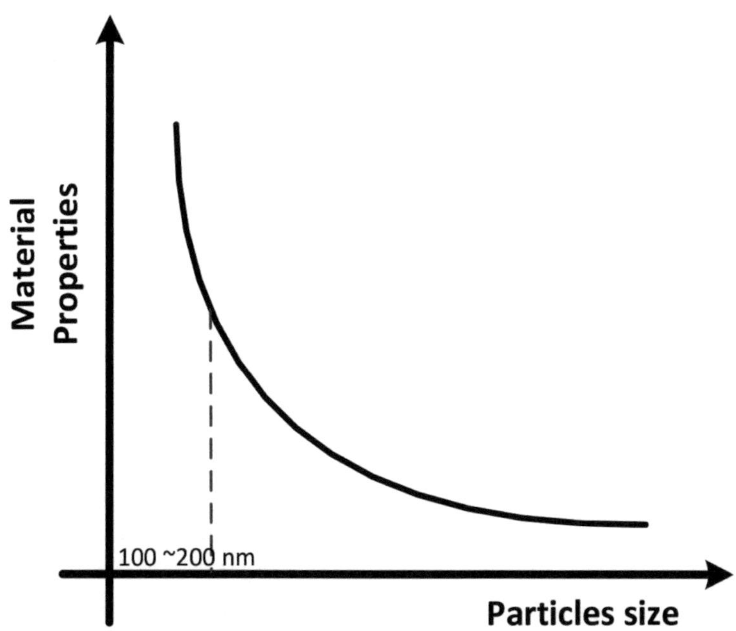

Figure 3.
Change in composite material properties with respect to particles' size.

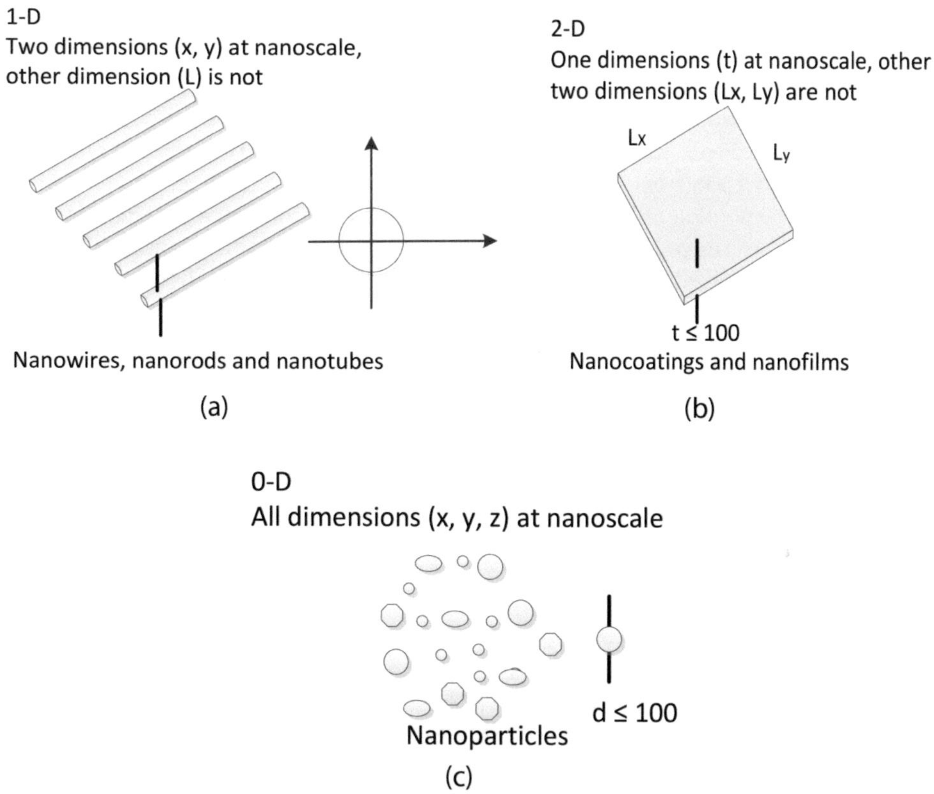

Figure 4.
Fillers' types, sizes, and shapes.

nanocomposites is the nanoparticle dispersion; they have heterogeneous surfaces, which cause variability in contact with the polyimide matrices. For better dispersion, several methods have been used such as the orientation of nanoparticles by applying an electric field, chemical treatment by using coupling agents, and plasma treatment. All these methods are adopted to solve the problems of heterogeneous agglomeration and the compatibility between the polyimide matrix and the nanoparticles. The nanoparticle mixing techniques such as mechanical milling, ultrasonication, high-speed stirrer, liquid dissolving, or heat melting can also affect the nanoparticle dispersion level. The size of particles and filler loading percentage can also influence the dispersion level [17]. Particles smaller than 80 nm tend to agglomerate and form more significant chunks of particles. The intermolecular forces keep nanoparticles together. The following particle parameters can influence the overall properties of polymer composites.

1. The size and shape of particles

2. The degree of particle dispersion

3. The surface modification of the particles

4. The particle-matrix interaction

1.3.4 Nanoparticles surface treatment

The interlinking of the PI and the nanoparticles depends on the functional groups and the surface energy of PI and nanoparticles. Some hydroxyl (OH) functional groups can be formed on nanoparticles after surface modification, which provide a better interface and tightly bound with the PI matrix. To modify nanoparticle surface, different methods have been presented in recent papers such as deposition reaction modification, chemical surface treatment modification, high energy such as plasma source modification, and intercalation modification [14]. Affinity and polarity compatibility can also be used to create materials with homogenous dispersion of nanoparticles. For better interlink between the silica nanoparticles and the polyimide matrix, the surface of polar silica nanoparticles is modified using the KH550 coupling agent. Silica nanoparticles are hydrophilic, and polyimide is nonpolar, which is not compatible with mixing; therefore, the surface of the silica is modified to make it hydrophobic with hydroxyl (OH) functional groups on its surface, which are easy to bond with aromatic polyimide functional groups [14]. This surface modification is adopted to ensure success in the application. The interphase region around nanoparticles can be controlled using surface treatment [14].

1.3.5 Coupling agents

KH-550 silane coupling agent chemically known as 3-(2,3-epoxypropoxy) propyl trimethoxy silane is selected for silica nanoparticle surface modification. It contains an organic functional group, a linker, a silicon, and a hydroxyl group. The organic functional group can bond to the organic aromatic polyimide ring. The general chemical formula for the silane group and the hydrolyzable functional group, typically alkoxy, amine, or chlorine, involved in the reaction with the inorganic silica substrate. The silanol groups of the nanosilica surface can react with the hydroxyl groups of the silane after hydroxylation through hydrogen bonding. The covalent bond improves interfacial adhesion between inorganic silica particles and organic monomer molecules.

1.3.6 Plasma treatment

Nonthermal plasma technique is also prevalent these days to modify the surface of nanoparticles. This plasma technique enhances the compatibility between nanoparticles and polymer by modifying the interfacial area of nanoparticles. The plasma is produced by using a dielectric barrier discharge, which generates ions and reactive species high-energy electrons that interact with the surface of nanoparticles to modify their surface characteristics [14]. Therefore, higher surface reactivity and stronger interactions between the nanoparticles and the surrounding polymers can be acquired, comparing with traditional coupling agent modification.

2. Synthesis of polyimide

When one monomer reacts with other monomers, it forms a carbon chain of the polymer. In case of polyimide, the monomers such as diamine (ODA) react with another monomer such as dianhydride (PMDA) to form polyamic acid solution (PAA). In this reaction, an oxygen atom of diamine reacts with the hydrogen atom of dianhydride, and the hydrogen of dianhydride reacts with the carbon of diamine to give a repeated unit of polyamic acid. To obtain polyimide film, thermal heat is applied for several hours to evaporate DMAC solvent. The chemistry and the properties of polyimides can vary due to the availability of several monomers.

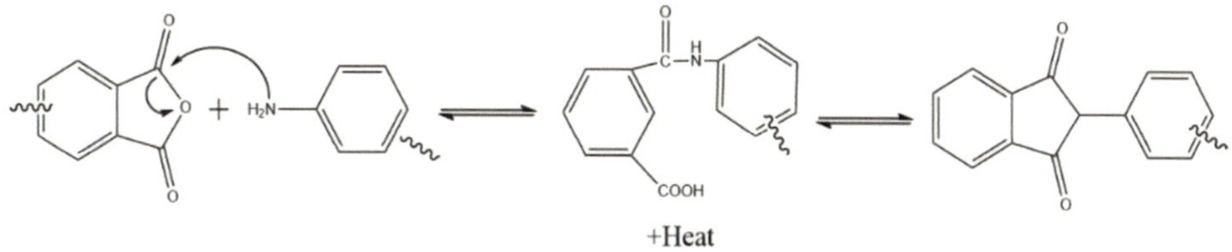

Figure 5.
Generalized reaction mechanism of aromatic imide formation.

The ratio of monomers can affect the molecular weight of PAA and change the molar mass of the final PI film [6, 18] (**Figure 5**).

2.1 Preparation mechanism for polyamic acid (PAA)

The aromatic PI film can only be synthesized from the solvent route using a two-step method. In the first step, PAA solution is synthesized using a reaction between dianhydride (pyromellitic dianhydride PMDA) and diamine (4,4'-oxydianiline ODA) at room temperature as shown in **Figure 6**. The dipolar aprotic solvents, such as N-methyl pyrrolidone (NMP) and N, N-dimethylacetamide (DMAc) are used to synthesize PAA solution. In the second step, PAA solution is converted into the final PI films after applying thermal imidization to evaporate solvent as explained in **Figure 7**. To make sure that PMDA did not absorb any moisture, it was heated for 2 h at 150°C. ODA was added into the beaker and mixed with DMAc for half an hour. Mechanical stirrer was used to stir the solution. PMDA was added into the solution in three parts, and the solution was further stirred for 24 h to get yellow color high molecular weight PAA solution as shown in **Figure 6**.

The reaction between the monomers to obtain viscous PAA solution is strongly dependent on the precise measurements. The following discussion will highlight important factors to select the monomers, solvents, and reaction conditions to get better results and avoid side reactions. The formation of the PAA solution may also involve some reversible reactions, which leads to the opening of the anhydride ring. Despite that, the forward rate of reaction is larger than the reverse rate depending on the purity of reagents. The molecular weight of the PAA product is also relying on the rate of difference between the forward and reverse reactions. Reverse reaction provokes when carboxyl group strikes to the adjacent polyacid group [6]. Some reagents can be used to stop the reverse reaction and provokes a forward reaction. The amino group basicity and PMDA electrophilicity in different solvents can change the equilibrium constant, and the reaction is exothermic; therefore, it should carry out at room temperature to minimize the equilibrium constant effect.

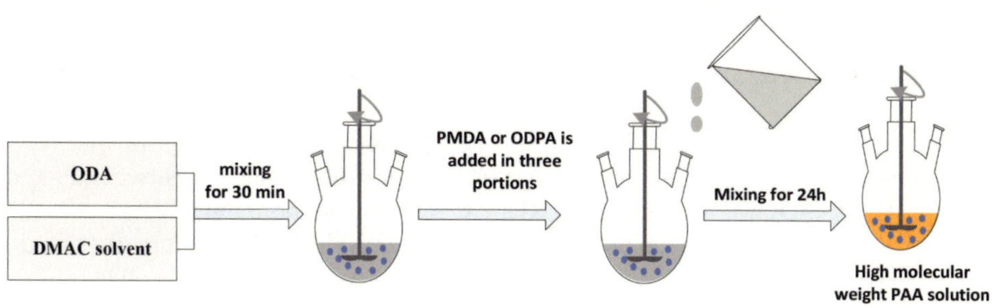

Figure 6.
Synthesis process of PAA solution.

2.2 Monomer reactivity conditions

As explained in the above section, the PAA solution is formed by nucleophilic substitution of carbon atom in carbonyl group of the dianhydride with a diamine. Therefore, electrophilicity of carbonyl group carbon atom and nucleophilicity of amino group nitrogen atom can be controlled using electron affinity measurement of both dianhydride and diamine. PMDA has the highest electron affinity and shows strong reactivity with diamines compared with other dianhydrides such as DSDA, BTDA, BPDA, and ODPA. The change in the structure of the diamines and dianhydride can affect the reaction significantly. These highest electron affinity dianhydrides can easily absorb moisture; therefore, they must keep moisture-free environment.

2.3 Factors involved in the molecular weight of PAA

The following factors are required to increase or decrease the molecular weight of PAA:

1. A slight increase in dianhydride quantity and the addition of solid dianhydride into the diamine solution can increase the molecular weight of PAA.

2. The order of the addition of the monomers, the solid mode of dianhydride addition into diamine solution, can avoid an immediate reaction and minimize the side reactions with water and impurities.

3. Minimize the side reactions to avoid dianhydride reaction with water and impurities.

4. Lesser quantity of solvent will help to reduce the impurities and water contents and increase the concentration of the monomer.

5. Storage of PAA solution for a long time can decrease the molecular weight. It can be due to the initiative of the hydrolysis process and chemical breakdown of a compound, and the other reason for this low viscosity can be the reverse reaction.

6. The solvents can also slightly affect molecular weight. The universal solvents are DMF, DMAC, and NMP, and the reaction is exothermic in which monomers are basic aromatic amines and protic anhydrides, and the final product is an acid. Therefore, the strong rate of the reaction is expected for more basic and more polar solvents.

2.4 Synthesis of PAA/nanoparticle composite solution

The PAA/SiO_2 nanocomposite solution was synthesized by applying the in-situ polymerization method, as elaborated in **Figure 7**. For the better link between PI and nanoparticles, SiO_2 was treated with a KH-550 coupling agent to modify its surface. Altered surface SiO_2 nanoparticles and dimethylacetamide (DMAC) were dispersed by applying ultrasonication and high-speed stirring, then oxy dianiline (ODA) was added into the solution and mixed for 1 h. After that pyromellitic dianhydride (PMDA) was added into two portions. For the first portion, almost 90% PMDA was added and mixed for 30 min, then the remaining 10% PMDA was added and mixed for 2–6 h until yellowish color high molecular weight PAA solution was obtained.

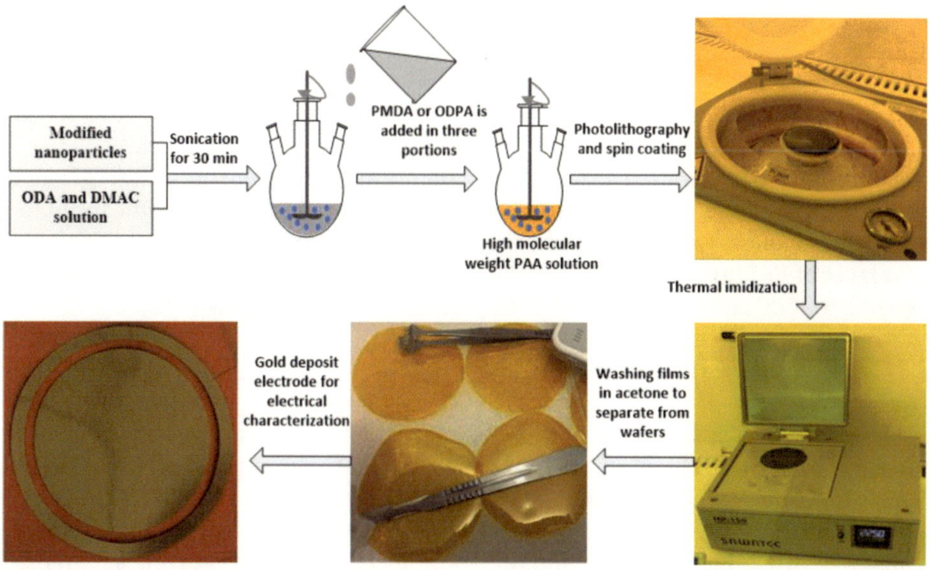

Figure 7.
Polyimide film synthesis process.

2.5 PI film casting using glass and brass substrate

Higher molecular weight polyamic acid solution dissolved in dimethylacetamide-based solvents is suitable for PI film casting using a variety of substrates, for example, teflon, alumina, glass, silicon wafers, brass, and copper. After coating the polyamic solution on the substrate, the precursor is thermally cured into an aromatic polyimide film. Both static and dynamic casting techniques can be used depending on the final product size and available tooling. In static casting, glass was used as a substrate to cast the PAA solution on it using a glass rod. Typically, this technique is easy, but it requires more material per substrate and difficult to control the thickness of PI films.

2.6 Casting PI films using a spin coating technique

Spin coating is the most prevalent technique to deposit PAA solution onto substrates such as silicon wafers and brass. The dynamic deposition technique uses less material, but it requires precision to control the operation. While depositing, it is important to pour PAA solution in the center of the wafer. Generally, we can achieve the maximum speed up to 6000 rpm, if the vacuum chunk has enough suction to hold the substrate while depositing PAA solution onto the surface of substrates. The PAA solution was homogenously distributed over the silicon wafer substrates during spinning. The acceleration of speed was programmed to start with low speed, slowly rise to maximum speed, and finally decrease the speed slowly to allow the coating flow across the substrate edges homogenously. Several spin speed steps can be used to control the flow of solution to cover more than 80% of the substrate before achieving the final speed.

2.7 Casting PI and nanocomposite/PI multilayer films

PI is obtained using the curing method in which a solvent is evaporated with the increase of temperature. During the PI film curing process from the polyamic acid (PAA) solution, there is a strong tendency of nanoparticles to get agglomerated. They may float on the surface or decant. Both effects can increase the chances of agglomeration and may influence the interface thickness and permittivity of samples and result in lower dielectric properties of PI films. Keeping this in mind, we prepared a multilayer (two and three layers) structure in which the top layer consists of a very thin PI/SiO_2 nanocomposite (NPI) layer, and the bottom is composed of pure PI layer [4]. By doing this, we are giving less space for nanoparticles to get agglomerated. In order to cast the PI/nanocomposite multilayer films, two different spin speeds were used for a two-layer structure. After calibrating the right speed to get right thickness, we select 20 s at speed of 500 rpm to get around 60 μm thick first base layer of PI film and soft bake for 30 min at 60 and 100°C temperature, then we use 30 s of 1000 rpm to get second nanocomposite/PI 20 μm thick layer.

2.8 Controlling the thickness

The thickness of the casted PI film depends on several factors such as the viscosity, molecular weight of the PAA solution, and the spin speed of spin coater. Varying PI film thickness in the range of 10–150 μm can be achieved by using a spin coating method. Two different viscosity solutions of PAA are used to obtain PI films as shown in **Figure 8**. The film thickness decreases with increasing spin rate. To avoid the bubbles and the accumulation of excessive blocks of PAA solution, spin cycle time should increase because prolonger spin times improve film coating uniformity, but at the same time, it can reduce the film thickness.

2.9 Thermal imidization of a polyamic acid (PAA)

The PI films can be cured of the PAA solution after applying thermal imidization. After casting the PAA solution onto the substrate, thermal steps with temperature from 100 to 350°C were used to evaporate the solvents. According to the literature, various thermal steps and different temperature ranges have been utilized to achieve 100% imidization from PAA to PI. Two main thermal imidization ways are as follows:

1. Baking films with slowly increasing in temperature up to 350°C according to the flexibility and T_g of the PI film.

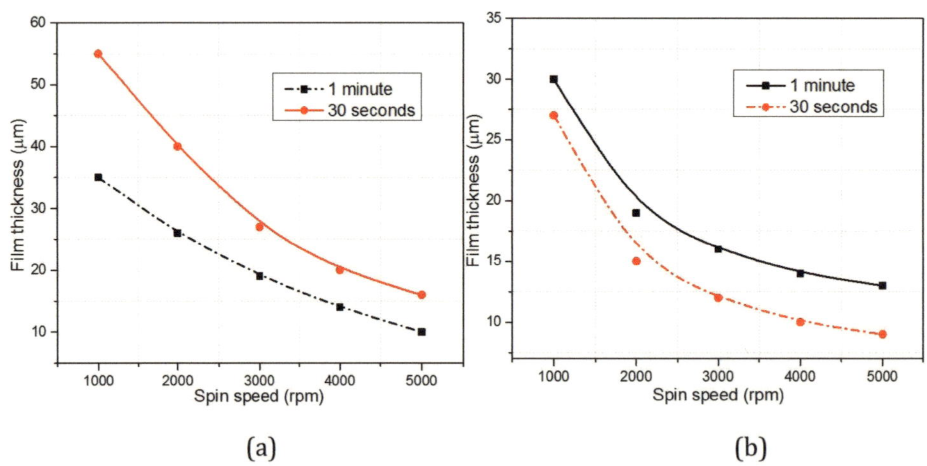

Figure 8.
Film thickness variation at different spin coater speeds (a) PAA solution high molecular weight and (b) PAA solution slight less molecular weight.

2. Start baking PAA with the temperature 100°C for 1 h, then baking at 200°C for 1 hour, then baking at 300°C and keeping for 1 h and slowly cool down to room temperature.

Several arduous factors are involved in the above simple-looking thermal imidization process to predispose the degree of imidization of PI films. During imidization, the remaining solvent at the later stages determines the stability of PI films. In early stages, the imidization process is faster, and several dynamic changing in physical properties happen due to the basicity of the amide solvent to accept protons and the amicable conformation of the amic acid to increase the mobility of the reacting functional groups [6]. At later stages, the rate of imidization slows down due to the decyclohydration of the open amic acid group into the closed imide ring, which decreases the chain mobility, and the T_g approaches the reaction temperature [6].

2.10 Determination of the degree of imidization using FTIR

When one monomer reacts with other monomer, it forms a carbon chain of the polymer. In our case, a single unit of two monomers such as diamine (ODA) and dianhydride (PMDA) reacts with each other to form a single unit of polyamic acid (PAA) solution. In this reaction, an oxygen atom of diamine reacts with the hydrogen atom of dianhydride, and the hydrogen atom of dianhydride reacts with the carbon atom of diamine to give us a repeated unit of PAA. After obtaining PAA, thermal imidization was applied to cure PI and PI/SiO_2 films.

Fourier transform infrared spectroscopy (FTIR) is a promising tool to determine the degree of imidization despite a minor error in measurement due to its sensitivity to chemical change. The bond composition of PI atomic structure was studied by FTIR spectroscopy to analyze the chemical bonds present in PI. The bands most frequently utilized are imide absorption bands near 1780 cm^{-1} (C=O asymmetrical stretching), 1380 cm^{-1} (C–N stretching), and 725 cm^{-1} (C=O bending). This study describes the chemical characterization of half cured and fully cured PI and PI/SiO_2 films. The range of wavelength used was from 650 to 4000 cm^{-1}. The PI subatomic structure is composed of a molecular chain containing the functional groups such as aromatic rings, amide rings, and some noncyclic ether rings. These functional groups have discrete chemical bonds such as C–N–C, C=O, benzene (C_6H_n), ether link (C–O–C), and –OCH_2–CH_2 deformation. The functional groups of PI polymer chains provide the level of linkage with subatoms as well as the linkage with silicon oxide nanoparticles. Imide group contains imide carbonyl in-phase and out-of-phase stretching, the C–N–C axial, transverse, and out-of-phase stretching. The absorbance peaks at 1371, 1112, and 721 cm^{-1} indicate the transverse and axial stretching of C–N–C bond, and the reasons of 1781 and 1720 cm^{-1} peaks are in-phase and out-of-phase stretching of C=O in imide ring. Aromatic ring is divided into tangential, radial skeletal, and out-of-plane vibrations. The peak at 1590 cm^{-1} relates to tangential C–C vibrations. The peak at 1280 cm^{-1} is tangential phenyl ring vibrations. The other 1480, 1168, 1087, 1011, 843, and 788 cm^{-1} peaks describe about the tangential, radial skeletal, and out-of-plane bending vibrations of C_6H_n in aromatic ring. The peaks at 1234 and 1454 cm^{-1} relate to the bond of ether link (C–O–C), and 1416 cm^{-1} peak corresponds to –OCH_2–CH_2 deformation. **Figure 9(a)** and **(b)** shows the FTIR spectra of changes in the bonds of half cured and fully cured PI films at different temperatures and time to determine the degree of imidization to obtain imide ring.

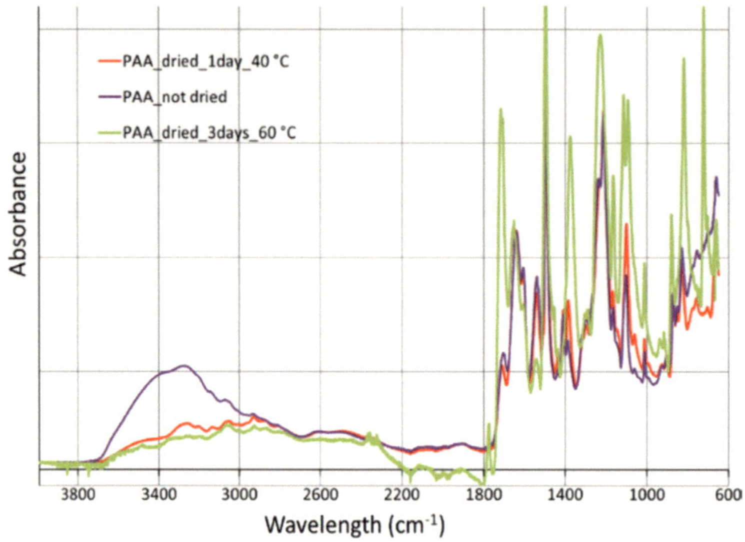

(a)

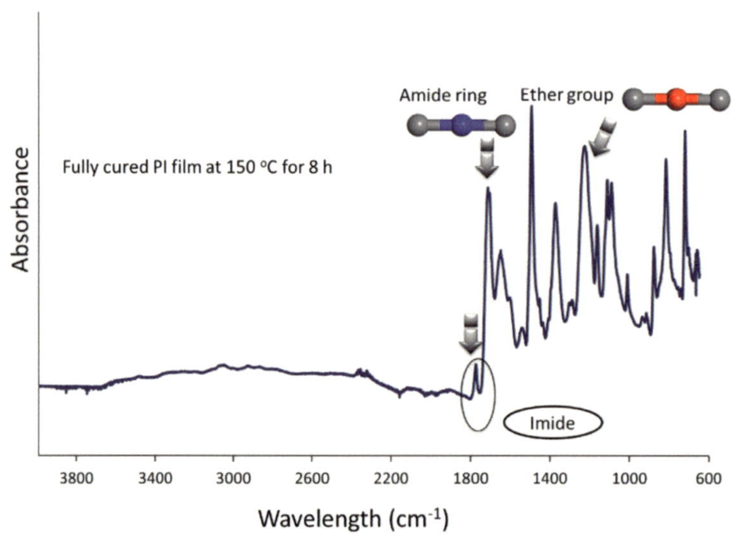

(b)

Figure 9.
(a) Half cured PI film at different temperature and time and (b) zoom version of main function group peaks.

In fully cured PI films, as shown from **Figure 9**, the bonds are grouped into amide rings, aromatic rings, and noncyclical stretching based on atom vibrations [18]. The intensity of the absorption band at 3463 cm^{-1} corresponding to the stretch vibration of –OH bond.

3. Dielectric properties of PI/nanocomposites and their modeling

To simulate the electrical field distribution in PI/nanocomposites, COMSOL-MATLAB Live link and image processing tools were used to build the PI/SIO_2 (NPI) 3D model. The boundary conditions and geometry of the model were obtained from TEM/SEM images of laboratory-produced samples, as shown in

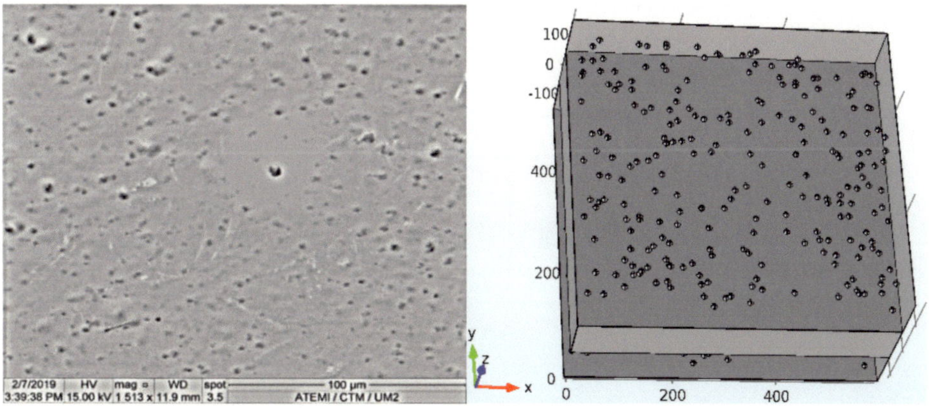

(a) (b)

Figure 10.
(a) TEM image of PI/SiO$_2$ and (b) model in COMSOL using MATLAB Live Link.

Figure 10a, and its respectfully constructed model, as shown in **Figure 10b**. The positions of nanoparticle coordinates (x, y) in the model were taken from TEM/SEM images after converting the image into binary. For z-axis information, there are two possible ways to estimate the z-axis parameter. Either we can assume color intensity level from SEM/TEM images of **Figure 10a** as the depth (z-axis) of nanoparticles to construct 3D model or we can use SEM cross-sectional view information (if particle size is viewable from SEM/TEM) to find the depth of nanoparticles inside the bulk of samples to build model. For our samples, we have used the color intensity technique to obtain the z-axis parameter and create a 3D model. To construct the model, we picked a clear TEM/SEM image, applied an image processing to remove background noise, and finally converted it into the binary. To select the required data from the binary image, we adjusted the image to binary conversion threshold, nanoparticles size, and circularity level. The data contained different information about the particle size and shape (x, y) coordinates. The z-axis related to the depth of each particle is obtained using a small code in MATLAB according to the intensity level of each particle. In this simulation model, the dielectric permittivity of the PI matrix is taken to be $\varepsilon_1 = 3.4$, and that of the silica nanoparticles is $\varepsilon_2 = 3.6$. The nanoparticles were assumed to have a spherical shape of 10 nm radiuses calculated after image processing. The top surface of the 3D model was applied by the constant electric potential of ($V = 10$ V), and the bottom surface was grounded to 0 V. The volume fraction of nanoparticles is constant and set at 1% for single and multilayer insulation structures. The details to construct the model are presented in **Figure 11**.

After constructing the continuum model in COMSOL, the finite element method (FEM) was utilized to observe the nonlinear electric field distribution, as shown in **Figure 12**. The nanodielectric modeling with FEM can be used to predict its dielectric strength based on electric field distribution. FEM was used to calculate the electric field enhancement factor and polarization charge density, as shown in **Figure 13a** and **b**, respectively. Single and multilayers of PI/silica nanocomposite models were simulated with altered nanoparticle permittivity and position. The analysis is focused on calculating the peak electric field enhancement factor (peak EFEF) in the sample from Eq. (3).

$$\text{Peak EFEF} = \frac{\text{maximum electricfield (kV/mm)}}{\text{applied electric field (kV/mm)}} \tag{3}$$

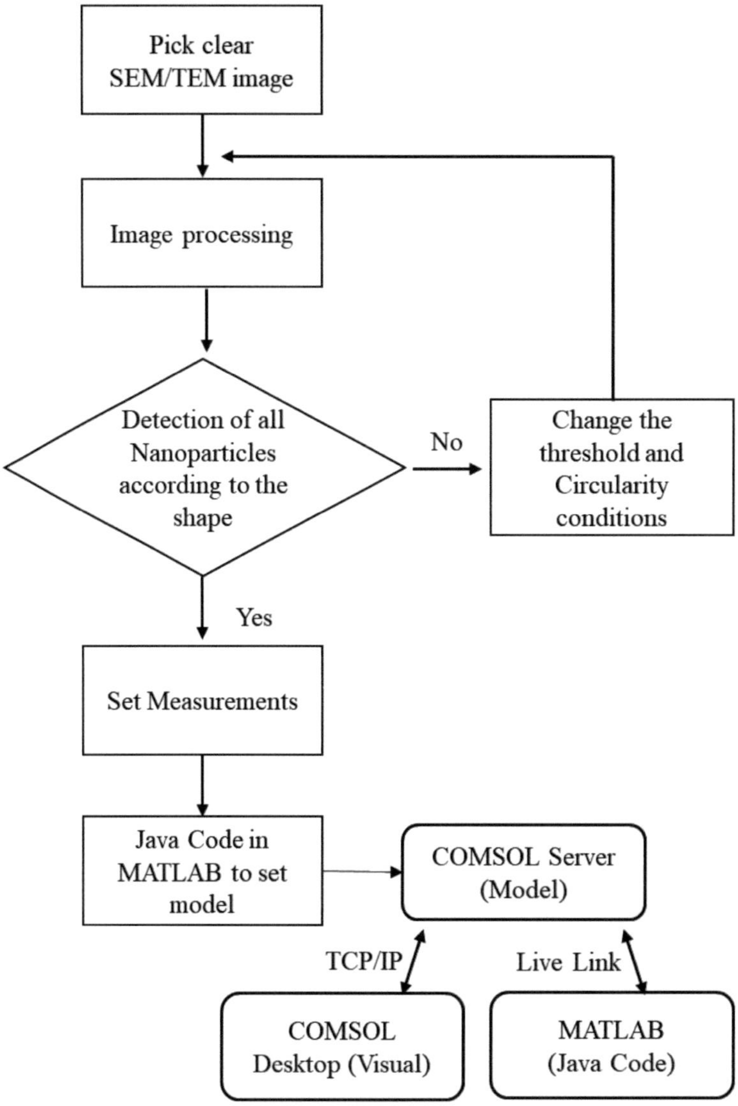

Figure 11.
Model building algorithm.

The simulation results from **Figure 13** show that the electric field enhancement factor (EFEF) and polarization charge density vary with the increase in nanoparticle permittivity, and lesser EFEF is observed in three-layer PI films compared with single-layer models such as top agglomeration and bottom agglomeration. As shown in the simulation solution of **Figure 12**, the maximum electric field exists at the surface between the nanoparticles and the PI matrix in the direction of the z-axis, while the linear electric field is observed in the remaining area of PI/nanocomposite model.

The uniform dispersion of nanoparticle models improved the electric field distribution and reduced the electric field localization as happened in single-layer agglomerated models. The highest electric field and polarization density were obtained in a single-layer top agglomeration model. One of the main reasons for this enhancement can be the change in equivalence capacitance of the sample due to the agglomeration of the nanoparticles, which influence the effective permittivity of the overall sample. The electric field is also influenced due to the shape of the outer surface of nanoparticles because if agglomeration happens, there is a chance that the

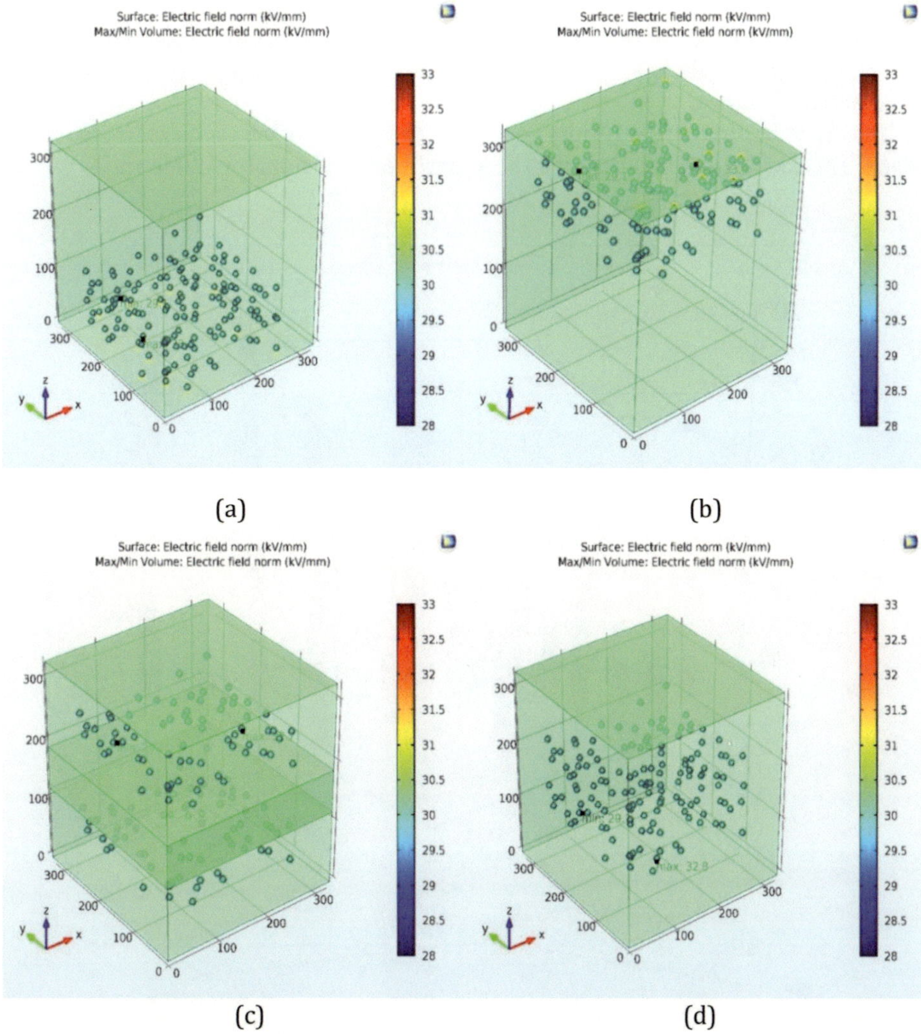

Figure 12.
(a) Top agglomeration, (b) bottom agglomeration, (c) three-layer homogenous dispersion, and (d) three-layer original homogenous distribution.

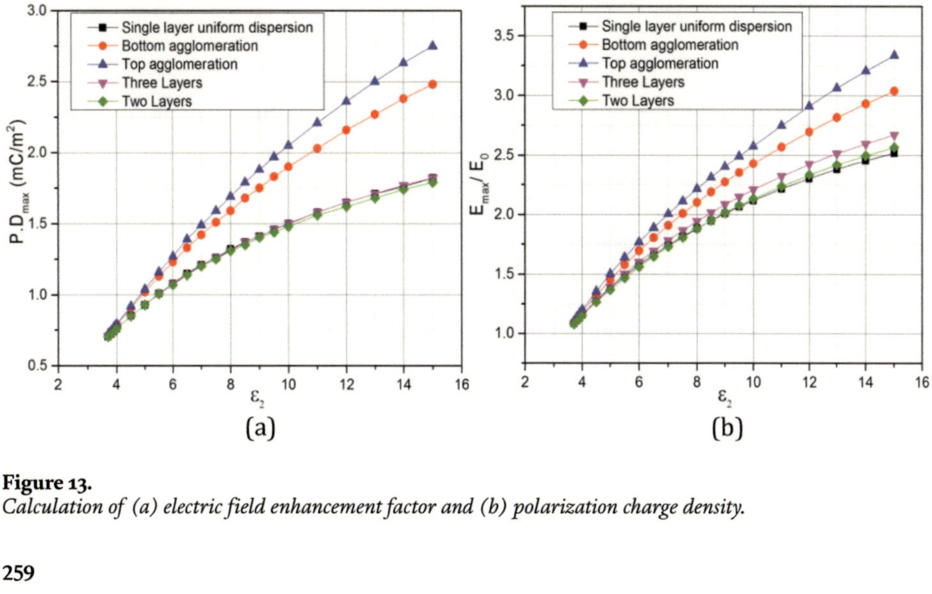

Figure 13.
Calculation of (a) electric field enhancement factor and (b) polarization charge density.

combined particles may change the circularity level and transform the shape into the sharp edges, which increase the local electric field around those edges. Another reason for higher EFEF in agglomeration models can be due to a decrease in the interparticle distance.

To verify the simulation results, electrical characterizations such as the space charge are also determined through the experimental results for single and

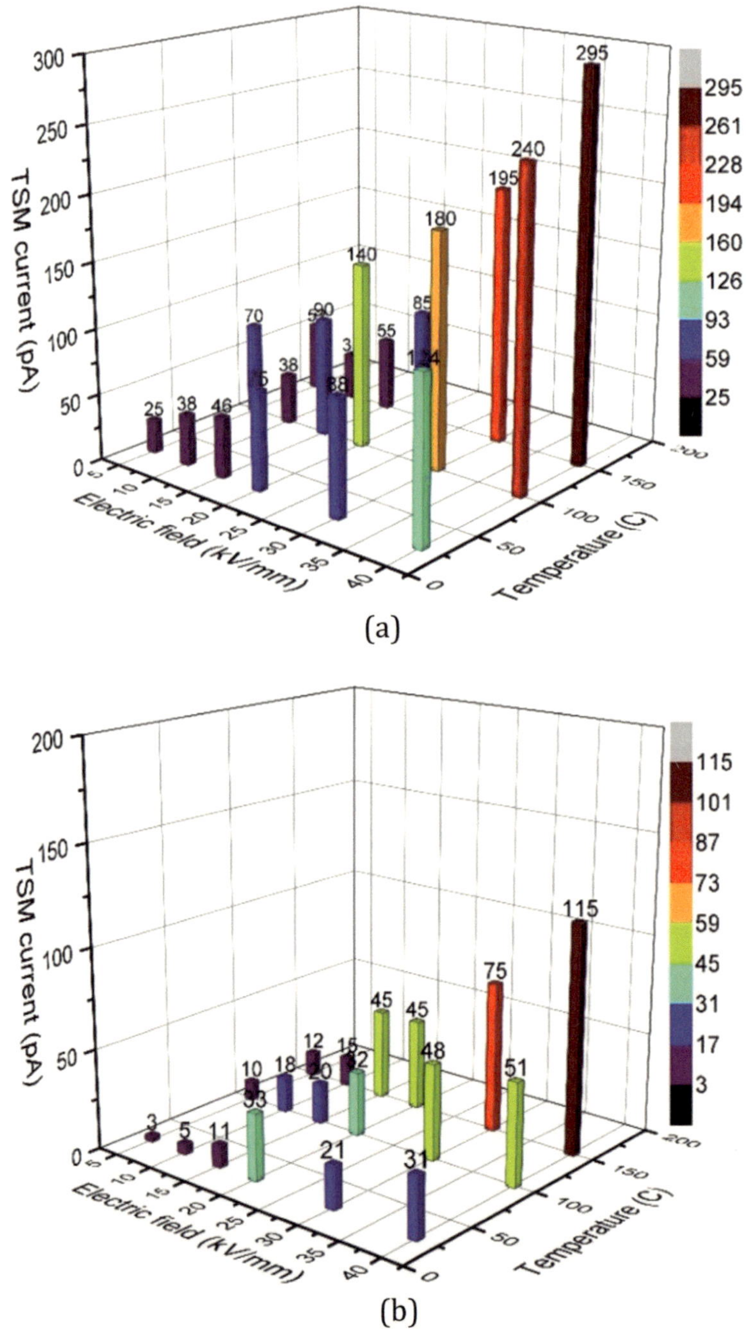

Figure 14.
TS current after applying different temperatures and electric fields: (a) single-layer NPI samples and (b) two-layer PI-NPI samples.

multilayer PI nanocomposite films [19]. For space charge measurements, the experiments are conducted using the thermal step method (TSM) to detect the capacity of these new insulating materials to accumulate/release space charges after thermoelectric poling, close to the practical applications. An electric field between 5 and 40 kV/mm is applied at 50, 100, and 150°C for 1 h [19]. The results presented in **Figure 14(a)** for single-layer NPI films and **Figure 14(b)** for double-layer PI-NPI films illustrate that the amplitude of the TSM current acquired for double-layer PI-NPI films is lesser than the single-layer NPI films at all electric fields and temperature conditions. The lower amplitude of TSM current is due to a lower level of space charge accumulation. Moreover, two opposite TSM current signs correspond to different dominant charges. The current sign changes from positive to negative from low temperature 50°C to higher temperature 150°C, especially for single-layer NPI samples. The sign of current corresponds to the same dominant charge injected from the electrode and shows an increase in accumulated space charge when the poling applied field increases. A thin layer of NPI on PI samples highlights lower TSM currents, and this behavior is more visible at higher electric fields because the signal-to-noise ratio is improved at the higher electric field. Thus, the coating of thin NPI on PI acts as a barrier and reduces the charge injection from electrodes. These coated layers are discharged resistive and increase the charge dissipation rate to reduce the space charge accumulation. Adding nanoparticles in the form of multistructure improves its dispersion by reducing the layer thickness and therefore leads to better dielectric properties, as presented in results.

4. Conclusions

A brief introduction of the physical properties of PI films and their derived nanoparticles was described in this chapter. The preparation of PI nanocomposite is a complex process with many variables involved. Several methods were probed before an optimal synthesis process was found. A detailed synthesis process optimization of multilayer PI nanocomposite films is described in this chapter to understand all variables, which can influence the dielectric properties of the final product. The polyimide/nanocomposite multilayer 3D model based on actual boundary conditions from SEM images of synthesized samples is also constructed and simulated in COMSOL multiphysics software. Effect of nanoparticle agglomeration in microstructures, with the impact of nanoparticle dispersion on the electric field enhancement, is explicitly described in this model. The dielectric properties such as space charge using thermal step method (TSM) technique were also measured to compare simulation results. Our results demonstrate that the chances of nanoparticle agglomeration are reduced by using a thin layer of PI/nanocomposite on PI film instead of using one single layer of PI/nanocomposite film. In consequence, less space charge and low electrical fields are observed in multilayer films. Our methods will help to reliably predict the dielectric strength of polymer/nanocomposite insulating materials. Additionally, the new synthesized multilayer PI/nanocomposite insulating material will ensure reliable operation for electric motors and increase its lifespan.

Acknowledgements

My deepest gratitude goes to my advisors, Prof. Jerome Castellon and Prof. Serge Agnel. Without their continuous support and guidance, I was not able to complete this task. I have benefited tremendously from their vision and technical insights.

I am also greatly indebted to my parents for their incredible love, endless encouragement, and heartfelt prayers. My special thanks are extended to Prof. Jean Pierre Habbas and his team in the Institute Charles Gerhardt Montpellier for providing his polymer synthesis laboratory and a friendly environment to synthesize the samples. He never let me stuck in the chemical synthesis process whenever I faced difficulty, and he was there to solve it and always come up with a solution. I also express my warmest gratitude to Prof. Zhou kai from the Sichuan University Chengdu China to provide moral and technical support to finish this task on time.

Conflict of interest

The book chapter was written with the contributions of all authors. All authors have approved the final version of the chapter. The authors declare no competing financial interest.

Author details

Shakeel Akram[1*], Jérôme Castellon[1*], Serge Agnel[1] and Jean-Pierre Habas[2]

1 Institut d'Electronique et des Systèmes (IES), CNRS, University of Montpellier, Montpellier, France

2 Institut Charles Gerhardt de Montpellier (ICGM), CNRS, Equipe Ingénierie et Architectures Macromoléculaires, University of Montpellier, Montpellier, France

*Address all correspondence to: akram@ies.univ-montp2.fr
and jerome.castellon@umontpellier.fr

References

[1] Yang Y, Jinliang H, Guangning W, Jun H. Thermal stabilization effect of Al_2O_3 nano-dopants improves the high-temperature dielectric performance of polyimide. Scientific Reports. 2015;**5**:16986. DOI: 10.1038/srep16986

[2] Shakeel A, Yang Y, Zhong X, Bhutta S, Wu G, Castellon J, et al. Influence of nano layer structure of polyimide film on space charge behavior and trap levels. IEEE Transactions on Dielectrics and Electrical Insulation. 2018;**25**:146-469. DOI: 10.1109/TDEI.2018.007506

[3] Diaham S, Zelmat S, Locatelli ML, Dinculescu S, Decup M, Lebey T. Dielectric breakdown of polyimide films: Area, thickness and temperature dependence. IEEE Transactions on Dielectrics and Electrical Insulation. 2010;**17**:18-27. DOI: 10.1109/TDEI.2010.5411997

[4] Akram S, Castellon J, Agnel S, Khan MZ. Space charge analysis of multi-structure polyimide films using TSM. In: Proceedings of the IEEE International Conference on Electrical Insulation and Dielectric Phenomena (CEIDP '18); 21-24 October 2018; Cancun. Mexico: IEEE; 2018. pp. 34-37

[5] Andritsch T, Fabiani D, Vazquez IR. Nanodielectrics—Examples of preparation and microstructure. IEEE Electrical Insulation Magazine. 2013;**29**:21-28. DOI: 10.1109/MEI.2013.6648750

[6] Debra LD. Synthesis and characterization of thermosetting polyimide oligomers for microelectronics packing [thesis]. Blacksburg Virginia: Virginia Polytechnic Institute and State University; 2000

[7] Simpson JO, Clair AK. Fundamental insight on developing low dielectric constant polyimides. Thin Solid Films. 1997;**308-309**:480-485. DOI: 10.1016/S0040-6090(97)00481-1

[8] Luo Y, Wu G, Liu J, Peng J, Zhu G, Guoqiang G. Investigation of temperature effects on voltage endurance for polyimide/Al_2O_3 nanodielectrics. IEEE Transactions on Dielectrics and Electrical Insulation. 2014;**21**:1824-1834. DOI: 10.1109/TDEI.2014.004305

[9] Jonscher AK. Dielectric relaxation in solids. Journal of Physics D: Applied Physics. 1999;**32**:57-70. DOI: 10.1088/0022-3727/32/14/201

[10] Shakeel A, Wu G, GuoQiang G, Liu Y. Cavity and interface effect of PI-film on charge accumulation and PD activity under bipolar pulse voltage. Journal of Electrical Engineering and Technology. 2015;**10**:2089-2098. DOI: 10.5370/JEET.2015.10.5.2089

[11] Hafiz ZA, Murat K. The changes in electrical and interfacial properties of polyimide exposed to dielectric barrier bischarge in SF6 medium. The Scientific World Journal. 2013;**2013**:1-7. DOI: 10.1155/2013/890454

[12] Starr W. Polymeric outdoor insulation. IEEE Transactions on Electrical Insulation. 1990;**25**:125-136. DOI: 10.1109/14.64507

[13] Frechette MF, Trudeau ML, Alamdar HD, Boily S. Introductory remarks on nanodielectrics. IEEE Transactions on Dielectrics and Electrical Insulation. 2004;**11**:808-818. DOI: 10.1109/TDEI.2004.1349786

[14] Nelson JK. Dielectric Polymer Nanocomposites. Springer; 2010. pp. 31-45. DOI: 10.1007/978-1-4419-1591-7

[15] Roman K. Thermal and electrical properties of nanocomposites,

including material properties [Thesis]. Astrophysical Journal Letters. 2012

[16] Mai Y-W, Yu Z-Z. Polymer Nanocomposites. Woodhead Publishing Ltd. [Book]; 2006

[17] Gaurav M, Suprakas Sinha R, Shofner ML, Shanfeng W, Jin Z. Polymer nanocomposites Processing, characterization, and applications. Journal of Nanomaterials. 2013;**2014**:403492. DOI: 10.1155/2014/403492

[18] Shakeel A, Guoqiang G, Yang L, Jian Z, Wu G. Degradation mechanism of $A1_2O_3$ nano filled polyimide film due to surface discharge under square impulse voltage. IEEE Transactions on Dielectrics and Electrical Insulation. 2015;**22**:3341-3349. DOI: 10.1109/TDEI.2015.005059

[19] Shakeel A, Jérôme C, Serge A, Jean-Pierre H, Zhou K, Tariq Nazir M. Impact of nanocomposite thin layer on nanoparticles dispersion and their dielectric properties. In: Proceedings of the IEEE International Conference on Electrical Insulation and Dielectric Phenomena (CEIDP '19); 20-23 October. Vol. 2019. Richland, Washington. USA: IEEE; 2019. pp. 381-384

Chapter 14

Effect of Molecular Structure Modification and Nano-Doping on Charge Transportation of Polyimide Films for Winding Insulation

Boxue Du, Ranran Xu, Jiwen Xing and Jin Li

Abstract

Polyimide (PI) is widely employed as winding insulation in high voltage devices, such as extra-high voltage electric reactor and inverter-fed motor. The injection and accumulation of charges on the surface of PI films will lead to electrical field distortion and reduced lifespan of winding insulation, especially for the operation environment of high temperature and high voltage. This chapter focuses on effects of surface molecular modification and nanoparticles on dynamic characteristics of surface charge and space charge of pure PI films, including three sections. The effect of molecular structure on the surface charge dynamics of PI films was studied firstly. The chapter investigated that how molecular structure affects surface charge of polyimide nanocomposite films. Furthermore, the effect of surface molecular modification on space charge characteristics of multilayer PI films was researched. The results illustrate that surface molecular modification and nanoparticles can comprehensively suppress space charge accumulation and improve dielectric property.

Keywords: polyimide, surface molecular modification, nanoparticles, surface charge, space charge

1. Introduction

Polyimide film (PI) is widely applied in reactor extra-high voltage (EHV) winding insulation, wind turbine winding, and variable frequency motor winding insulation due to its excellent physical, chemical, and heat resistance properties, with the wide range of applications in electronics and aerospace. During the operation of electrical equipment, the corona and partial discharge will inevitably happen and be accompanied by temperature rise [1], making insulation materials of EHV equipment in a complex environment where high temperature, high voltage, corona, and partial discharge exist together for a long time. A large amount of surface and space charge will inject into insulation materials, leading to the accumulation of local charges in the insulation and electrical field distortion, which will reduce the life of equipment. Therefore, it is vital to study the surface charge and space charge transport mechanism of polyimide to improve the electrical properties of the film.

Nanocomposite dielectrics have shown excellent insulation performance due to the special size effect of nanoparticles, especially in corona resistance. Plenty of researchers have added inorganic nanoparticles to the polymer matrix to prepare nanocomposite dielectrics and achieved some results in practical applications [2–8]. Zhong added the silane coupling agent when preparing the SiO_2/PI composite films, the results showed that the coupling agent improved the dispersion of the nanoparticles and the interface morphology [9]. Zha Junwei et al. prepared PI/ZnO composite films via in-situ polymerization, with studying the material's electrical resistance, volume resistivity, dielectric constant, and dielectric loss, the better corona resistance of the composite films is acquired. However, both the dielectric loss and the dielectric constant increase with the growth of the ZnO content [10]. The Al_2O_3/PI composite films were prepared using in-situ polymerization. The corona resistance testing results showed that the corona resistance improved when the mass fraction of nano-Al_2O_3 increased, and the corona resistance reaches the peak when in 20 wt%. Nevertheless, the agglomeration phenomenon becomes more and more obvious with the mass fraction gradually starts from 5 wt%, observed by the electron microscope [11]. Based on the above studies, it can be found that some achievements have been obtained in the nano-doped polyimide films, but its electrical resistance is not ideal. The research status of nano-composite polyimide films is still lingering in pursuing higher content of inorganic nanoparticles, better compatibility, and the better physical and chemical structure of nanocomposites. The surface and space charges have not been studied.

On the other hand, the electrical properties of fluoropolymers have not attracted the attention of researchers. In fact, the molecular structure modification (fluorination), as a kind of polymer surface chemical structure that changes the chemical composition of the polymer surface layer, has mature applications in the chemical industry [12, 13]. A Zhenlian found that the fluorocarbon layer is formed on the surface of polyethylene by fluorination, and the space charge characteristics have been studied [14]. It was found that the fluorinated layer can suppress the space charge injection. However, the current research has not carried out on the dynamic characteristics of surface and space charge accumulation as well as dissipation and on the effect of the introduction of fluorine on the depth of dielectric traps. So far, the application of surface fluorination is mainly used to improve the adhesion properties, barrier properties, and anti-permeability properties of polymers. There are few reports on its application in electrical insulation [15]. In the chapter, the molecular structure modification is used to control the surface structure of polyimide film and nano-composite polyimide film, and the surface and space charge dynamic characteristics of the film are studied to develop new polymer dielectrics, which is important in the field of engineering dielectric.

2. Experiments

2.1 The sample preparation

The chapter uses the two-step method to prepare the polyimide films. Pyromellitic dianhydride (PMDA) and 4,4′-diaminodiphenyl ether (ODA) are added into dimethylacetamide (DMAc). The polyamic acid (PAA) was prepared by polymerization reaction in the process. The PAA was coated on the glass plate, placed in a vacuum environment to remove air inside PAA, which was placed in the environment where the temperature was 60°C (1 h), 120°C (2 h), 150°C (1 h),

200°C (1 h), 250°C (1 h) and 300°C (1 h). After the heating was stopped, waited for 6 hours, and the PI film with a certain thickness was prepared.

In addition, the nano-composite polyimide film was prepared by in-situ polymerization. Firstly, the nano-particles Al_2O_3 (particle size is less than 20 nm) and surface modifier KH550 were placed in an organic solvent DMAc for ultrasonic process for 1 hour to ensure that the nanoparticles are uniformly dispersed in the organic solvent. Then the appropriate amount of ODA was added into solution and stir for 1 hour. The PMDA was added to the suspension in batches under the water bath environment at a constant temperature of 40°C, and mechanical stirring was continued until the viscosity of the entire reaction solution suddenly increased. After degassing and imidization, PI films doping with different contents of Al_2O_3 (0, 1, 3, 5, 7 wt%) with a certain thickness were obtained. The scanning electron microscope (SEM) was used to observe the presence of nanoparticles in distribution in the PI matrix.

The surface molecular modification is achieved in terms of surface fluorination. The PI sample is placed in a closed reactor, the air inside the reactor is purged, and the reactor is filled with a certain proportion of fluorine gas and nitrogen gas mixture (12.5 and 20%). The internal reaction temperature was adjusted by the temperature control device (reaction temperature is room temperature and 55°C, respectively), the air pressure was 500 mbar, and the reaction time was 15, 30, 45, and 60 min. The sample after surface molecular modification was used to study the effect of different reaction time on the surface and space charge of the PI sample. The SEM and infrared spectrum analysis tester were used to verify the effect of fluorination.

The multilayer PI sample is a composite of pure single-layer PI film with the same fluorination time.

2.2 Testing method

2.2.1 Surface charge dynamic characteristics experiment

The surface charge measurement system is shown in **Figure 1(a)**. The entire corona measuring device is in a closed transparent container to ensure the same temperature and humidity. The needle plate electrode was used, the plate electrode is below the needle electrode, the distance between the needle tip and the plate electrode is 5 mm, and the distance between the plate and the surface of sample is

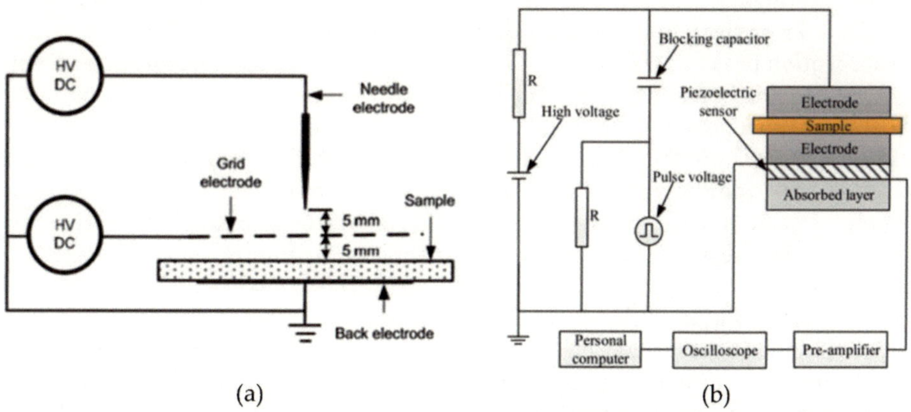

Figure 1.
(a) Surface charge measurement system and (b) space charge measurement system.

also 5 mm. The sample is placed on the ground electrode. Turn on the voltage power supply, and the corona time is 10, 20, and 30 min. When the corona time is reached, turn off the high voltage power supply, quickly transfer the sample to the electrostatic potentiometer and record its surface potential for 4 hours.

According to the measured results, the surface charge density can be calculated, and the calculation formula is

$$\sigma = \frac{\varepsilon_0 \varepsilon_r}{d} V_{(t_0)} \tag{1}$$

Among them, σ is the required surface charge density, d is the thickness of the sample, V is the sample surface potential, ε_0 and ε_r are the vacuum dielectric constant and the relative dielectric constant [16].

2.2.2 Space charge dynamic characteristics experiment

This chapter uses the pulsed electro-acoustic (PEA) method to test and study the space charge dynamic distribution in the PI sample. The measurement device is shown in **Figure 1(b)**, which mainly is made up of high voltage DC power supply, pulse generator, electrode system, preamplifier, computer, and oscilloscope. The measuring principle is that a pulse voltage is applied across the sample, and this pulse will generate an electric field inside the sample, causing space charge vibration in the sample. This vibration is propagated outward by means of sound waves. The amplitude of sound waves stands for the amount of charge, and the time when the acoustic wave reaches the piezoelectric sensor reflects the position of the space charge in the sample. During the test, the applied DC voltage was 5000 V, the measuring time was 3600 s, and data were recorded per 600 s. In addition, the ambient temperature was 25°C, and the relative humidity was ~40%.

3. Results and discussion

3.1 The effect of molecular structure on the surface charge dynamics of PI films

The ATR-FTIR spectrum of the PI films before and after fluorination is shown in **Figure 2**. In **Figure 2(a)**, the infrared spectrum of the original polyimide film was described. From the graph, we can find the typical characteristics of PI films, with the absorption peaks at 1780 and 1720 cm^{-1}. Furthermore, the absorption peaks of the C=C double bond of the benzene ring are at 1500 and 1100 cm^{-1}, the vibrational absorption peaks of the C—N bond are at 1373 cm^{-1}, and the absorption peaks at 810 cm^{-1} are the vibration of benzene-H bond. The absorption peak at 1237 and 3200 cm^{-1} indicates that a small amount of ODA remained during the molecular polymerization reaction. The infrared spectrum of the sample after fluorination is shown in **Figure 2(b)**. As can be seen that the PI film after the surface molecular structure has the obvious C—F, $C—F_2$ and $C—F_3$ absorption peak is in the range of 950–1340 cm^{-1}. **Figure 2** shows that the C—F bond is the strongest bond in the polymer, higher a great than the C—H bond. So, the surface fluorination will make the fluorine element replace the hydrogen element on the PI films surface, which results in the reduction and even disappearance of C—H bonds in the molecular structure of the surface layer and accompanied by the formation of C—F, $C—F_2$, and $C—F_3$. There is a dense C—F layer on the PI film surface layer.

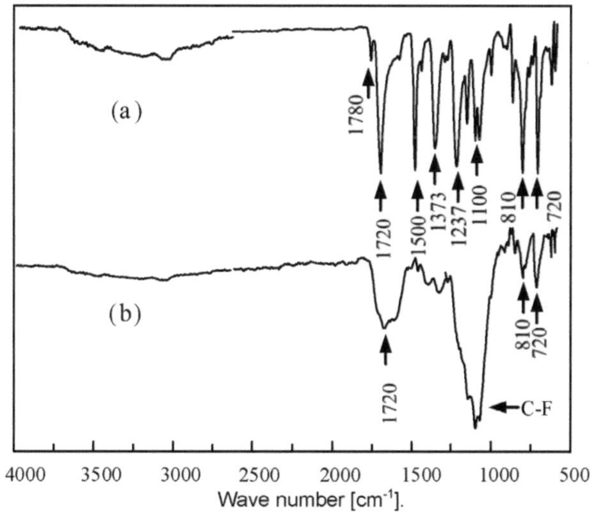

Figure 2.
The infrared spectrum of (a) original film and (b) fluorinated film.

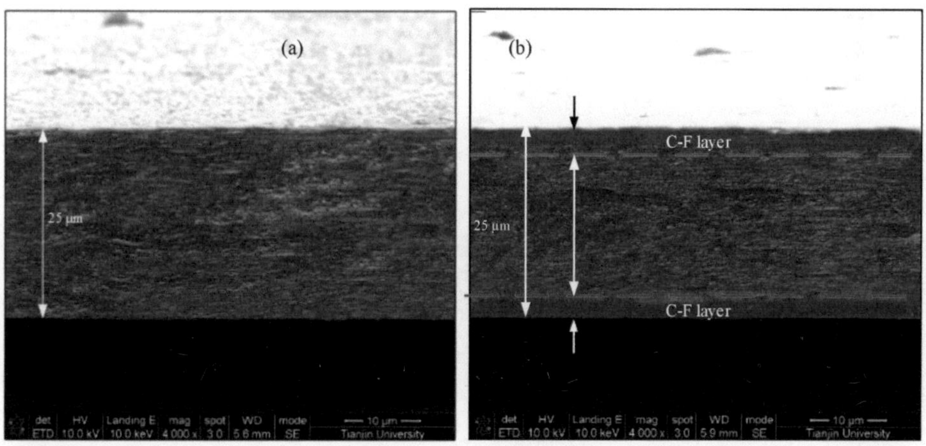

Figure 3.
The SEM of (a) original film and (b) fluorinated film of PI.

Figure 3 describes the microstructure of original PI film without fluorination and the fluorination film for 60 min by the SEM, respectively. As can be seen from the figure, a dense fluorinated layer is formed on the surface of the sample. The thickness of the sample is about 2.6 μm when the reaction time is 60 min. It is obvious that the surface molecular modification can change the chemical structure of the surface of the sample and form a C-F layer on the surface when combining infrared spectral analysis with the SEM results comprehensively.

The surface charge density of the PI sample was measured using the test system of **Figure 1(a)**. **Figure 4** shows the change of the surface charge density of the original sample and the sample fluorinated for 45 min under different corona times. It is noticeable that the surface charge density dissipates rapidly at the beginning, and then the dissipation speed gradually slows down and remains a steady trend. The positive and negative charges have similar trends. There are three main ways for surface charges to dissipate: (1) migrating to the ground electrode on the back, (2) migrating along the surface by the tangential electric field and entering the earth

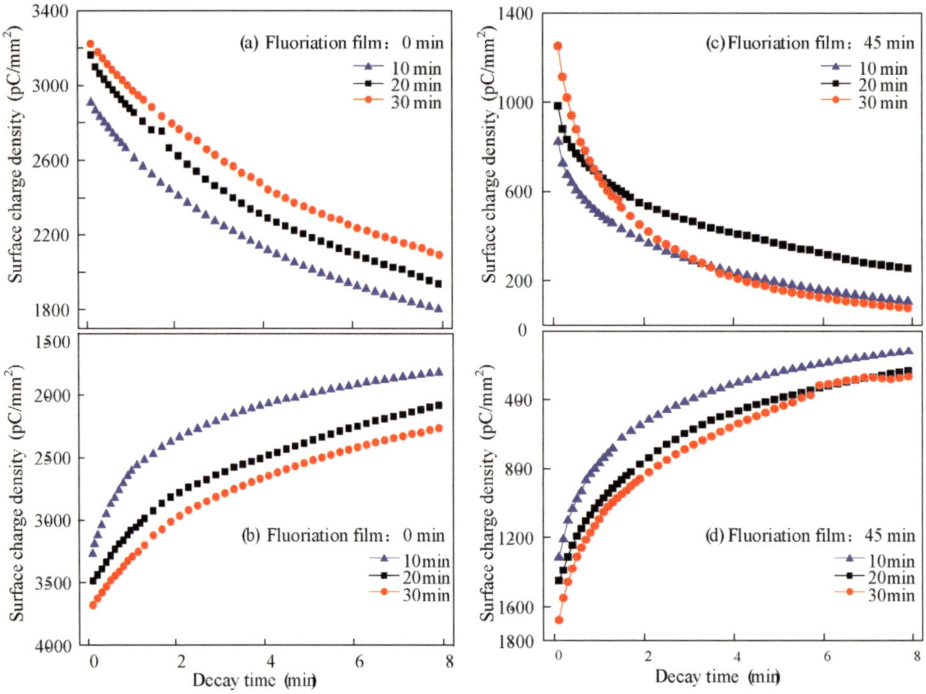

Figure 4.
The surface charge density of sample without fluorination under the (a) positive voltage, (b) negative voltage and the surface charge density of sample with fluorination under the (c) positive voltage, (d) negative voltage.

through the ground electrode, and (3) neutralizing with heterogeneous charges in the air. Which dissipating path dominates depends on various factors such as the surface characteristics of the solid medium, the gas atmosphere, and the electrode structure [17, 18]. Since the normal electric field on the film surface is much larger than the tangential electric field in this experiment. The most possible way for the surface charge to dissipate is to migrate to the ground electrode on the back and neutralize with heterogeneous charges in the air.

Figure 4(a) and **(b)** is the surface charge density of the sample without surface molecular modification. It is clear that the surface charge density gradually increases from 2900 to 3200 pC/mm^2 with the corona time rising from 10 to 30 min. During the dissipation process, the surface charge density tends to stabilize as the corona time increases, which is due to the energy required to restrain the injected charge increases with the corona time growing. The electric field formed by the charge which has been injected suppresses the large amount of the original charge transfer, alleviating the charge dissipation process [19].

According to the graph, the initial surface charge density increases with the increase of the corona time, but the dynamic of the charge is different from the dynamic of the original sample. The surface charge density of the sample under the voltage for 20 min is larger than those for 10 min and the attenuation curve is flatter, while the surface charge density of the sample under voltage for 30 min decays faster than the previous two samples. Referring to **Figure 4(c)** and **(d)**, the surface charges of the samples have the similar trend after 8 min for the samples that are under voltage for 30 and 20 min. The reason may be that the surface layer after fluorination has the fluorinated layer on the surface of the sample, which can effectively suppress the injection of charge. As the corona time increases, a large amount of charge accumulates in the fluorinated layer on the surface. When the power is turned off, the charge neutralizes and dissipates into the sample body and the air, and most of change accumulated on the surface charge neutralizes with opposite charge in the air. Therefore, although the initial surface charge density increases when the voltage time is 30 min, most of change accumulates on the surface fluorinated layer and does not enter the deep traps inside sample, which lead to the similar trend after 8 min.

The influence of different fluorination times on the surface charge dissipation of the sample was evaluated by the dissipation time. The dissipation time is regarded as the time from the charge starting dissipates to the remaining 10% of the premier surface charge. The longer time stands for the more slowly charge dissipation. **Figure 5** shows the surface charge dissipation time under different conditions. **Figure 5(a)** is the dissipation time of the sample fluorinated 30 min under different corona times, and **Figure 5(b)** is the dissipation time of sample with fluorination time when the corona time is 10 min.

From **Figure 5(a)**, the dissipation time gradually increases as the corona time increases, and the dissipation time of the negative charge is shorter than that of the positive charge.

According to **Figure 5(b)**, the positive charge dissipation time of original sample is 80 min, whereas the negative charge is 100 min, both which are more than 1 h. After fluorination, the total dissipation time is less than 18 min, and the charge dissipation time of the sample fluorinated for 45 min is the shortest. For the original sample, the dissipation time of the negative charge is longer than the positive charge. However, the dissipation time of the negative charge of the sample after fluorination is lower than the positive charge, due to the strong electronegativity of fluorine element which can absorb electrons to form a shielding layer on the surface layer under the negative corona, thereby improving the dissipating speed.

This section uses polymer trap theory to further study the surface charge transport mechanism of polyimide materials. According to the trap theory, the trap energy level and density in the polymer can be calculated by the following formula (2) and (3).

$$\Delta E = kT \ln (\nu t) \tag{2}$$

$$N(E) = \frac{4\varepsilon_0 \varepsilon_r}{qLkTL} \frac{tdV}{dt} \tag{3}$$

ΔE is the trap energy level, T is the absolute temperature, k is the Boltzmann constant, t is the time, $N\ (E)$ is the trap density, L is the sample thickness, ν is the escape frequency, q is the basic charge, and V is the surface potential.

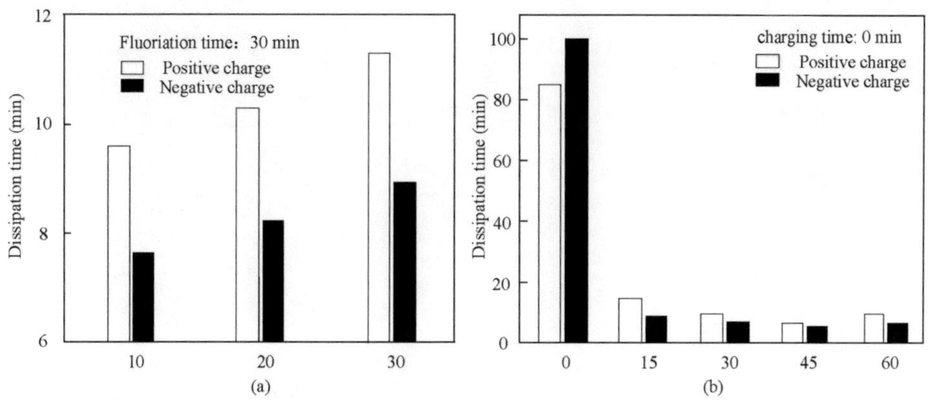

Figure 5.
The surface charge density of sample without fluorination under the (a) positive voltage and (b) negative voltage.

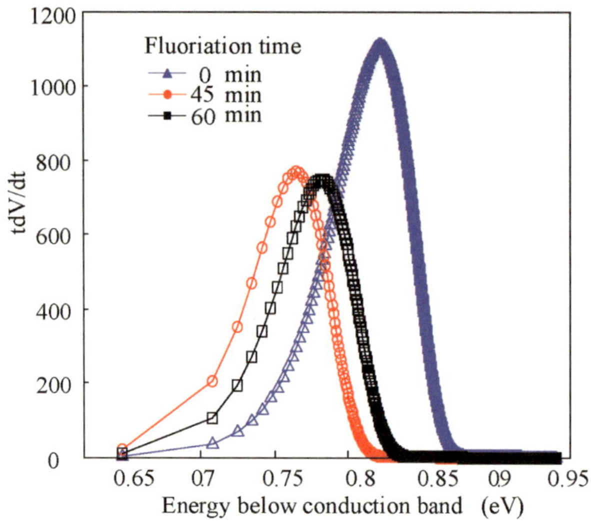

Figure 6.
The trap distributions of sample under various fluorination times.

The trap parameters of the polyimide film under different fluorination times are calculated. As shown in **Figure 6**, the energy level represents the depth of the trap. For the sample without fluorination, the trap level is most distributed at 0.84 eV, while the samples fluorinated for 45 and 60 min are most distributed at 0.76 and 0.79 eV. The results indicate that the trap level of sample with fluorination was shallower than original sample. The polyimide after fluorination makes up for some defects on the surface of the sample, thereby reducing the trap depth of the sample and causing surface charge difficult to accumulate, and dissipates faster, which is in line with the former conclusion.

3.2 The effect of molecular structure on surface charge dynamics of PI nanocomposites

Figure 7 is the SEM of the PI nanocomposite film doping with 3 and 5 wt% of Al_2O_3. The nanoparticles are dispersed in the PI matrix uniform when the mass fraction is 3 wt%, but large-size agglomerates appear with a particle size above

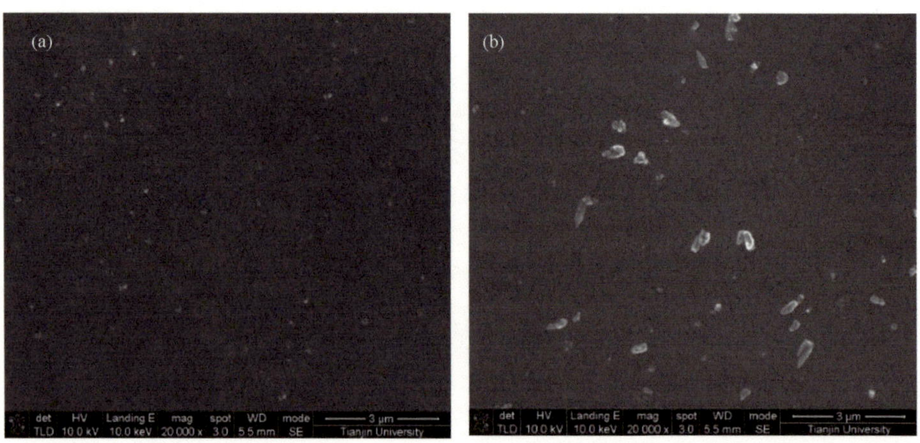

Figure 7.
The SEM images of PI films doping with (a) 3 wt% and (b) 5 wt% of Al_2O_3.

500 nm when the mass fraction is 5 wt%. Nano-composite PI films with different mass amounts are divided into two groups, one of which is subjected to fluorinate for 30 min and the other is original films.

The experiment used the surface charge dynamic measurement system to study surface charge dynamics. The conditions were to maintain relative humidity at ~40%, room temperature, grid voltages to ±3 kV, and corona times to 5, 10, and 15 min. The result is shown in **Figure 8**.

According to **Figure 8(a)**, the surface charge density gradually increases as the corona time increases, and the initial surface charge density of the negative charge is higher than the positive charge. Comparing **Figure 8(a)** and **(c)**, it is found that the surface charge density of the sample after fluorination decreased more and dissipate faster. Referring to **Figure 8**, it can be found that the fluorination has stronger effect on the PI film than the nanocomposite PI film. Comparing **Figure 8 (c)** and **(d)** with **Figure 4(c)** and **(d)**, the surface charge density of the sample with nanoparticles is higher than sample without nanoparticles and the charge dissipation is slower. The surface charge density of fluorinated PI film with Al_2O_3 decays below ±200 pC/mm^2 in a short time. But after the Al_2O_3 are added, the sample charge density is 300 pC/mm^2 after 35 min, which indicates that the sample with nanoparticles has a stronger ability to capture surface charges. These studies show that nanoparticles and surface molecular modification have opposite effect on the surface charge dynamics of polyimide films. The latter can make sample traps shallower and the former makes the surface charge agglomerate and dissipate slowly.

Figure 9 describes the surface charge dissipation of the nano-composite PI film. It is clear that the surface charge dissipating time of the fluorinated sample is

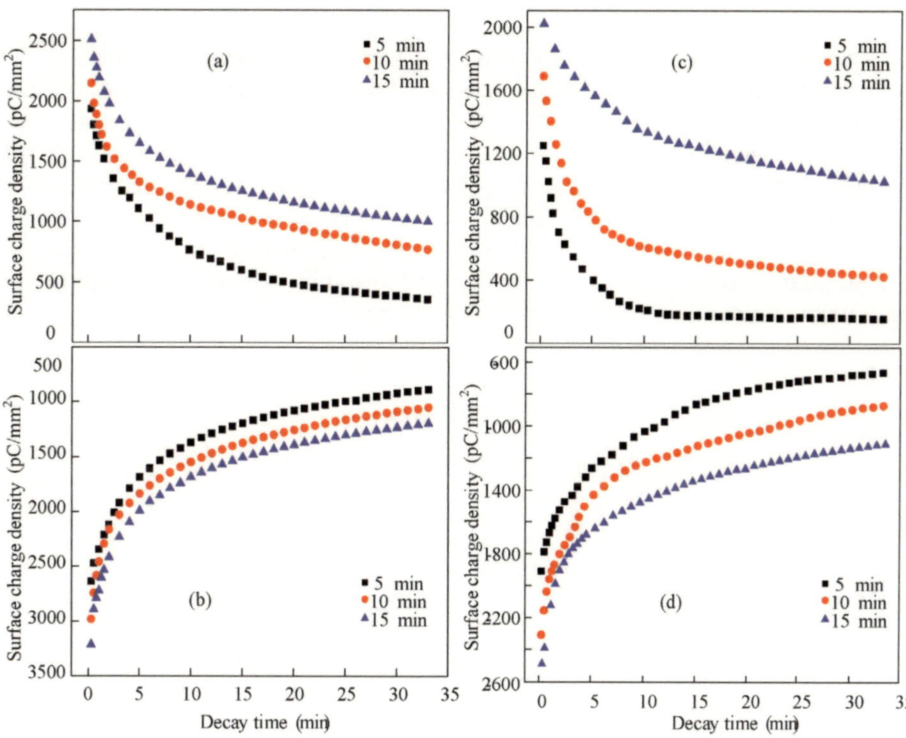

Figure 8.
The surface charge density of original sample doping with 3 wt% Al_2O_3 under the (a) positive voltage, (b) negative voltage and the surface charge density of fluorination sample doping with 3 wt% Al_2O_3 under the (c) positive voltage, (d) negative voltage.

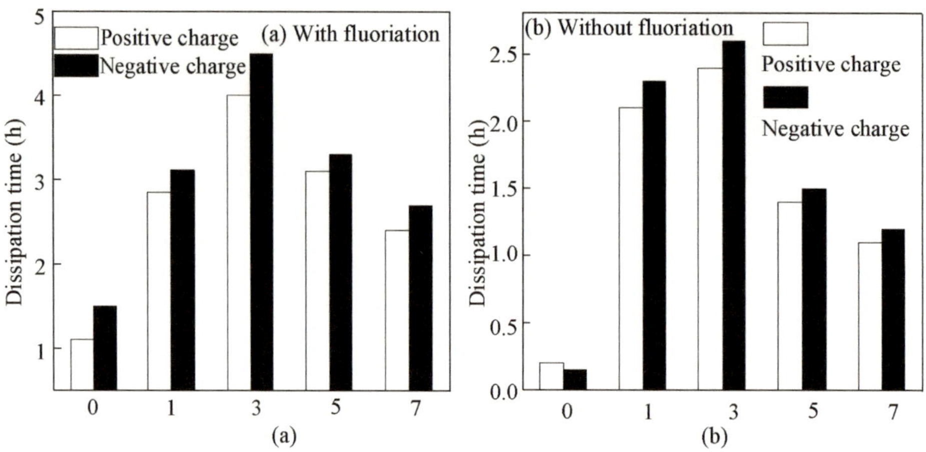

Figure 9.
The surface charge dissipation time of the nano-composite PI film (a) with fluorination (wt%) and (b) without fluorination (wt%).

shorter than the sample without fluorination, indicating that fluorination improves the surface charge dissipation. The dissipation time increases firstly and then decreases with the increase of the Al_2O_3 content for all samples. The surface charge dissipates slowly when Al_2O_3 content is 3 wt%.

In order to further study the effect of nanoparticles on the transport mechanism of the surface charge of the PI film, the trap properties are analyzed. It can be seen from **Figure 10(a)** that the nano-composite film with 3 wt% Al_2O_3 has the highest trap energy level and the most trap density, the trap energy level of the film without nanoparticles is the lowest. The trap energy level is lower when the Al_2O_3 content is 7 wt% than 3 wt%. This indicates that the interface which formed between the nanoparticles and matrix makes the trap level of nanocomposite PI deeper, and the deep trap inhibits the injection of the charge. As the content of Al_2O_3 changes, the distribution and quantity of the interface will change. When the content is high, the agglomeration will occur and affect trap depth of the sample and the accumulation as well as dissipation characteristics. This is also the reason why the dissipation time is longest when the content is 3 wt%. This chapter finds that the nanoparticles will deepen the traps of the polyimide film, increase the ability of surface charges to accumulate on the surface, and suppress the dissipation.

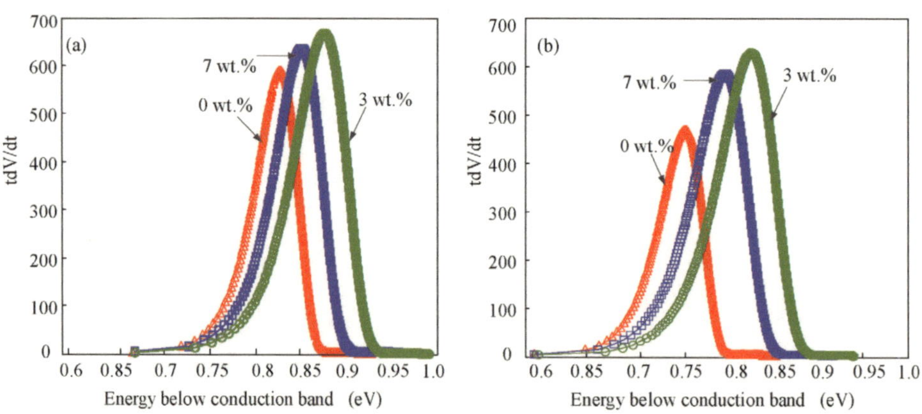

Figure 10.
The trap distributions of sample doping with various mass contents of Al_2O_3 (a) without fluorination, (b) with fluorination.

Combining the conclusion of fluorination with nanocomposites, it is clear that comprehensive application of surface molecular structure modification and nanoparticles can improve the electrical resistance while improving its surface charge dissipation, optimizing the trap distribution of the sample, and reducing the surface energy.

3.3 The effect of surface molecular modification on space charge characteristics of multilayer PI films

The samples used in the section are multilayer PI films without nanoparticles and fluorinated for 15, 30, 45, and 60 min, respectively, which are named as PI15, PI30, PI45, and PI60. In addition, the samples without fluorination named PI10 are

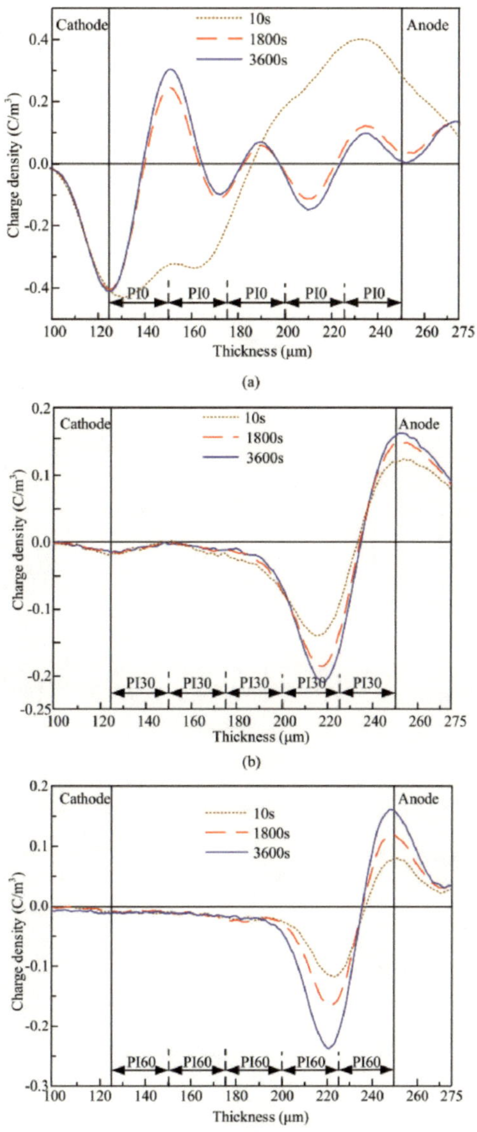

Figure 11.
The space charge distribution of multilayered PI films by stacking (a) five samples without fluorination, (b) five samples fluorinated for 30 min, (c) five samples fluorinated for 60 min.

used as the contrast group. The testing system is the PEA method in **Figure 1(b)**. During the measurement, five samples with the same fluorination time were stacked into single multilayer film, and the silicone oil was applied between the interfaces to remove air. The applied voltage time was 3600 s, and the amplitude was 5 kV.

The space charge distribution of multilayered PI films is shown in **Figure 11**. In the figure, the thickness of each layer is 25 μm. Each layer of the sample is marked with PI0, PI30, and PI60. It can be seen from **Figure 11(a)** that when the voltage time is 10 s, a certain number of homosexual charges appear near the anode and cathode. As the voltage time increases, the space charge distribution changes. When the voltage time is 1800 and 3600 s, a lot of positive charge occur near cathode, followed by the negative charge accumulating near the next interface, and then the

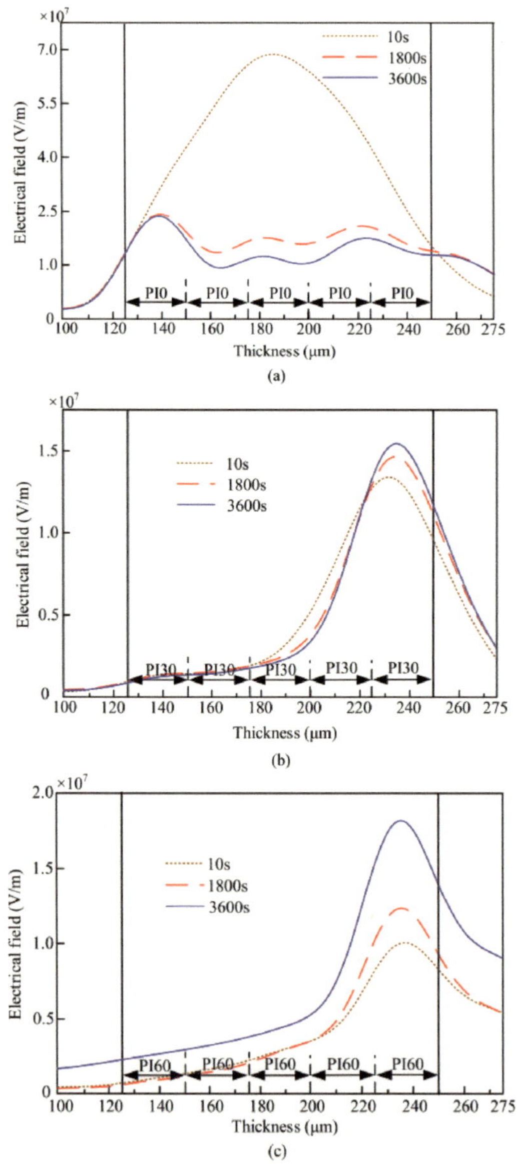

Figure 12.
The electrical distribution of multilayered PI films by stacking (a) five sample without fluorination, (b) five sample with fluorinated for 30 min, and (c) five sample with fluorinated for 60 min.

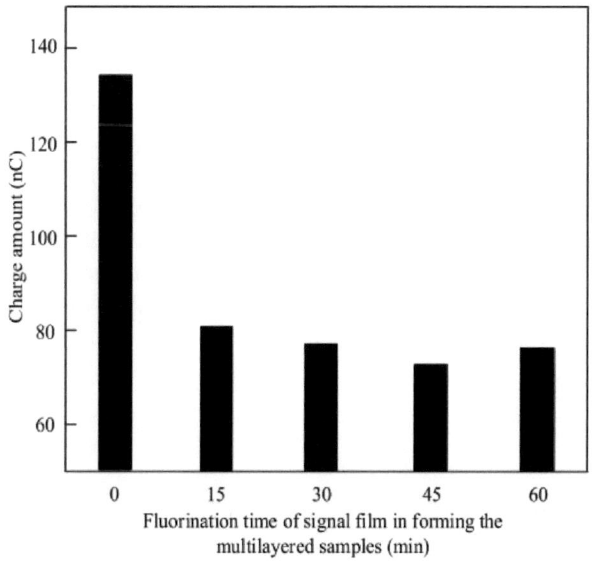

Figure 13.
Relationship between the charge of multilayer samples and the fluorination time.

positive and negative charges alternately appear, until the small amount of positive charge exists near the anode. The largest space charge density amplitude near the cathode is 0.3 C/m^3.

According to **Figure 11(b)**, the charge distribution is very different from the space charge distribution of the untreated samples. The space charge distribution of the five layers of PI30 has the same trend from 10 to 3600 s. A certain amount of positive charge accumulates near the anode, and a large amount of negative charge accumulates at 210–220 μm near the interface near the anode. The maximum space charge density amplitude is 0.21 C/m^3. In **Figure 11(c)**, the space charge distribution trend of the five-layer PI60 sample is similar with the five-layer PI30 film. In addition, the space charge distribution of the five-layer PI15 and PI45 samples has the similar trend with five-layer PI30 and PI60 samples. To avoid repetition, the space charge distribution of the two is not given in this chapter.

Figure 12 shows the distribution of the electric field of the sample after different fluorination times. In **Figure 12(a)**, the sample accumulates a large number of homosexual charges near the cathode and anode during the initial stage of applying voltage, resulting in the high electric field strength with 7.0×10^7 V/m. As the voltage time increases, the charge is transferred and accumulated inside the sample, leading to the change of electrical distribution. **Figure 11(b)** and **(c)** shows the internal electric field strength of the five-layer PI30 and PI60 samples. It can be found that the electric field strength of PI30 is much lower than PI10, which means surface molecular modification can optimize the electric field. The highest internal electric field strength of PI30 is 1.6×10^7 V/m, and PI60 is 1.8×10^7 V/m.

The studies in **Figures 11** and **12** show that the surface structure and internal trap energy level of the sample after surface molecular modification is adjusted can make the sample effectively inhibit the injection of charge into the sample, thereby suppressing the space charge accumulation in the sample.

In order to analyze the amount of space charge injected into the internal of sample, the total amount is calculated according to formula (4):

$$Q = \int_{d_2}^{d_1} |\rho(x)| S dx \quad (4)$$

In the formula, S is the effective area, and d_1 and d_2 are the thickness start and end positions of the sample. $\rho(x)$ is the charge density at x.

The charge amount of the sample with different fluorination times is shown in **Figure 13**. It can be seen that the sample without fluorination has a charge of 135 nC, and the amount of space charge of all sample with fluorination is below 80 nC. With the increase of the fluorination time, the amount of charge decreases slightly and then increases again. The charge amount of sample fluorinated for 45 min is the lowest (75 nC) among all the fluorinated samples.

4. Conclusions

In this chapter, nano-composite polyimide films with different amounts of Al_2O_3 are prepared by in-situ polymerization. PI films with and without surface molecular modification are prepared. Based on the DC corona method and PEA method, the surface charge and space charge dynamic of the sample were analyzed. The main research results obtained are as follows:

1. The infrared spectroscopy and SEM images show that surface molecular modification can effectively change the chemical structure of the sample surface and form a dense CF layer on the surface. The surface charge test shows that the number of shallow traps on the surface of the PI film after surface molecular modification increases, making it difficult to accumulate charges on the surface and improving charge dissipation, which can effectively solve the problem of surface charge accumulation for polymer. And in this chapter, the PI films fluorinated for 45 min has the lowest trap energy level and the lest charge accumulation.

2. For nano-composite PI film, the nanoparticles increase the surface trap level and density of the PI film, generating charge accumulation and block the charge dissipation. The dissipation time of films increases first and then decreases with Al_2O_3 content increases. The PI film doping with 3 wt% Al_2O_3 has the highest trap energy level and the largest trap density, which brings the most serious charge accumulation and the longest dissipation time.

3. The amount of space charge of the multilayer PI film decreased from 135 75 nC with the time of surface molecular modification rising from 0 to 45 min, but the amount of charge decreases slightly when the time increases from 45 to 60 min.

4. This chapter finds that the comprehensive application of the surface molecular modification and nanoparticles can effectively improve the polyimide's electrical properties, regulating its surface charge distribution and the trap level of the sample.

Author details

Boxue Du, Ranran Xu, Jiwen Xing and Jin Li*
Key Laboratory of Smart Grid of Education Ministry, School of Electrical and Information Engineering, Tianjin University, Tianjin, China

*Address all correspondence to: lijin@tju.edu.cn

References

[1] Kirkici H, Serkan M, Koppisetty K. Nano/micro dielectric surface flashover in partial vacuum. IEEE Transactions on Dielectrics and Electrical Insulation (TDEI). 2007;**14**(4):790-795

[2] Maity P, Basu S, Parameswaran V, et al. Degradation of polymer dielectrics with nanometric metal-oxide fillers due to surface discharges. IEEE Transactions on Dielectrics and Electrical Insulation (TDEI). 2008;**15**(1):52-62

[3] Bin M, Gubanski SM, et al. Dielectric properties and resistance to corona and ozone of epoxy compositions filled with micro-and nano-fillers. Annual Report Conference on Electrical Insulation and Dielectric Phenomena; 2009. pp. 672-677

[4] Venkatesulu B, Joy Thomas M. Corona aging studies on silicone rubber nanocomposites. IEEE Transactions on Dielectrics and Electrical Insulation (TDEI). 2010;**17**(2):625-634

[5] Venkatesulu B, Joy Thomas M. Erosion resistance of alumina-filled silicone rubber nanocomposites. IEEE Transactions on Dielectrics and Electrical Insulation (TDEI). 2010;**17**(2):615-624

[6] Iyer G, Gorur RS, Krivda A. Corona resistance of epoxy nanocomposites: Experimental results and modeling. IEEE Transactions on Dielectrics and Electrical Insulation (TDEI). 2012;**19**(1): 118-125

[7] Du BX, Zhang JW, Liu Y. Effect of concentration on tracking failure of epoxy/TiO2 nanocomposites under dc voltage. IEEE Transactions on Dielectrics and Electrical Insulation (TDEI). 2012;**19**(5):1750-1759

[8] Tanaka T, Mulhaupt GCMR. Polymer nanocomposites as dielectrics and electrical insulation-perspectives for processing technologies, material characterization and future applications. IEEE Transactions on Dielectrics and Electrical Insulation (TDEI). 2004;**11**(5):763-784

[9] Kharitonov AP. Direct fluorination of polymers-from fundamental research to industrial applications. Progress in Organic Coating. 2008;**61**:192-204

[10] Tressaud A, Durand E, Labrugere C, et al. Modification of surface properties of Carbonbased and polymeric materials through fluorination routes: From fundamental research to industrial applications. Journal of Fluorine Chemistry. 2007;**128**:378-391

[11] Zhong JH. Research on Corona resistance of hybrid polyimide film. Engineering Plastic. 2005;**33**(9):13-15

[12] Junwei Z. Insulation characteristics of PI/ZnO corona-resistant hybrid film. Chinese Society for Electrical Engineering. 2009;**29**(34):122-127

[13] Hongyan L. Dielectric properties of polyimide/Al_2O_3 composite film. Chinese Society for Electrical Engineering. 2006;**26**(20):166-169

[14] Jiang Y, An Z, Liu C, et al. Influence of oxyfluorination time on space charge behavior in polyethylene. IEEE Transactions on Dielectrics and Electrical Insulation (TDEI). 2010;**17**(6):1814-1823

[15] Kharitonov AP, Kharitonova LN. Surface modification of polymers by direct fluorination: A convenient approach to improve commercial properties of polymeric articles. Pure and Applied Chemistry. 2009;**81**: 451-471

[16] Hosono T, Kato K, Morita A, et al. Surface charges on alumina in vacuum with varying surface roughness and

electric field distribution. IEEE Transactions on Dielectrics and Electrical Insulation (TDEI). 2007;**14** (3):627-633

[17] Das-Gupta DK. Electrical properties of surfaces of polymeric insulators. IEEE Transactions on Dielectrics and Electrical Insulation. 1992;**27**(5):909-923

[18] Fu M, Chen G, Dissado L, et al. The effect of gamma irradiation on space charge behaviour and dielectric spectroscopy of low-density polyethylene. In: International Conference on Solid Dielectrics; 2007. pp. 442-445

[19] Wang FP, Xia ZF, Qiu XL, et al. Piezoelectric properties and charge dynamics in poly(vinylidene fluoride-hexafluoropropylene) copolymer films with different content of HFP. IEEE Transactions on Dielectrics and Electrical Insulation (TDEI). 2006;**13** (5):1132-1139

Section 8

New Sustainable and Eco-Friendly Polyimides

Chapter 15

New High-Performance Materials: Bio-Based, Eco-Friendly Polyimides

Radu Dan Rusu and Marc J.M. Abadie

Abstract

The development of high-performance bio-based polyimides (PIs) seems a difficult task due to the incompatibility between petrochemical-derived, aromatic monomers and renewable, natural resources. Moreover, their production usually implies less eco-friendly experimental conditions, especially in terms of solvents and thermal conditions. In this chapter, we touch some of the most significant research endeavors that were devoted in the last decade to engineering naturally derived PI building blocks based on nontoxic, bio-renewable feedstocks. In most cases, the structural motifs of natural products are modified toward amine functionalities that are then used in classical or nonconventional methods for PI synthesis. We follow their evolution as viable alternatives to traditional starting compounds and prove they are able to generate eco-friendly PI materials that retain a combination of high-performance characteristics, or even bring some novel, enhanced features to the field. At the same time, serious progress has been made in the field of nonconventional synthetic and processing options for the development of PI-based materials. Greener experimental conditions such as ionic liquids, supercritical fluids, microwaves, and geothermal techniques represent feasible routes and reduce the negative environmental footprint of PIs' development. We also approach some insights regarding the sustainability, degradation, and recycling of PI-based materials.

Keywords: bio-based polyimides, renewable monomers, polyimide recycling, ionic liquids, supercritical fluids, microwave synthesis, geothermal synthesis

1. Introduction

Long-established, commercial or innovative, tailor-made, functional polyimides (PIs) represent a benchmark for high-performance polymers and demonstrate many advantages, which include excellent thermo-oxidative stability, chemical inertness, high mechanical resistance, high dielectric strength, as well as a remarkable combination of thermal, mechanical, and electrical insulating properties [1–4]. Therefore, these heterocyclic polymers have found applications in many advanced technology industries, such as aviation, spaceflight, microelectronics (printed and integrated circuit board, flexible chip carriers) [5], composite materials [6], automotive, packaging industries, or separation membranes [7]. Moreover, PIs' biocompatibility was heavily explored in the last two decades and polyimide-based materials entered the high-demanding area of biomedical applications, such as cell substrates, retina stimulation implants, cortical recordings, neural stimulation devices, and others [8].

This outstanding collection of traits mainly comes from their highly symmetrical backbones, extreme structural rigidity, and intra- and inter-chain hydrogen bonding. However, these features also imply some intrinsic weaknesses such as poor solubility in organic media, low thermal coefficient, poor corona resistance, relatively high thermal expansivity and, very important from the industrial point of view, difficult and expensive processability, which therefore restrict some of their potential applications [9–11].

PIs usually consist of linear, planar, (fully or partially) aromatic architectures that contain several flexible (bulky or pendant), heteroaliphatic structural elements which can provide additional features [12–14]. Completely aliphatic and cross-linked versions are also available.

Traditionally, linear PIs are prepared by a classical, two-step polycondensation method via a soluble poly(amic acid) (PAA) intermediate [15, 16].

The overall process displays a stepwise mechanism (a polyaddition alternative is also scarcely used in some labs) and runs at a rate dictated by the reactivity of the starting compounds. The formation of the PAA precursor is usually performed in common dipolar amidic solvents such as DMF (N, N-dimethylformamide), DMAc (N, N-dimethylacetamide), NMP (N-methyl-2-pyrrolidinone), TMU (1,1,3,3-tetramethylurea) which act as Lewis "bases," or in other nonamidic solvents such as DMSO (dimethyl sulfoxide) and m-cresol at low temperatures (from room temperature up to 80–100°C, depending on the basicity of the aromatic amines).

The cyclodehydration is conducted either by thermal or chemical imidization, both pathways being effective for either soluble or insoluble PIs. The former implies the stepwise heating of PAAs at various temperatures (usually from 100°C up to 250°C, depending on the polyimide's structure and the solvent used during synthesis) for various periods (mainly 1 h at each temperature). The latter involves the use of a chemical dehydrating agent of acidic (acetic anhydride, trifluoroacetic anhydride, and formic acid) or aminic (pyridine and trialkyl amines) nature to promote ring closure at temperatures between 20 and 80°C [17].

A one-step synthetic pathway is also used for the small-scale, in-house preparation of soluble polyimidic materials. This is based on the polycondensation of the monomers at high temperatures without the isolation of the PAA precursor [18].

Besides the widely employed reaction of (at least) a dianhydride with (at least) a diamine, an alternative that uses a dianhydride and diisocyanate is also known [19–21].

In the case of cross-linked PIs, the synthesis follows the same routes as above, with the usual additional presence of a diamine or triamine as a cross-linker in the PAA preparation stage [22].

Other functions may be also used in the development of cross-linked PIs, such as a diphenylethynylene structure along the main chain [23], or the use of a benzoxazine-containing monomer [24]. However, it must be noted that the three-dimensional, cross-linked structures cannot be recycled.

Given the experimental conditions employed in the synthesis of heterocyclic polymers in general, PIs included, it is difficult to envisage synthetic pathways or experimental conditions obeying the strict criteria of green chemistry, ecological synthesis, or eco-friendly polymers.

However, due to their specific macromolecular architectures, linearity, and rigid arrangement, PIs can be already considered as long-life, and therefore sustainable materials, that are however not produced by eco-friendly methods, especially in terms of solvents (some of them are not so green) and thermal conditions (which imply a high energy intake).

At this point, a special comment must be inscribed to the starting building blocks. The majority of dianhydride- and diamine-type monomers used in the synthesis of PIs are produced using petroleum-based chemicals as the raw materials. Furthermore, some of these monomers (especially some diamines like methylene dianiline, or the less used isocyanates) and their intermediates are highly toxic, carcinogenic, or endocrine-disrupting [25]. Therefore, considerable research endeavors have been devoted to achieving benign polyimide building blocks based on nontoxic, bio-renewable feedstocks. At the same time, long-term efforts were (and still are, as it will be detailed later on) dedicated to finding less toxic experimental conditions and synthetic pathways.

Considering all of the above, it seems difficult at first to come close to the Holy Grail of bio-based and/or green polyimide chemistry. However, in this chapter, we seek to demonstrate that there are some viable, strong alternatives to the traditional polyimide building blocks or classical experimental conditions. Moreover, these can unlock more eco-friendly, bio-based PI materials (even by using green chemistry routes) which maintain the same combination of high-performance characteristics and even bring some new or enhanced features to the field. We will mainly focus on bio-based polyimides developed in the last 10 years from natural, renewable resources and will also touch the basics of some less harmful synthetic platforms like synthesis in ionic liquids, supercritical fluids, microwave conditions, and other nonconventional methods. Some interesting insights regarding the sustainability, degradation, and recycling of PI-based materials are also provided.

2. Bio-based polyimides

A plethora of reliable studies and publications on bio-based, green polymers have been extensively produced in the last 10 years, some of the most solid reviews on the topic being cited herein [26–28].

Various natural, eco-friendly chemical structures have been produced and used after several modifications of their chemical function to access sustainable and eco-friendly polymers. These include vanillin [29] or bio-succinic acid [30] for semi-aromatic polyesters, lignins for phenolic oligomers [31], epoxidized castor oils for thermosetting architectures [32], catechol-based moieties for highly adhesive polymers [33], biodegradable copolymers from SO2, renewable eugenol for biodegradable materials [34] or cashew nut shell liquid for various systems [35].

The development of high-performance bio-based polymers such as PIs is currently indispensable to establish the sustainable and green objective of the research society. However, its accomplishment proves a difficult task due to the incompatibility of their aromatic monomers with renewable, natural resources. PIs were built for a long time from petrochemical-derived monomers, but some recent attempts to synthesize them from bio-based monomers are chancing this trend. Among eco-polymers, PIs' synthesis starting from natural and renewable resources is not to be outdone. As in the case of already well-developed eco-friendly polymers, the chemical functions present in natural products have been modified to obtain, in the majority of cases, amine functions. Several attempts have been made to prepare partially or completely biopolyimides using bio-based aromatic diamines and various dianhydrides.

After thoroughly sweeping the available studies, we were able to find a handful of biopolyimide systems (bioPIs) based on different green or eco-friendly diamines. Some of the most important ones are described below.

2.1 PIs derived from natural phenols and polyols

2.1.1 PIs derived from cardanol

Cardanol is a phenolic lipid obtained from an anacardic acid, the main component of cashew nutshell liquid, and the starting point in the development of many interesting organic building blocks [36].

A partially bio-based aromatic diamine, namely 1,1-bis(4-aminophenyl)-3-pentadecylcyclohexane, was synthesized starting from cardanol via a 3-pentadecylcyclohexanone intermediate in the presence of aniline and aniline hydrochloride under reflux conditions.

This bulky, unsymmetrical diamine was used in the development of processable PIs with reasonably high molar masses in the range of 15–32 kDa [36].

The obtained polyimides and copolyimides were soluble in common organic solvents and maintained high thermal stability, with a glass transition temperature (Tg) in the 160–250°C framework (depending on the dianhydride partner) and a 10% mass loss temperature (Td10) recorded above 500°C.

2.1.2 PIs derived from isomannide

- Another approach used biomass isomannide to develop both bio-based diamine and dianhydride monomers and optically transparent bioPIs therefrom [37]. The two hydroxylic groups of isomannide were used as the starting points of various chemical transformations to obtain one cycloaliphatic and two semi-aromatic diamines, together with a flexible dianhydride.

The diamines were used in combination with the bio-based isomannide anhydride or with the commercial 4,4-oxydiphthalic anhydride (ODPA) to produce completely or partially bio-based PIs through the classic two-step method. The ordered arrangement of the isomannide heterocycle conveyed a certain degree of crystallization in the PI framework and afforded materials only soluble in common, high boiling point solvents.

The relatively rigid alicyclic isomannide imparted good optical transparency (transmittances above 80% at 450 nm), reasonably high thermal resistance (Tg between 227 and 264°C, Td10 greater than 430°C in nitrogen) and outstanding mechanical features (tensile strength above 90 MPa, elongation at break over 6%). Surprisingly, the use of the fully alicyclic monomer generated PIs of superior thermal stability when compared to the ones containing the semi-aromatic bio-based diamines.

Two of the isomannide-based diamines were further used in both their isomeric forms in combination with other six commercial dianhydrides. As expected, similar results were obtained: soluble, processable bioPIs with thermo-mechanical stability comparable with those of analogous petroleum-based PIs. They also provide the additional features of optical transparency and optical activity which qualify them for applications like liquid crystal alignment and solar cells [38, 39].

2.1.3 PIs derived from myo-inositol (MI)

Myo-inositol or cyclohexanehexol is a naturally occurring cyclohexane decorated with six hydroxylic groups that is widely encountered in animals and plants. One of the many stereoisomers of inositol, it is the most commonly used biologically active isomer [40].

Until now, a large number of bioactive molecules bearing inositol-derived components were used as building blocks in developing macromolecular backbones containing diols, like polyurethanes [41] or polyamides [42]. It was a matter of time until it would have been used to prepare polyimidic architectures through its incorporation in diamines [15, 16, 28, 40].

For example, myo-inositol was treated with 1,1-dimethoxycyclohexane to yield a heterocyclic compound with three alicyclic units, the central one bearing two free hydroxyl groups [42]. These hydroxyl moieties are further used to introduce two aromatic amine functions through substitution and subsequent reduction reactions [40].

The target diamine readily underwent polycondensation with widely used, commercial dianhydrides in microwave conditions [40].

The resulting bioPIs, whose main chains inherited the bulky 6-5-6-5-6 polyalicyclic system from the diamine monomers, are soluble in common organic solvents at room temperature, even in chloroform, dichloromethane, and acetone. They displayed quite respectable molar masses, between 40 and 99 kDa, and relatively narrow polydispersities, with PDI values below 1.6. All bioPI films displayed characteristic thermal stability and outstanding mechanical features (tensile modulus above 3.5 GPa), even higher than those of most petroleum-based PIs. Their most important traits are the optical ones (light coloration and high transmittance values [for a PI], above 82%) and are also derived from the voluminous cycloaliphatic segment and its ability to hamper the charge-transfer complex usually found in PI materials.

2.2 PIs derived from anethole

Anethole is a natural styrene analog that is widely found in various in essential oils from plants and can be used in a peculiar preparation protocol (polymerization of Wagner-Jauregg type monomers containing preformed imide rings) of bio-based PIs [43].

Starting from this phenylpropene derivative, a new multicyclic monomer containing Wagner-Jauregg type imide motifs was synthesized via cascade, double Diels-Alder reaction. The polycondensation reaction of this diphenol-type monomer between and decafluorobiphenyl yielded a colorless and transparent (86% transmittance at 450 nm) poly(ether imide) film.

The partial bioPI showed reasonably high molar mass (72 kDa, 1.6 dispersity), sound thermal stability (5% weight loss [Td5] starting above 410°C, Tg higher than 360°C) and above reasonable mechanical characteristics (1.90 GPa tensile strength, 53.3 MPa elastic modulus, 5.4% elongation). These traits make them eligible for various optoelectronic applications in harsh conditions.

2.3 PIs derived from vanillin and its derivatives

Vanillin is one of the few commercially available, bio-based, aromatic compounds, and therefore, it is widely used in the polymer community dealing with bio-based macromolecular materials, PIs included [44, 45].

Vanillin was used as the starting point in the preparation of two dimers further used in two synthetic pathways for the development of bio-based aromatic diamines.

The first one involves three steps: phenol alkylation, oxidation of divanillin's aldehyde moieties, and subsequent reduction of the obtained oxime toward a methylated divanillylamine [46].

The second route uses the Curtius rearrangement and employs the synthesis of an acyl azide intermediate from divanillic acid, its transformation into a diisocyanate, and further hydrolysis toward another divanillin derivative decorated with two amino groups.

Another sequential synthetic scheme was used to prepare a diisocyanate starting from vanillic acid and its methylated dimer.

Therefore, these accessible synthetic transformations afford the preparation of a broad range of bio-based PI building blocks and corresponding polymers. One particularity of these procedures is that they unlock bio-based isocyanates which can be further used in combinations with dianhydrides to obtain bioPIs [45].

This series of partially bioPIs shows moderate molar masses (Mw from 49.5 to 75.8 kDa), characteristic polydispersity (2.0–2.5), acceptable solubility, and good thermal stability (Tg between 260 and 330°C, Td10 around 460°C).

The ability of vanillic acid to act as a versatile building block was explored even further together with another lignin-derivative, syringic acid, in the development of diacylhydrazides. These were then used in a two-step polycondensation procedure with aromatic anhydrides to generate partially bio-based poly(amide imide)s. The new materials provided transparent, flexible, and tough films with high thermal stability [47].

A simpler synthetic pathway can be also used to transform vanillin into an asymmetric diamine.

The aminic building block was then used to build partially bio-based PIs with four conventional, commercial dianhydrides and the results were compared with an analogous series based on a symmetric aromatic amine.

As expected, the use of the asymmetric structural moiety hampers the formation of charge-transfer complexes while vanillin's aromatic structure maintains backbone rigidity. This is translated into improved solubility, optical transparency, and hydrophobicity, without any negative impact upon thermal or mechanical stability [48].

The vanillin route was also pursued to develop a bio-based diamine containing aromatic, pyridine, and aliphatic structural moieties. This was used in combination with an alicyclic dianhydride, Epiclon, to obtain a semi-aromatic bioPI through the two-stage route.

The resulting flexible PI backbone afforded easily processable films with improved solubility and proper chemical and thermal stability (Td10 around 317°C). The PAA precursor was also employed to prepare a series of Ag nanocomposites by sonication. The casted films displayed a homogeneous dispersion of Ag nanoparticles within the bioPI matrix due to the high compatibility of the composites' elements. This resulted in Ag-induced, partial crystallinity, improved thermal properties (Td10 from 317°C up to 357°C, depending on the Ag amount), and antibacterial activity against *E. coli* [49].

2.4 PIs derived from camphor

Another bio-renewable, natural resource, camphor, a waxy, alicyclic forestry product was used in its (D) form to produce two diamine building blocks via an accessible synthetic route [50, 51]. The diamine was further condensed with two aromatic dianhydrides to prepare semi-alicyclic PIs with high solubility and optical transparency by the conventional two-step method.

The partially bio-based PIs were comparable to their fully aromatic counterparts (based on the combination of the similar rigid diamines with the same dianhydrides) in terms of solubility, optical features, thermal, and mechanical resilience. They provided transmittance values between 75 and 81% at 500 nm, Td10 from 390 to 519°C, tensile strength above 110 MPa, and elastic modulus in the range 2.4–3.24 GPa.

2.5 PIs derived from *Escherichia coli*

Bio-available aromatic diamines were derived from genetically manipulated *Escherichia coli*, through a photodimer of an aromatic amino acid, namely

4-aminocinnamic acid. This was produced by the fermentation and bioconversion of the genetically engineered microorganism from glucose via 4-aminophenylalanine [52].

The diamine displaying an alicyclic structure sandwiched between two aromatic rings was synthesized via a 4,4′-diamino-α-truxillic acid dihydrochloride, obtained through a [2+2] photocycloaddition of the microorganism-derived monoamine's hydrochloride salt.

This new building block was polycondensed afterward with different common dianhydrides following the two-stage procedure.

The bio-based PAAs displayed high molar masses, between 250 and 400 kDa, while the inadequate solubility of the thermally cyclized PIs precluded molecular mass determination. These bio-based PI films showed ultrahigh thermal resistance with Td10 values over 425°C and no Tg values below 350°C, which are the highest thermal features of all bio-based plastics reported thus far. They also showed high tensile strength and Young's moduli, excellent transparency, and high refractive indices, and adequate cell compatibility [53].

The transparent bioPI based on 1,2,3,4-cyclobutanetetracarboxylic dianhydride showed electrical insulative properties comparable to those of Kapton®, the most used PI dielectric. As in the case of the commercial PI, the volume resistivity proved to be directly connected to the annealing time and water uptake [54].

The precursor of this bioPI was used to prepare silica hybrids by sol-gel polycondensation with silicon alkoxide and in vacuo thermal annealing. The method generated transparent, thermo-mechanically robust films with excellent electrical stability [55].

The applicative potential of this bioPI was explored even further, by developing bionanohybrids through the sputtering of ITO nanolayers on the functionalized, reactive surface of a bioPI. The obtained materials displayed thermal, mechanical, electrical, optical, and adhesions performances comparable to or superior (especially in terms of optical transparency, ITO adherence, and device resistivity) to the extensively employed Kapton® PI film. This allowed the development of flexible, robust, and transparent electrodes for high tech electronic devices [56].

Other cycloaliphatic dianhydrides were also polymerized with the bio-based diamine coming from 4-aminocinnamic acid to draw a correlation between the PIs' solubility and the structural motifs belonging to the dianhydride.

The study concluded that a lower cycloaliphatic ring strain determines PI microstructures with improved flexibility and reduced Tg. All semi-aromatic bioPIs maintained relatively high molar masses (50–80 kDa) and displayed improved solubility and processability while preserving high thermal stability (Td10 temperatures above 375°C) [57, 58]. One of them was also used to develop highly transparent, flexible TiO2 and ZrO2 hybrid films that display the basic features of memory devices with tunable memory properties.

Ductile bioPI films were obtained from the same renewable semi-aromatic diamine by copolymerization with different binary mixtures of the aforementioned dianhydrides [59]. The copolymers displayed high-performance features comparable to those of the Kapton® film. For example, they maintained a high thermal resistance (Td10 values above 406°C and Tg over 208°C), improved tensile strength and elongation at break, and a Young's modulus around 4 GPa. Also, the copolymers showed adhesion strength to the carbon plate in the range of 0.22–4.47 MPa, which is similar to that of cyanoacrylate-based superglues [59].

The same bio-based, exotic building block, 4-aminocinnamic acid, was employed to develop two other diamines, 4, 4′-diaminostylbene and its hydrogenated version, by using Grubb's olefin metathesis as a key step.

The two diamines were further polymerized via two-stage polycondensation with the same dianhydrides as the photodimer ester coming from 4-aminocinnamic acid. High molecular masses PAAs were obtained both as films and fibrils and their thermal imidization afforded partially bio-based PIs with high thermal stability and mechanical properties superior to Kapton® [60].

2.6 PIs derived from peptides and amino acids

A monoamine-substituted version of the above-mentioned, microorganism-derived building tool was used to develop an aromatic amino acid, 4-amino-L-phenylalanine. This was used as a core to build a cyclic dipeptide, 3,6-di(4-aminophenylmethyl)-2,5-diketopiperazine through iterative protection/coupling/deprotection protocols [61].

Several bioPIs incorporating the diketopiperazine heterocyclic structure in the backbone were prepared by polycondensation of the peptide with commercially available aromatic dianhydrides.

Through a judicious design, the cyclic dipeptide monomer contains a centro-symmetric amide functionality in the diketopiperazine ring adjoined by two aromatic rings through a methylenic bridge. This type of architecture is able to induce auto-aggregation of the corresponding polymer chains through hydrogen bonding and π-π interactions. As a consequence, these bioPIs and their PAAs can self-assemble into nanosized conglomerates of various shapes (spheres, spiky balls, flakes, and rods), as observed by scanning electron microscopy (SEM) investigation. This particular feature (and its combination with the innate characteristics of any PI) enables these PIs suitable for applications such as fillers, heat-resistant superhydrophobic coatings, and ultralow-dielectric-constant films.

In addition, the peptide-based PIs display a high thermal resistance, particularly the one derived from pyromellitic dianhydride (PMDA), which showed the highest Td10 (around 432°C) and a Tg value above the decomposition temperature.

The literature provides few examples of other amino acids used as building blocks for bioPIs. For instance, isoleucine, valine, methionine, and phenylalanine were used together with benzimidazole pendant units to build four chiral diamines.

The diamines were polymerized with a particular dianhydride containing a pendant trifluoromethyl segment to attain optically active, aromatic bioPIs.

This resulted in amorphous, versatile PIs with improved solubility and molar masses above 124 kDa. The materials showed accessible Tg (142–165°C), suitable mechanical resilience (97 MPa ultimate strength), and a rather lower thermal stability (Td10 around 255°C) as compared to the PIs presented herein. They were further involved in sol-gel procedures to produce Ti bionanocomposites with improved UV absorption and enhanced gas permeability [62].

2.7 PIs derived from bio-based adenine

Adenine, as an amine-substituted purine, is one of the four building blocks of the DNA's double helix supramolecular structure and therefore an important naturally abundant structural framework for multiple hydrogen bonds [63].

An adenine-containing diamine was synthesized through the nucleophilic substitution of biomass adenine with 4-chloronitrobenzene and the subsequent reduction of the nitro groups. The adenine-containing diamine was then polycondensed in a single-step reaction with the widely used 4,4′-(hexafluoroisopropylidene) diphthalic anhydride (6FDA) to obtain a new bioPI.

The unique conjugated heterocyclic structure of adenine and its propensity toward hydrogen-bonding interactions generated a bioPI with excellent solubility and outstanding combined (thermal and mechanical) performance. The Tg value was higher than 350°C and the Td10 over 500°C (both in nitrogen and air). The tensile strength was up to 144 MPa, and the elastic modulus exceeded 4.1 GPa. A low dielectric constant of 2.8 (measured at 10 MHz) adds an important feature to this bio-based high-performance material. Similar results were obtained when the diamine was combined with another widely employed commercial dianhydride, ODPA [64].

The same adenine-based diamine was used together with PMDA to obtain a new bioPI and study its thermal expansion conduct along the in-plane direction. The PI films proved a rare in-plane thermal contraction feature which maintained its negative nature even above Tg. Infrared spectroscopy studies showed a mutation of the adenine-powered hydrogen bonding from a purine type to carbonyl-based interactions [65].

2.8 PIs derived from fatty diamines

Many of the PIs' characteristics are an outcome of the hard blocks coming from both the dianhydride and diamine segment. By replacing one of them with a softer block of natural origin, bio-based PIs with interesting traits are obtained. For example, a long, branched, cycloaliphatic, fatty diamine was employed as the soft block together with several, flexible or rigid, hard aromatic dianhydrides in the construction of some PIs, and the effect of the soft-hard combination was investigated [66].

The integration of flexible (or bulky) hard segments containing various spacers provided amorphous PIs, while the use of a highly rigid dianhydride segment yielded a semicrystalline polymer with separated nanophases after annealing.

The soft-hard combination resulted in highly soluble, processable, thermoplastic PIs of moderate molecular mass, with Tg values near room temperature, yet relatively thermal stability. They display superior processability as compared to most polyimides and are also available for recycling without significant changes in their outstanding mechanical features.

The four materials displayed high elongations (especially the one containing the highly rigid hard segment) and qualify as trustworthy candidates for applications that require high damping characteristics in ambient conditions (shock absorbers, noise, or vibration insulating materials, sealants).

Further studies revealed a self-healing capacity of these materials at room temperature. Rheological and tensile investigations proved the two-step kinetics and physical characteristic of the healing process and its dependency on the particular relaxation behavior of the PIs during stepwise healing. Mechanical integrity is maintained during healing due to the interplay of primary and secondary interactions between the soft and hard structural motifs [67, 68].

Another fatty diamine, the commercial Priamine™ 1074 coming from vegetable extracts (tall oil and soybean oil), was used as a starting material in the synthesis of partially bio-based fluorinated PIs.

The dimer containing fully renewable C36 enabled the preparation of highly soluble, thermally resistant PIs with a high content of biomass (up to 48.9%). During the two-stage polycondensation process, additional building blocks were used to ensure double-bond terminal groups which were further employed in the preparation of UV-cross-linked coatings. The obtained materials can satisfy some key requirements of microelectronics photoresists, like high optical transparency, adequate thermal stability, low water absorption, and suitable adhesion [69].

3. Green and nonconventional synthesis and processing of polyimides

As mentioned earlier, PIs are not usually produced by eco-friendly synthetic methods, especially in terms of solvents and thermal conditions. As detailed in the previous section, the production of some bioPIs starting from renewable building blocks involves the use of heavy-duty, rather toxic solvents (cresols in particular), with a negative environmental impact. This is a drawback which needs further research dedicated to greener experimental conditions.

On the other side, a large number of research studies were dedicated in the last two decades to the progress of nonconventional synthetic platforms or experimental conditions which can be used with some undeniable success in the development of PI-based materials. These come quite close to the strict criteria of green or sustainable chemistry based on ecological synthesis.

In this section, we will focus on some less harmful experimental conditions like synthesis (and processing) in ionic liquids, supercritical fluids, microwave conditions, and other nonconventional methods, as depicted by the scientific literature of the last 10 years.

3.1 Synthesis and processing of PIs in supercritical fluids

A supercritical fluid can be largely defined as a substance whose temperature and pressure are above its critical values. Above the critical temperature and pressure, a substance has the density of a liquid and the rheological properties of a gas. The characteristic density of a liquid allows the supercritical fluid to dissolve the substances, while the flowing features of the gas offer the advantage of lower reaction times [70].

This unique combination of physical traits offers several benefits in the use of supercritical fluids, both in macromolecular chemistry and materials science, but also in areas such as fine organic synthesis, catalysis, coordinative chemistry, and biochemistry.

Supercritical fluids and especially carbon dioxide (scCO2) have been successfully used in the last decade as solvents, antisolvents, or plasticizers in the synthesis and processing of PI-type materials. The synthetic procedure offers some chemical, ecological and economic conveniences. scCO2 is inert, nontoxic, nonflammable, relatively inexpensive, and, as an ambient gas, the solvent's removal after usage and depressurization is quite easy and hampers any ecological drawbacks. In some cases, the improved quality of the obtained products is an important factor in choosing supercritical liquids to the detriment of conventional organic solvents [70–78].

On the other hand, the use of supercritical fluids requires relatively high pressures and special equipment, and these considerations must be carefully balanced with the perceived advantages for a particular application.

The behavior of PI systems in the presence of scCO2 strongly depends on several parameters: the structural elements of the polymer, its physico-chemical characteristics (Tg, degree of crystallinity, cross-linking), the properties of pure scCO2 (molecular structure, critical points), the nature of the interactions between scCO2 and the polymer, and, obviously, the external temperature and pressure [72, 73].

Solubility is the crucial factor in the synthesis of polymers in scCO2. While scCO2 is a good solvent for low molecular mass, polar and nonpolar molecules, it is a very weak solvent for most high molecular mass polymers in mild conditions (below 100°C and 1000 bar). In some cases, a mixture of scCO2 and a common organic solvent provides satisfying results [74, 75].

The majority of research dealing with the synthesis of PIs in scCO2 focuses on the polycondensation of common, commercial monomers. 6FDA or similar fluorine-containing monomers are the preferred building blocks in this regard because most fluorinated polyimides seem to have a higher solubility in scCO2 [76–78].

Most studies report reactions with good yields and products with variable inherent viscosities, depending on the monomers' structure and reaction conditions. The molecular mass of the resulting PIs is influenced both by the concentration of monomers, by the reaction time and temperature, and by the scCO2 pressure. Some articles report a catalytic effect of scCO2, since the reaction rate increases in the presence of traces of water. Subsequently, the small water amounts can lead to the formation of cyclization products of lower molecular mass. At the same time, the employment of scCO2 in the synthesis of polymers leads to a small decrease in the Tg and dielectric constant values, due to a plasticizing effect. This can be an important advantage when it comes to the processability possibilities of the final material [79–82].

An increasing trend regarding the processing of polyimide materials using scCO2 was also observed in the last decade. The scCO2 dissolved in a condensed phase causes various changes in the microstructure of the polymer and a considerable reduction of its viscosity due to the increase in the free volume of the polymer matrix. This leads to severe modifications in the transport properties of the polymer, the permeability, and selectivity coefficients growing several times [83, 84]. As a consequence, scCO2 is largely used in the production of nanoparticles, foams, aerogels, and membranes based on linear [85–87] or cross-linked [88–92] PIs. These materials have interesting features (high porosity, tunable shrinkage, nano- and micro-cavities, high surface area, ultralow dielectric constant, and mechanical resilience) that open up new possibilities in the applicability of PI-based systems such as catalytic systems, separation membranes, aviation, aerospace, construction, and others [93–97].

For a rational design and optimization of these processes, it is essential to know the solubility, diffusibility, density, and permeability of the imidic macromolecular compound.

Nevertheless, despite the advantages and promising results attained so far, supercritical fluid technology has little chance to replace conventional methods for the time being. This is due to the limitations imposed by the solubility issue, the viscosity and molecular mass of the reaction products, as well as by the special equipment required by the supercritical conditions.

3.2 Synthesis and processing of PIs in ionic liquids

Ionic liquids had a cutting edge impact in the last two decades in polymer chemistry and materials science. They are widely used in both academia and industrial research and are considered a new class of ecologically compatible, green solvents, which can reduce or even replace the often dangerous, polluting, classic organic solvents.

The ionic liquids term generally implies different types of salts with a structure similar to ordinary salts like NaCl. However, while common salts melt at high temperatures, most ionic liquids remain liquid in a temperature range between room temperature and 200°C. They are usually formed by highly polar combinations of organic, voluminous, and asymmetric cations, decorated with aliphatic residues, and inorganic anions of symmetrical, regular shape, this chemical composition resulting in a low melting point. Poor packaging and poor coordination of ions in the structure of ionic liquids is the reason why they remain liquid at room temperature [98–100].

This peculiar structural composition unlocks a unique range of physicochemical features that make them environmentally compatible and suitable for various applications. They can dissolve different types of organic, inorganic, and organometallic materials, and their solvent properties can be modified to satisfy a particular application, by varying the combinations of anions and cations. They have high thermal conductivity, are liquid on a broad temperature range, and thermally stable up to 300°C, which allows the kinetic control of a reaction within wide limits. In addition, they possess extremely low vapor pressures (and hence do not evaporate) and are immiscible with many organic solvents [98, 101].

The benefits of ionic liquids were widely applied at first in the synthesis of classical PIs by the two-stage polycondensation reaction of common commercial diamines and dianhydrides [102]. Nevertheless, the solubilizing capacity of ionic liquids enables their usage in the polycondensation of more exotic building blocks with special features [103–107]. The method usually results in satisfying yields, without further addition of other catalysts. The obtained compounds show high molecular masses and inherent viscosities, enhanced thermal stability, and proper mechanical resistance.

Some studies also evidence an activating or catalytic effect of ionic liquids that boost the monomers' reactivity and unlock PIs with high molecular mass [102, 108].

The polycondensation of said monomers is usually performed in hydrophobic ionic liquids. Otherwise, the retained water will cause a decrease in the monomers' reactivity and hamper the formation of cyclization products with high molecular masses. The procedure generally involves easy work-up and solvent reusability and provides the potential for further scale-up attempts.

The innate immiscibility of ionic liquids with organic solvents was also explored in the green, interfacial polymerization of cross-linked PIs to obtain composite osmotic membranes [109].

In the early stages of their usage, ionic liquids were criticized for "sticking" to the PI material, even after various purification attempts. A small quantity of ionic liquid within the microstructure of a PI film or membrane would interfere with its thermo-mechanical features.

In the last decade, this issue was transformed into a benefit, through the advantageous combination of the versatility of both PIs and ionic liquids. The synthesis of these materials is usually performed by polycondensation of classical PI building blocks with ionic monomers (especially diamines) or via commercial PI modification. The resulting macromolecular species is called polyimidic ionenes or imidic poly(ionic liquid)s. The strategy uses ionic liquids' strong affinity to CO2 with the high-performance characteristics of PIs for the development of new materials for gas membrane separation [110–113].

3.3 Synthesis of PIs in microwave conditions

As it can be easily observed from the previous sections of this chapter, classical heating remains the primary means of stimulating chemical reactions that are difficult to carry out under ambient conditions, such as those used in PIs chemistry. In recent years, this technique is strongly rivaled by several modern heating techniques, the most important being microwave heating [114].

The main advantages of using microwaves in the synthesis of PIs are the rapid completion of polycondensation, high purity of final products, uniform temperature rise, and overall greener energy balance. Some small drawbacks refer to the less accessible upscale and the need to re-dissolve the final product to prepare PI films or membranes [115, 116].

The outstanding downsizing of reaction time is the strongest point of this experimental technique. Microwave-assisted polycondensation reactions proceed very rapidly, requiring only a few minutes of irradiation to obtain binary or ternary soluble polyimides with more than acceptable inherent viscosity (up to 1.2 dL/g) and molecular masses (up to 200 kDa) [117]. The irradiation time is a key parameter that can easily turn from a benefit to a drawback. Too long irradiation times or too high irradiation powers are many times translated into PIs of lower molecular mass, due to the partial degradation of the polymer [118].

The polycondensation proceeds through a one-stage mechanism, the imidization being performed employing microwave heating [119] or with the help of additional chemical initiators [120]. At the same time, microwaves can be used only to perform the cyclization of the PAA precursors obtained by conventional methods. This procedure is able to generate PI films with improved mechanical features as compared to their thermally cyclized counterparts [121].

The reaction can be easily extended to a wide range of building blocks (common monomers' salts included [122, 123]) and even to the direct production of various composites [114, 116, 124]. The viscosity value of the final products usually depends on the solubility of the final compound in the reaction medium. The initial amount of solvent or overall monomer concentration thus plays an important role in the characteristics of the final material.

The most used solvents in microwave-assisted PI synthesis are the polar ones, with a high dielectric constant and high boiling point like NMP, DMF, and alike. Under microwave irradiation, these solvents increase their temperature extremely fast and reach the boiling point in a short time. Although nonpolar solvents do not absorb microwaves, they can be still be used in combination with small amounts of polar solvent or salts.

Some reactions are solvent-free and only require pertinent microwave-absorbing monomers [125] or additives (CuO is an efficient example) [126]. The latter are used to quickly raise the reaction temperature to the melting of (at least one of) the monomers, thus empowering melt polycondensation. Such a strategy eliminates the usual, tedious washing process and the use of any (potentially) toxic organic reagents, further increasing the green character of the technique.

The one-stage polycondensation mechanism, combined with a small heating time and reduced (if any) contact with solvent usually translates in higher optical transparencies as compared to PI analogs obtained by the classical method [118].

Several studies compared the classical, thermal-, and microwave-assisted polycondensation procedures applied on the same common starting compounds in terms of order degree within polymeric backbones, imidization level, thermal and mechanical properties of resulting films. Microwaves assured a higher order level of the final products and proved to be more efficient and, of course, faster in reaching imidization [115, 122]. The imidization degree attained by microwaves was double the size of the one obtained by common thermal cyclization at 200°C and was almost complete at 250°C. Moreover, the PIs obtained by microwaves displayed higher thermal stability and superior mechanical resilience (even up to 30% higher) as compared to their thermally imidized counterparts [115, 117, 127].

Even if the method comes with some technological limitations regarding upscale and industrial use, the results obtained so far are quite impressive and, together with their obvious green character, require extensive research on the topic.

3.4 Other green and non-conventional synthetic procedures

A new, alternative method for PI synthesis appeared roughly 25 years ago, involving the use of the ultimate green solvent, water, in special temperature, and pressure conditions [128, 129]. The method is now generally known as hydrothermal polymerization (HTP) due to its resemblance to the geomimetic conditions of silicates' forming by condensation in the hydrothermal veins of the Earth's crust [129, 130].

HTP uses the same building blocks as the conventional PI synthesis: a dianhydride (usually, the original, more stable, tetracarboxylic acid is employed since the dianhydride will automatically hydrolyze in water) and a diamine. They form a nylon-type AB monomer salt intermediate which is subjected to polycondensation in water, at elevated temperatures and pressures accessible through a steel pressure vessel [131–133].

Since the monomer salt formation was undeniably proven by separate studies [128, 131, 132], one version of the method begins with the separate synthesis and purification (washing and filtration) of the monomer salt as the starting building block. Although not mandatory, this will provide the (almost) perfect stoichiometry required by any polycondensation reaction to ensure high yields.

Typical HTP experimental conditions for high monomer conversion are a 200°C temperature, a 16.7 bar pressure, and a reaction time between 6 and 24 h. The imidic product precipitates on cooling in aqueous residues of low toxicity. There are small amounts of residual solvent and no volatile impurities trapped in the final product.

At first glance, the reaction seems remarkable or even paradoxical, since it appears to contradict Le Chatelier's principle and the formation of PIs by polycondensation. A reversible reaction, with water as a side product, leading to products that cannot be (classically) obtained systematically unless firm absence or removal of water, feels inappropriate to be carried in water. Nevertheless, it works with noteworthy results. First, the salt formation increases hydrolytic stability, prevents the reverse reaction, and leads to increased imidization rates. Second, the aforementioned principle is still followed, the (theoretically) reversible polycondensation reaction being generally driven by the innate insolubility of the synthesized imide products.

The reaction follows a classical stepwise polymerization with a three-fold mechanism strongly depending on the temperature (TR) (reaction time and monomer concentration must be also considered) [132]. In the sub-hydrothermal regime, sHTP, the polymerization takes place in solution and leads to amorphous, short, low oligomeric PIS of a zwitterionic nature that coexists with unreacted, less soluble monomers. Between 100 and 130°C, longer macromolecular chains start to form in an ordered fashion, leading to semicrystalline PIs. When the reaction temperature comes to close to 200°C, the order degree and overall morphological homogeneity rise considerably and crystalline imidic products are generated. Solid-state polymerization (SSP) takes place in the high-temperature regime and is correlated to the polymerization temperature of any given monomer. If sHTP and SSP are mostly suppressed by judicious selection of starting compounds and reaction parameters, (almost) completely cyclized, highly crystalline PIs with particular morphological features are accessible by this method.

The potential usage of the method was recently broadened by the successful PI synthesis in various protic polar solvents (ethanol, isopropanol, and glycerin) or several aqueous mixtures therefrom. This unlocks the employment of new building blocks and fine tuning of reaction conditions toward desired PI morphologies [134].

Most HTP reactions performed so far used various aromatic dianhydrides/tetracarboxylic acids and commercial diamines and reproduced the structure of common or commercially available polyimides. The obtained molecular masses were in most cases significantly close to those of the PIs prepared by the classical method. Other basic properties (aspect, optical transparency, thermal, and mechanical features) are strongly dependent on the starting monomers and crystallinity degree of the polycondensation product.

Nevertheless, the method allows the synthesis of new PI structures [135] and composites [136], which can also be assisted by microwaves [137] or extended toward the production of PI fibers by green electrospinning or application-driven materials [133, 138].

Although only in its infancy and not easily accessible due to the special reaction setup, HTP delivers interesting results which require extension toward new building blocks and structures. Most importantly, the method is close to achieving all the strict criteria imposed by an ideal green chemical industry: high efficiency, economy, low (if any) toxicity, and benign environmental impact.

4. Recycling of polyimides

Polymer recycling is (or should be) a vital criterion to be taken into account in the life cycle of a product, especially when it comes to PIs. They represent key, reliable, long-life materials for a wide range of high-tech applications, but their recycling is a major challenge. Several technological and economic hurdles hamper the progress in this field.

On one side, recycling should enable the reintegration of discarded PI materials in the technological cycle or the production of new PI materials from recycled ones (especially since PIs represent materials of high-end value). On the other, it implies the quest for real sustainable solutions for the PI wastes issue, even if they represent a very small percent of the total polymer volume in use/to be recycled. Last but not least, it also means cutting down the dependence on oil-based raw materials, this matter being (at least partially) treated in the previous sections.

As in the case of most materials, mechanical and chemical recycling and combustion are the most common polymers' recycling methods, PIs included.

When it comes to the recycling or reuse of PI-based materials, two options are available, depending on whether you are dealing with a thermoplastic or thermoset. Important constraints are met in both cases.

Only a low percent (between 5 and 8%) of thermoplastic PIs is prone to recycling by reuse. There is no clear, general answer to the dilemma regarding the faith of the more than 92% left.

In the case of PI thermosets, the problem is even more complex, since the three-dimensional structure impedes easy changes in the physical state, especially to the molten state. Cross-linked PIs cannot be melted, molded, or extruded and therefore suffer from severe limitations in terms of recycling or require complicated or energy-intensive recycling methods. One easy example would be their reuse as additives in composites. If such, the problem of the matrix-reinforcement interface will arise from the homogeneity of the composite.

4.1 Degradation of polyimides

At this point, a separate discussion must be dedicated to the degradation of PI materials. Degradation (commonly known as aging) usually encompasses any modification of the initial polymeric material's features due to one or several external factors like light, heat, or chemicals. This is a major concern when it comes to the application of PI-based materials. Moreover, degradation is a key process when it comes to the reuse or recycling of a PI-based product to meet green environmental criteria. Its importance is also crucial when considering the final disposal or destruction/decomposition of the envisaged material and its environmental impact.

Assisted degradation of polymeric materials is an intricate procedure due to multifaceted mechanisms involving highly reactive macro- and micro-radicals and a plethora of degradation and decomposition products. These mechanisms are strongly dependent on the environment, more precisely on the presence or absence of oxygen. In the particular case of the highly stable PIs, there are also strongly time-dependent mechanisms [139–141].

The degradation of PI-based materials can be the subject of various pathways that will be briefly approached below.

4.1.1 Thermal (oxidative) degradation

As repeatedly mentioned in this chapter, PIs and PI-based materials display outstanding thermal stability, both in air or inert atmosphere. Degradation starts at very high temperatures, usually above 350°C for aromatic PIs, depending on the polymer's structure. Naturally, decomposition follows at much higher temperature regimes.

The thermal degradation of the widely used Kapton® film mainly generates CO2 from various sources (initial hydrolysis of the imidic units, followed by decarboxylation of the resulting acidic group), CO, and other volatiles from further transformations of imidic and aromatic moieties (a total of five degradation steps with variable activation energies) [141–144].

Several studies investigated the exact origin and nature of the volatiles resulting from the thermally induced degradation of common thermoplastic PIs: CO2, HCN, NH3, N2, H2O, CH4, and HCs. The presence of F- or S-rich structural motifs determines other, usually toxic volatiles. The decomposition process follows several pathways: depolymerization, pyrolytic reformation, successive homolytic and hydrolytic cleavages, hydrogen ablation, progressive molecular rearrangements, and loss of organic functionality through radical scission [140, 145, 146]. In every pathway, CO2 remains the main degradation product, while the nature of the other volatiles is dictated by the chemical structure of the polymer.

The resulting carbon-rich materials display improved compatibility between the organic (PI) and inorganic (carbon) parts which are mixed at the molecular level and enable several features that can be tuned by thermal treatment. Such materials can find various applications as fillers for gas purification membranes, dielectric composite films, coatings for electronic devices, and others.

Naturally, the degradation of cross-linked PIs has a more puzzling nature due to the three-dimensional architecture of the materials and requires different temperature and activation energy ranges [8]. Nevertheless, the pyrolysis of such thermosets can be used to form new reticulated structures. For example, a PI bearing cross-links of anhydride nature formed new reticulation points during thermal treatment at 430°C, which finally resulted in microporous membranes with potential gas transport properties [147].

In the presence of oxygen, the mechanisms of thermal polymer degradation become even more complex due to the formation of highly reactive peroxide macroradicals which enable a cascade of degradation reactions [148–150].

4.1.2 Hydrolytic degradation

Several experimental and computational studies were also dedicated to the chemical (especially hydrolytic) degradation or aging of common or less conventional PIs and model compounds [143, 151–153]. Two main mechanisms were evidenced: hydrolytic degradation and water-induced plasticization, both of them being thermally activated and severely time-dependent (long exposure times [month scale] are usually involved).

In its early stages, plasticization can be cautiously reversed with no significant impact upon thermal or mechanical features. Further on, water induces irreversible hydrothermal defects like blistering and/or delamination [153]. The process is stronger when high temperatures or pressures are involved [152, 154].

The mechanism is usually based on the hydrolysis-activated chain scission of PI macromolecules through the attack of water molecules on carbonyl groups or other humidity labile structural motifs [155, 156].

4.1.3 Radiation degradation

Several studies on this issue were performed due to the employment of Kapton® and other commercial PIs in aircraft and aerospace applications. Kapton® in particular is renowned for its outstanding resilience to ionizing irradiation, as compared to other widely used PIs.

The photodegradation of PIs results in macro- and micro-cracks coming from the scission of macromolecular backbones and low molecular mass volatiles such as CO and CO2 (and CHF3 in the case of 6FDA-containing PIs). The resulting oligomeric or polymeric chains display phenol, amine, isocyanate, and carboxylic acid terminal groups [157, 158]. Most UV irradiation experiments showed biexponential degradation kinetics [159].

In the case of beam-induced degradation processes, CO and CO2 are the major volatile fragments, with an additional significant contribution of short hydrocarbons like CxHy [160–162]. The degradation requires heavy energetic conditions and can be inhibited by the addition of various fillers [157]. Additionally, the process is accelerated by air exposure, which quickly determines a significant loss of radicals.

When the aging experiments mimicked low earth orbit, space-flight conditions (temperatures between 10 and 300 K, ultra-high vacuum [10–11 Torr], and high concentration of electronically excited atoms [108 atoms•cm$-^{3}$ of atomic O]), the degradation proceeded through chain scission and generated H2, CO, and CO2 [163, 164]. Similar experiments mimicking geostationary orbit conditions (involving high energy [90 keV] electrons) showed a board range of radiation-induced damage (breakage of chemical bonds and formation of new ones) and a strong effect on optical (lower transmittance) and charge transport (higher conductivity) features [165].

Most ion- or radiation-induced degradation studies have been performed at room temperature, and very little is known about degradation effects determined by irradiation experiments performed at extreme (very low [cryogenic] or high) temperatures [160].

4.1.4 Biodegradation

PIs are generally known to be biostable and bioinert, a very useful trait when it comes to biomedical applications (implantable devices, especially) [8]. However, the biodegradation of some PI-based materials would prove an advantageous feature in terms of recyclability and environmental impact. However, this requires specially designed building blocks and very strict control of the overall synthetic procedure.

One successful example in this regard is the two-stage synthesis of a biodegradable PI starting from an aliphatic amine derived from poly(propylene fumarate) and two commercial aromatic dianhydrides [166].

Both the PAA precursor and the cyclized PI proved biodegradable in a buffer solution, with a weight loss between 20 and 40% after 60 days of exposure, depending on the amount of amine used.

This represents a good starting point for a new research topic in the field of bioPIs. Similar studies are currently considering some of the partially or completely bio-based PIs described in the second section of this chapter.

4.2 Recycling of polyimides

Generally, two recycling technologies apply to PI materials: chemical and mechanical. The chemical recycling is a viable solution since it allows the maximum output of the final products (starting monomers, partially imidized powders) with a high degree of purity. Its main disadvantages are the long duration (multistage process) and energetic intensity.

In recent years, mechano-chemistry (e.g., ball milling) has proved its advantages over conventional chemical methods (fly ash modification, rubber, and plastic recycling), which lead to its intensive application in polymer recycling, PIs included. The process is simpler, economically favorable, and more eco-friendly (no solvents or intermediate fusion are used). Moreover, it allows attaining final products in a metastable state, which is not easily accessible by other conventional recycling methods.

Solid-state mecano-chemistry can be successfully applied in the recycling of PI film wastes with the purpose to develop thermostable blends and multicomponent tribocompositions. To gain applicative potential, these can be prepared in combination with other high-performance polymers, carbon black, diamond powder, and Al-Cu-Fe crystals [167].

The recycling treatment method depends on the physical form of the PI waste. PI-based powder compositions are obtained through low-energy planetary ball mill, PI film wastes are treated by high-energy planetary ball mill, while bulk samples are obtained by compression molding.

Multicomponent systems with a "multiscaled structure" based on PI waste and fluorinated ethylene propylene were obtained by the above methods. The resulting composites displayed an increased wear resistance and a smaller friction coefficient as compared to the raw propylenic material.

As mentioned before, a disadvantage of recycled PI usage in the development of composite materials is the lack of control of the interface and cohesion between composite elements. This results in a composite with unsatisfying integrity and poor mechanical features.

One solution to this issue is the use of polymer wastes of the same nature and proper tuning of the milling process. For example, common PMDA-ODA (pyromellitic dianhydride-4,4′-oxydianiline) PI films were grinded to powders and mixed with another commercial PI powder (BTDA [3,3′,4,4′-benzophenonetetracarboxylic dianhydride]-ODA) and a liquid, BTDA-based, commercial PI resin [168].

Several adjustments were performed to the high energy ball milling process, especially in terms of duration. The milling time strongly influences the average size and the particle-size distribution of the PI powder and the corresponding mechanical features of the resulting PI composites. Finally, bulk samples were obtained by compression molding at 400°C. The technique provided access to homogeneous materials with a flexural strength of 87 MPa, deformation at failure around 4.2%, and linear elastic modulus between 50and 350°C.

The milling duration also influences the chemical structure of the resulting recycled PI and its cohesion with the other two components of the composite material. Shorter milling times (45 min and 65 min) create larger particles that are strongly reinforced through links provided by the PI resin during compression molding. On the other side, longer milling times (90 min and 180 min) provide smaller particles that are more compatible with and reinforced by the other powdered component (BTDA-ODA commercial PI).

Another mechanical recycling method is the application of uniaxial stretching on PI films to control and reduce the coefficient of thermal expansion of the recycled material [169]. The resulting PI material is stretchable at a certain operating temperature regime and draw ratios. Increased stretching stress (performed in the machine's direction) leads to recycled films with higher birefringence and Young's modulus and lower thermal expansion, mostly independent of temperature.

A very useful recycling procedure was proposed for gelled PAAs, a common issue for researchers dealing with PI synthesis. PAAs frequently form gels during synthesis or storage due to various intermolecular and intramolecular interactions, the entanglement of long, high mass chains, partial imidization, or even undesired cross-linking. The formed PAA gels are usually discarded both in research and industry, which translates into time and economic loss and, more importantly, to negative environmental impact.

The recycling method was applied for a broad range of conventional PI building blocks and is based on the conversion of PAA gels to homogeneous solutions by using common microwave irradiation for a short time at room temperature. In some random cases, similar results can be obtained by less green conditions: heating the gelled PAA at 135°C for at least an hour [170].

The resulting PAA solutions can be successfully converted to the corresponding PIs. The resulting materials maintain the original film-forming features and show superior thermal (Tg values included), mechanical, and dielectric features as compared to the original PIs (obtained from the ordinary, homogeneous PAA solutions).

Recyclability can be also attained by finding new applications for already established, commercial PI materials, as to ensure a longer product lifetime. For example, the Kapton PI film was used to build an efficient, active particulate matter air filter which, very important, is also recyclable. Patterned through-holes were developed by ion etching on the common film (15 μm hole diameter; 30 μm interhole distance) [171]. These holes are combined with the strong electrostatic forces coming from Kapton®'s high work function to capture particulate matter. The device was tested under real working conditions and proved efficient in long-term filtration of dust particles with dimensions ranging from 0.3 up to 10 μm.

Moreover, the highly flexible, thermally stable filter is easily recyclable and reusable by simple washing (with tap water), which makes it suitable for various air filter-based applications (air purifiers, air conditioners, humidifiers, and industrial filtration systems).

4.3 Recyclable PIs by design

Another strategic approach in the field of recycling PI-based materials is to ensure recyclability through judicious design, a tactic inspired by applications based on cyclic operations.

One example in this direction is the development of highly hydrophilic, composite recyclable PI adsorbents for wastewater treatment and removal of heavy metal ions [172]. These can be easily attained by the common sol-gel process which leads to hybrid PI/silica materials with Cu2+ adsorption yields comparable to those of common activated carbon adsorbents. The innate resilience of PIs affords adsorbents with stable adsorption performance over 50 recycling processes.

A more complex concept for hybrid, recyclable PIs is the development of intricate, green, poly(imide imine) thermosets that combine the mechanical resilience of rigid PIs with the on-demand degradation and recyclability of flexible polyimines through dynamic covalent chemistry [173]. The greenness of the approach comes from the ability of imine units to provide heat- or water-triggered reversibility in the absence of expensive, environmentally unfriendly (transition metal) catalysts. The hybrid architecture is based on various ratios of an aromatic bisimide developed from widely used, commercial PI building blocks, a common aldehyde, and some aliphatic triamines or tetraamines as cross-linkers. The strategy affords thermosets with thermal and mechanical features comparable to common polyimides.

The key characteristic of these hybrid structures resides in the use of the primary amine as a catalyst to generate interchain imine exchange reactions at the interface of ruptured film strips. This feature leads to the conversation of mechanical properties and promotes a high healing output. The mechanical features of the (re) healed material are comparable or slightly superior to the original material up to the third generation.

The dynamic covalent chemistry concept unlocks novel pathways in the design of smart, high-performance PI-based materials able to lay out rehealability, repairability, and recyclability.

Another pathway to develop recyclable PIs is based on the introduction of flexible spacer units within and pendant to the macromolecular backbone. For example, the highly flexible 4,4′-diamino-3,3′-dimethyldiphenyl methane was polycondensed with BTDA to generate recyclable, PI-based, nanofiltration membranes. These are fabricated by phase inversion and retain a skinned asymmetric architecture [174].

The optimized membranes are efficient in rejecting Rose Bengal, PEG 1000, and typical salts with yields above 91%. They successfully corroborated their recyclability and display stable, longtime performance.

5. Conclusion

The development of high-performance, PI-based materials by following the rigid prerequisites of environmentally benign, green, or sustainable chemistry may still seem an arduous or eccentric task to the research society, especially to the industry-related part of it. The illusive incompatibility with the petrochemical-derived starting compounds and the usually harsh experimental conditions involved in their synthesis contribute to this stance. Nevertheless, judging by the firm advances made in the field in the last decade, the concept of bio-based or green PIs is not to be outdone.

Significant research endeavors have been devoted to engineer naturally derived building blocks based on nontoxic, bio-renewable feedstocks. These have evolved as viable and accessible alternatives to the traditional starting compounds and can be subjected to the classical experimental conditions of PI synthesis or even to greener ones. In most cases, the structural motifs of natural products are tailored to attain amine functionalities, while bio-based dianhydrides are still in their infancy.

Moreover, these building blocks can unlock eco-friendly, bio-based PI materials that maintain the same combination of high-performance characteristics and even bring some new or enhanced features to the field. There is a handful of high molar mass, soluble bioPIs with thermal stability, and mechanical features comparable with (and in a couple of cases even higher than) those of analogous petroleum-based PIs. They can be processed both as films and fibrils and provide industry-appealing features such as high optical transparency, optical activity, stable dielectric character, or tailorable hydrophilicity, without any negative impact upon thermal or mechanical resilience. Therefore, green polyimide chemistry exceeds the fading trend aura and already has the solid starting points of a long-term strategy. Based on the plethora of yet unexploited, renewable reagents available, the Holy Grail of high-performance, fully bio-based PIs is not unreachable.

At the same time, serious progress has been made in the field of nonconventional synthetic and processing platforms for the development of PI-based materials. Less harmful experimental conditions such as ionic liquids, supercritical fluids, microwaves, and geothermal-inspired setups represent workable options and further reduce the negative environmental footprint of PIs' production. While still undeveloped and with some technological limitations regarding upscale and industrial use, these methods deliver sound results and provide access to novel, impressive PI materials such as nanoparticles, foams, aerogels, and membranes.

Last, but not least, the status quo has slightly shifted from the constant investigation of the outstanding stability of PIs to the evaluation of several recycling strategies. The mechano-chemical pathway is the lead runner in the field, since it is economically favorable, more eco-friendly, and paves the way for new, interesting materials. Therefore, this represents the third valid point of departure for new research topics in the field of high-performance, bio-based PIs.

Author details

Radu Dan Rusu[1]* and Marc J.M. Abadie[1,2]

1 Petru Poni Institute of Macromolecular Chemistry, Romanian Academy, Iasi, Romania

2 Institute Charles Gerhardt Montpellier, University of Montpellier, Montpellier, France

*Address all correspondence to: radu.rusu@icmpp.ro

References

[1] Abadie MJM, editor. High Performance Polymers—Polyimides Based. From Chemistry to Applications. Rijeka: Intech Europe; 2012. p. 256. DOI: 10.5772/2834

[2] Bruma M, Damaceanu MD, Rusu RD. Study of thin films made from aromatic polymers containing six-member imide rings. High Performance Polymers. 2012;**24**:31-39. DOI: 10.1177/0954008311431115

[3] Damaceanu MD, Rusu RD, Musteata VE, Bruma M. Dielectric behavior of thin films made from poly (oxadiazole-naphthylimide)s. Soft Materials. 2011;**9**:44-63. DOI: 10.1080/1539445x.2010.525445

[4] Rusu RD, Damaceanu MD, Bruma M, Ronova IA. Effect of conformational parameters on thermal properties of some poly(oxadiazole-naphthylimide)s. Iranian Polymer Journal. 2011;**20**:29-40

[5] Shin HI, Chang JH. Transparent polyimide/organoclay nanocomposite films containing different diamine monomers. Polymers. 2020;**12**:135-153. DOI: 10.3390/polym12010135

[6] Nalwa HS, editor. Supramolecular Photosensitive and Electroactive Materials. San Diego: Academic Press; 2001. p. 970

[7] Escorihuela S, Tena A, Shishatskiy S, Escolá Stico S, Brinkmann T, Serra JM, et al. Gas separation properties of polyimide thin films on ceramic supports for high temperature applications. Membranes. 2018;**8**:16. DOI: 10.3390/membranes8010016

[8] Constantin CP, Aflori M, Damian RF, Rusu RD. Biocompatibility of polyimides: A mini-review. Materials. 2019;**12**:3166. DOI: 10.3390/ma12193166

[9] Rusu RD, Damaceanu MD, Bruma M. Aromatic copolyimides containing perylene units. Macromolecular Symposia. 2010;**296**:399-406. DOI: 10.1002/masy.201051055

[10] Damaceanu MD, Rusu RD, Bruma M. Six-member polyimides incorporating redox chromophores. Journal of Materials Science. 2012;**47**: 6179-6188. DOI: 10.1007/s10853-012-6542-8

[11] Rusu RD, Damaceanu MD, Bruma M. Comparative study of soluble poly(keto-naphthylimide)s. Revue Roumaine de Chimie. 2009;**54**: 1015-1022

[12] Damaceanu MD, Rusu RD, Bruma M, Rusanov AL. New thermally stable and organosoluble heterocyclic poly(naphthaleneimide)s. Polymers for Advanced Technologies. 2011;**22**: 420-429. DOI: 10.1002/pat.1519

[13] Damaceanu MD, Rusu RD, Bruma M. Copolyimides containing perylene and hexafluoroisopropylidene moieties. High Performance Polymers. 2012;**24**:50-57. DOI: 10.1177/0954008311431112

[14] Belomoina NM, Bulycheva EG, Rusanov AL, Rusu RD, Damaceanu MD, Bruma M. Polynaphthylimides based on isomeric 2,5-bis[(aminophenoxy) phenylen]-1,3,4-oxadiazoles. Polymer Science, Series B. 2011;**22**:420-429. DOI: 10.1134/S156009041009006X

[15] Sroog CE. Polyimides. Progress in Polymer Science. 1991;**16**:561-694. DOI: 10.1016/0079-6700(91)90010-I

[16] Rusu RD, Damaceanu MD, Bruma M. Copoly(1,3,4-oxadiazole-naphthylimide)s containing siloxane units in the main chain: Synthesis and properties. High Performance Polymers.

2011;**23**:384-393. DOI: 10.1177/0954008311409784

[17] Kricheldorf HR, editor. Progress in Polyimide Chemistry I. Berlin Heidelberg: Springer-Verlag; 1999. DOI: 10.1007/3-540-49815-X

[18] Damaceanu MD, Rusu RD, Musteata VE, Bruma M. Insulating polyimide films containing n-type perylenediimide moieties. Polymer International. 2012;**61**:1582-1591. DOI: 10.1002/pi.4253

[19] Barikani M. Polyimides derived from diisocyanates. Iranian Polymer Journal. 2002;**11**:215-236

[20] Yu F, Wang K, Liu XY, Zhan MS. Preparation and properties of rigid polyimide foams derived from dianhydride and isocyanate. Journal of Applied Polymer Science. 2013;**127**: 5075-5081. DOI: 10.1002/app.38121

[21] Paraskevopoulou P, Chriti D, Raptopoulos G, Anyfantis GC. Synthetic polymer aerogels in particulate form. Materials. 2019;**12**:1543. DOI: 10.3390/ma12091543

[22] Deligoz H, Ozgumus S, Yalcınyuva T, Yıldırım S, Deger D, Ulutas K. A novel cross-linked polyimide film: Synthesis and dielectric properties. Polymer. 2005;**46**: 3720-3729. DOI: 10.1016/j.polymer.2005.02.097

[23] Ando S, Harada M, Okada T, Ishige R. Effective reduction of volumetric thermal expansion of aromatic polyimide films by incorporating interchain crosslinking. Polymers. 2018;**10**:761

[24] Zhang K, Liu J, Ishida H. High performance crosslinked polyimide based on main-chain type polybenzoxazine. RSC Advances. 2014; **4**:62550-62556. DOI: 10.1039/c4ra12015a

[25] Pohanish RP. Sitting's Handbook of Toxic and Hazardous Chemicals and Carcinogens. 6th ed. Norwich: William Andrew Publishing; 2012. pp. 873-875

[26] Zhu Y, Romain C, Williams CK. Sustainable polymers from renewable resources. Nature. 2016;**540**:354-362. DOI: 10.1038/nature21001

[27] Schneiderman DK, Hillmyer MA. 50th anniversary perspective: There is a great future in sustainable polymers. Macromolecules. 2017;**50**:3733-3749. DOI: 10.1021/acs.macromol.7b00293

[28] Nguyen HTH, Qi PX, Rostagno M, Feteha A, Miller SA. The quest for high glass transition temperature bioplastics. Journal of Materials Chemistry A. 2018; **6**:9298-9331. DOI: 10.1039/c8ta00377g

[29] Llevot A, Grau E, Carlotti S, Grelier S, Cramail H. Renewable (semi) aromatic polyesters from symmetrical vanillin-based dimers. Polymer Chemistry. 2015;**6**:6058-6066. DOI: 10.1039/c5py00824g

[30] Short GN, Nguyen HTH, Scheurle PI, Miller SA. Aromatic polyesters from biosuccinic acid. Polymer Chemistry. 2018;**9**:4113-4119. DOI: 10.1039/c8py00862k

[31] Llevot A, Grau E, Carlotti S, Grelier S, Cramail H. Selective laccase-catalyzed dimerization of phenolic compounds derived from lignin: Towards originalsymmetrical bio-based (bis) aromatic monomers. Journal of Molecular Catalysis B: Enzymatic. 2016; **125**:34-41. DOI: 10.1016/j.molcatb.2015.12.006

[32] Noe C, Malburet S, Bouvet-Marchand A, Graillot A, Sangermano M. Cationic photopolymerization of bio-renewable epoxidized monomers. Progress in Organic Coating. 2019;**133**: 131-138. DOI: 10.1016/j.porgcoat.2019.03.054

[33] Faure E, Falentin-Daudré C, Jerome C, Lyskawa J, Fournier D, Woisel P, et al. Catechols as versatile platforms in polymer chemistry. Progress in Polymer Science. 2014;**38**: 236-270. DOI: 10.1016/j.progpolymsci.2012.06.004

[34] Liu LM, Ni YH, Zhi YF, Zhao WB, Pudukudy M, Jia QM, et al. Sustainable and biodegradable copolymers from SO2 and renewable eugenol: A novel urea fertilizer coating material with superior slow release performance. Macromolecules. 2020;**53**:936-945. DOI: 10.1021/acs.macromol.9b02202

[35] Chatterjee D, Sadavarte NV, Shingte RD, More AS, Tawade BV, Kulkarni AD, et al. Step-growth polymers from cashew nut shell liquid (CNSL)-based aromatic difunctional monomers. In: Anilkumar P, editor. Cashew Nut Shell Liquid: A Goldfield for Functional Materials. 1st ed. New York: Springer; 2017. pp. 163-214. DOI: 10.1007/978-3-319-47455-7_9

[36] Shingte RD, Tawade BV, Wadgaonkar PP. Partially bio-based processable polyimides based on aromatic diamine derived from cardanol. Green Materials. 2017;**5**:74-84. DOI: 10.1680/jgrma.17.00010

[37] Yang G, Zhang R, Huang H, Liu L, Wang L, Chen Y. Synthesis of novel biobased polyimides derived from isomannide with good optical transparency, solubility and thermal stability. RSC Advances. 2015;**5**: 67574-67582. DOI: 10.1039/C5RA14526K

[38] Ji XD, Yan JL, Liu XJ, Wang ZK, Wang Z. Synthesis and properties of polyimides derived from bis(4-aminophenyl)isohexides. High Performance Polymers. 2017;**29**: 197-204. DOI: 10.1177/0954008316634443

[39] Ji XD, Wang ZK, Yan JL, Wang Z. Partially bio-based polyimides from isohexide-derived diamines. Polymer. 2015;**74**:38-45. DOI: 10.1016/j.polymer.2015.07.051

[40] Zhang R, Li T, Zhou H, Huang H, Chen Y. Biobased transparent polyimides with excellent solubility and mechanical properties using myo-inositol derived diamines. Reactive and Functional Polymers. 2018;**128**:91-96. DOI: 10.1016/j.reactfunctpolym.2018.05.006

[41] Okamoto S, Onou S, Kobayashi M, Sudo A. Rigid triol and diol with adamantane-like core derived from naturally occurring myo-inositol and their polyaddition with diisocyanates. Journal of Polymer Science, Part A: Polymer Chemistry. 2014;**52**:3498-3505. DOI: 10.1002/pola.27414

[42] Sudo A, Sugita S. A highly rigid diamine monomer derived from naturally occurring myo-inositol and its use for polyamide synthesis. Journal of Polymer Science, Part A: Polymer Chemistry. 2016;**54**:3436-3443. DOI: 10.1002/pola.28231

[43] Wang C, Sun J, Tao Y, Fang L, Zhou J, Dai M, et al. Biomass materials derived from anethole: Conversion and application. Polymer Chemistry. 2020; **11**:954-963. DOI: 10.1039/c9py01513b

[44] Kuhire S, Ichake A, Grau E, Cramail H, Wadgaonkar P. Synthesis and characterization of partially bio-based polyimides based on biphenylene-containing diisocyanate derived from vanillic acid. European Polymer Journal. 2018;**109**:257-264. DOI: 10.1016/j.eurpolymj.2018.09.054

[45] Fache M, Boutevin B, Caillol S. Vanillin production from lignin and its use as a renewable chemical. ACS Sustainable Chemistry & Engineering. 2016;**4**:35-46. DOI: 10.1021/acssuschemeng.5b01344

[46] Savonnet E, Le Croz C, Grau E, Grelier S, Defoort B, Cramail H.

Divanillin-based aromatic amines: Synthesis and use as curing agents for fully vanillin-based epoxy thermosets. Frontiers in Chemistry. 2019;7:606. DOI: 10.3389/fchem.2019.00606

[47] Kuhire SS, Sharma P, Chakrabarty S, Wandgaonkar PP. Partially bio-based poly(amide imide)s by polycondensation of aromatic diacylhydrazides based on lignin-derived phenolic acids and aromatic dianhydrides: Synthesis, characterization, and computational studies. Journal of Polymer Science, Part A: Polymer Chemistry. 2017;**55**: 3636-3645. DOI: 10.1002/pola.28748

[48] Wang ZK, Li YJ, Zhu TS, Xiong L, Liu F, Qi HX. Conversion of renewable vanillin into high performance polyimides via an asymmetric aromatic diamine derivation. Polymer Degradation and Stability. 2019;**167**:67-76. DOI: 10.1016/j.polymdegradstab. 2019.06.002

[49] Mirzakhanian Z, Faghihi K, Geravi HA, Mahdies M. Semi-aromatic polyimide/Ag nanocomposite derived from vanillin. Journal of Applied Polymer Science. 2016;**133**:44001. DOI: 10.1002/app.44001

[50] Wu Y, Tian Z, Liu Z, Xiong Z, Liu F. Semi-alicyclic polyimides from bio-renewable natural-(D)-camphor with high solubility and optical transparency. In: Abadie MJM, editor. Proceedings of the 11th Polyimides & High Performance Polymers Conference, STEPI 11; 2–5 June 2019. Montpellier: U. Montpellier; 2020. pp. 26-42

[51] Zhang HS, Li J, Tian ZL, Liu F. Synthesis and properties of novel alicyclic-functionalized polyimides prepared from natural-(D)-camphor. Journal of Applied Polymer Science. 2013;**129**:3333-3340. DOI: 10.1002/app.39053

[52] Minakawa H, Masuo S, Takaya N. Fermentation and purification of microbial monomer 4-amminocinnamic acid to produce ultra-high performance bioplastics. Process Biochemistry. 2019; **77**:100-105. DOI: 10.1016/j.procbio.2018.11.021

[53] Suvannasara P, Tateyama S, Miyasato A, Matsumura K, Shimoda T, Ito T, et al. Bio-based polyimides from 4-aminocinnamic acid photodimer. Macromolecules. 2014;**47**:1586-1593. DOI: 10.1021/ma402499m

[54] Kato S, Yusof FAA, Harimoto T, Takada K, Kaneko T, Kawai M, et al. Electric volume resistivity for biopolyimide using 4,4′-diamino-α-truxillic acid and 1,2,3,4-cyclobutanetetracarboxylic dianhydride. Polymers. 2019;**11**:1552. DOI: 10.3390/polym11101552

[55] Dwivedi S, Sakamoto S, Kato S, Mitsumata T, Kaneko T. Effects of biopolyimide molecular design on their silica hybrids thermo-mechanical, optical and electrical properties. RSC Advances. 2018;**8**:14009-14016. DOI: 10.1039/c8ra01965g

[56] Dwivedi S, Kaneko T. Robustification of ITO nanolayer by surface-functionalization of transparent biopolyimide substrates. Journal of Applied Polymer Science. 2018;**135**: 46709. DOI: 10.1002/app.46709

[57] Dwivedi S, Kaneko T. Molecular design of soluble biopolyimide with high rigidity. Polymers. 2018;**10**:368. DOI: 10.3390/polym10040368

[58] Huang TT, Tsai CL, Tateyama S, Kaneko T, Liou GS. Highly transparent and flexible bio-based polyimide.TiO2 and ZrO2 hybrid films with tunable refractive index, Abbe number, and memory properties. Nanoscale. 2016;**8**: 12793-12802. DOI: 10.1039/c6nr03963d

[59] Shin H, Wang S, Tateyama S, Kaneko D, Kaneko T. Preparation of a ductile biopolyimide film by

copolymerization. Industrial and Engineering Chemistry Research. 2016; **55**:8761-8766. DOI: 10.1021/acs.iecr.6b02221

[60] Kumar A, Tateyama S, Yasaki K, Ali MA, Takaya N, Singh R, et al. Ultrahigh performance bio-based polyimides from 4,4′-diaminostilbene. Polymer. 2016;**83**:182-189. DOI: 10.1016/j.polymer.2015.12.008

[61] Hirayama T, Kumar A, Takada K, Kaneko T. Morphology-controlled self-assembly and synthesis of biopolyimide particles from 4-amino-l-phenylalanine. ACS Omega. 2020;**5**:2187-2195. DOI: 10.1021/acsomega.9b03231

[62] Ahmadizadegan H, Ranjbar M. Synthesis and characterization of green membranes polyimide/titania bionanocomposites containing amino acid and benzimidazole moieties for gas transport properties. Polymer Engineering and Science. 2018;**58**: 1666-1677. DOI: 10.1002/pen.24757

[63] Hu J, Wang Z, Lu Z, Chen C, Shi M, Wang J, et al. Bio-based adenine-containing high performance polyimide. Polymer. 2017;**119**:59-65. DOI: 10.1016/j.polymer.2017.05.012

[64] Li RK, Lu Z, Liu Y, Zeng K, Hu JH, Yang G. The retarding effects and structural evolution of a bio-based high-performance polyimide during thermal imidization. Journal of Applied Polymer Science. 2019;**136**:46953. DOI: 10.1002/app.46953

[65] Hu JH, Li RK, Chen C, Lu Z, Zeng K, Yang G. New insights into mechanism of negative in-plane CTE based on bio-based adenine-containing polyimide film. Polymer. 2018;**146**:133-141. DOI: 10.1016/j.polymer.2018.05.001

[66] Susa A, Bijleveld J, Hernandez M, Garcia SJ. Understanding the effect of the dianhydride structure on the properties of semiaromatic polyimides containing a bio-based fatty diamine. ACS Sustainable Chemistry & Engineering. 2018;**6**:668-678. DOI: 10.1021/acssuschemeng.7b03026

[67] Susa A, Mordvinkin A, Saalwachter K, van der Zwaag S, Garcia SJ. Identifying the role of primary and secondary interactions on the mechanical properties and healing of densely branched polyimides. Macromolecules. 2018;**51**:8333-8345. DOI: 10.1021/acs.macromol.8b01396

[68] Susa A, Bose RK, Grande AM, van der Zwaag S, Garcia SJ. Effect of the dianhydride/branched diamine ratio on the architecture and room temperature healing behavior of polyetherimides. ACS Applied Materials & Interfaces. 2016;**8**:34068-34079. DOI: 10.1021/acsami.6b10433

[69] Liu JC, Wang K, Xie YZ, Gao F, Zeng QT, Yuan Y, et al. Novel partially bio-based fluorinated polyimides from dimer fatty diamine for UV-cured coating. Journal of Coating Technology and Research. 2017;**14**:1325-1334. DOI: 10.1007/s11998-017-9931-8

[70] Kiran E. Supercritical fluids and polymers. The year in review: 2014. Journal of Supercritical Fluids. 2016;**110**: 126-153. DOI: 10.1016/j.supflu.2015.11.011

[71] Kratochvil AM, Koros WJ. Effects of supercritical CO2 conditioning on cross-linked polyimide membranes. Macromolecules. 2010;**43**:4679-4687. DOI: 10.1021/ma100535h

[72] Haruki M, Hasegawa Y, Fukui N, Kihara S, Takishima S. Deposition of aromatic polyimide thin films in supercritical carbon dioxide. Journal of Supercritical Fluids. 2014;**94**:147-153. DOI: 10.1016/j.supflu.2014.07.010

[73] Ishizaka T, Ishigaki A, Suzuki TM, Chatterjee M, Kawanami H. Green synthetic method and simple

size-control of polyimide nanoparticles in scCO2. Chemistry Letters. 2011;**40**: 849-851. DOI: 10.1246/cl.2011.849

[74] Haruki M, Fukui N, Kihara S, Takishima S. Measurement and prediction of the solubilities of aromatic polyimide monomers in supercritical carbon dioxide with acetone. The Journal of Chemical Thermodynamics. 2012;**54**:261-265. DOI: 10.1016/j.jct.2012.04.013

[75] Haruki M, Fukui N, Kobayashi F, Kihara S, Takishima S. Entrainer effect of N,N-dimethylformamide on solubility of monomers for aromatic polyimide in supercritical carbon dioxide. Industrial and Engineering Chemistry Research. 2011;**50**: 11942-11949. DOI: 10.1021/ie2013289

[76] Haruki M, Wasada A, Hosokama Y, Kihara S, Takishima S. Deposition of PMDA-TFDB type polyimide onto microscale trenches patterned on silicon wafer using supercritical carbon dioxide. Journal of Supercritical Fluids. 2017;**127**:200-207. DOI: 10.1016/j.supflu.2017.03.008

[77] Haruki M, Oda A, Wasada A, Kihara S, Takishima S. Deposition of fluorinated polyimide consisting of 6FDA and TFDB into microscale trenches using supercritical carbon dioxide. Journal of Supercritical Fluids. 2017;**119**:238-244. DOI: 10.1016/j.supflu.2016.10.001

[78] Haruki M, Oda A, Wasada A, Hasegawa Y, Kihara S, Takishima S. Phase appearance during polymerization of fluorinated polyimide monomers and deposition into the microscopic-scale trenches in supercritical carbon dioxide. Journal of Applied Polymer Science. 2016;**133**: 43334. DOI: 10.1002/APP.43334

[79] Ronova IA, Sinitsyna OV, Abramchuck SS, Nikolaev AY, Chisca S, Sava I, et al. Study of porous structure of polyimide films resulting by using various methods. Journal of Supercritical Fluids. 2012;**70**:146-155. DOI: 10.1016/j.supflu.2012.06.018

[80] Ronova IA, Bruma M, Kuznetsov AA, Nikolaev AY. Lowering the dielectric constant of polyimide thin films by swelling with supercritical carbon dioxide. Polymers for Advanced Technologies. 2013;**24**:615-622. DOI: 10.1002/pat.3123

[81] Li L, Chang KC, Lin XN, Zhang R, Lou JH. Insulating property improvement of polyimide in devices by low-temperature supercritical fluids. Advanced Electronic Materials. 2019;**5**: 1900580. DOI: 10.1002/aelm.201900580

[82] Haruki M, Hasegawa Y, Fukui N, Kihara S, Takishima S. Production of polyamic acid in supercritical carbon dioxide with N, N-dimethylformamide. Journal of Applied Polymer Science. 2014;**131**:39878. DOI: 10.1002/app.39878

[83] Ronova IA, Alentiev AY, Chisca S, Sava I, Bruma M, Nikolaev AY, et al. Change of microstructure of polyimide thin films under the action of supercritical carbon dioxide and its influence on the transport properties. Structural Chemistry. 2014;**25**:301-310. DOI: 10.1007/s11224-013-0282-5

[84] Scholes CA, Kanehashi S. Polymeric membrane gas separation performance improvements through supercritical CO2 treatment. Journal of Membrane Science. 2018;**566**:239-248. DOI: 10.1016/j.memsci.2018.09.014

[85] Lin HW, Tan CS. Preparation of polyamic acid and polyimide nanoparticles by compressed fluid antisolvent and thermal imidization. Journal of Supercritical Fluids. 2015;**99**: 103-111. DOI: 10.1016/j.supflu.2015.01.028

[86] Zhai WT, Feng WW, Ling JQ, Zheng WG. Fabrication of lightweight microcellular polyimide foams with three-dimensional shape by CO2 foaming and compression molding. Industrial and Engineering Chemistry Research. 2012;**51**:12827-12834. DOI: 10.1021/ie3017658

[87] Wu S, Du A, Huang SM, Sun W, Xiang YL, Zhou B. Solution-processable polyimide aerogels with high hydrophobicity. Materials Letters. 2016; **176**:118-121. DOI: 10.1016/j. matlet.2016.04.099

[88] Kuiper S, Embrechts A, Every HA, de Vries T, de Smet LCPM. Matrimid aerogels by temperature-controlled, solution-based crosslinking. Macromolecular Materials and Engineering. 2013;**298**:868-875. DOI: 10.1002/mame.201200335

[89] Fei ZF, Yang ZC, Chen GB, Li KF. Preparation of tetraethoxysilane-based silica aerogels with polyimide cross-linking from 3, 3′, 4, 4′-biphenyltetracarboxylic dianhydride and 4, 4′-oxydianiline. Journal of Sol-Gel Science and Technology. 2018;**3**: 506-513. DOI: 10.1007/s10971-017-4566-x

[90] Kim M, Eo K, Lim HJ, Kwon YK. Low shrinkage, mechanically strong polyimide hybrid aerogels containing hollow mesoporous silica nanospheres. Composites Science and Technology. 2018;**165**:355-361. DOI: 10.1016/j. compscitech.2018.07.021

[91] Zhang TY, Zhao Y, Wang K. Polyimide aerogels cross-linked with aminated Ag nanowires: Mechanically strong and tough. Polymers. 2017;**9**:530. DOI: 10.3390/polym9100530

[92] Teo N, Gu ZP, Jana SC. Polyimide-based aerogel foams via emulsion-templating. Polymer. 2018;**157**:95-102. DOI: 10.1016/j.polymer.2018.10.030

[93] Zhai CH, Jana SC. Tuning porous networks in polyimide aerogels for airborne nanoparticle filtration. ACS Applied Materials & Interfaces. 2017;**9**: 34068-34079. DOI: 10.1021/acsami .7b09345

[94] Fanf GQ, Li H, Liu JG, Ni HJ, Yang HX, Yang SY. Intrinsically atomic-oxygen-resistant POSS-containing polyimide aerogels: Synthesis and characterization. Chemistry Letters. 2015;**44**:1083-1085. DOI: 10.1246/ cl.150396

[95] Shen DX, Liu FG, Yang HX, Yang SY. Intrinsically highly hydrophobic semi-alicyclic fluorinated polyimide aerogel with ultralow dielectric constants. Chemistry Letters. 2013;**42**:1230-1232. DOI: 10.1246/cl.130623

[96] Zhang TY, Zhao Y, Muhetaer M, Wang K. Silver nanoparticles cross-linked polyimide aerogels with improved high temperature microstructure stabilities and high mechanical performances. Microporous and Mesoporous Materials. 2020;**297**: 110035. DOI: 10.1016/j.micromeso. 2020.110035

[97] Feng JZ, Wang X, Jiang YG, Du DX, Feng J. Study on thermal conductivities of aromatic polyimide aerogels. ACS Applied Materials & Interfaces. 2016;**8**: 12992-12996. DOI: 10.1021/acsami. 6b02183

[98] Singh SK, Savoy AW. Ionic liquids synthesis and applications. An overview. Journal of Molecular Liquids. 2020;**297**:112038. DOI: 10.1016/j. molliq.2019.112038

[99] Villa R, Alvarez E, Porcar R, Garcia-Verdugo E, Luis SV, Lozano P. Ionic liquids as an enabling tool to integrate reaction and separation processes. Green Chemistry. 2019;**21**:6527-6544. DOI: 10.1039/c9gc02553g

[100] Sinha AK, Singh R, Kumar R. Towards an understanding of the

ambiphilic character of ionic liquids for green synthesis of chemically diverse architectures. Asian Journal of Organic Chemistry. 2020;**9**:706-720. DOI: 10.1002/ajoc.202000065

[101] Xing DY, Chan SY, Sung TS. Fabrication of porous and interconnected PBI/P84 ultrafiltration membranes using [EMIM]OAc as the green solvent. Chemical Engineering Science. 2013;**87**:194-203. DOI: 10.1016/j.ces.2012.10.004

[102] Chen M, Wang SL. Non-thermal polyimidization reaction using base-ionic liquid medium as a dual catalyst-solvent. RSC Advances. 2016; **6**:96914-96917. DOI: 10.1039/c6ra23162d

[103] Mallakpour S, Zadehnazari A. Tailored synthesis of nanostructured polymer thin films from optically active and thermally stable poly(amide-co-imide)s containing hydroxyl pendant groups in a green ionic solvent. Polymer-Plastics Technology and Engineering. 2012;**51**:1097-1105. DOI: 10.1080/03602559.2012.689060

[104] Mallakpour S, Zarei M. Novel, thermally stable and chiral poly(amide-imide)s derived from a new diamine containing pyridine ring and various amino acid-based diacids: Fabrication and characterization. High Performance Polymers. 2013;**25**:245-253. DOI: 10.1177/0954008312461658

[105] Gharekhani E, Mehdipour-Ataei S, Taghi-Ganji M. Soluble and thermally stable new pyridine-based poly(ether imide)s synthesized through conventional and ionic liquid methods. Iranian Polymer Journal. 2011;**20**: 491-502

[106] Mansoori Y, Atghia SV, Sanaei SS, Zamanloo MR, Imanzadeh G, Eskandari H. New, organo-soluble, thermally stable aromatic polyimides and poly(amide-imide) based on 2-[5-(3,5-dinitrophenyl)-1,3, 4-oxadiazole-2-yl] pyridine. Polymer International. 2012;**61**: 1213-1220. DOI: 10.1002/pi.4234

[107] Akbarian-Feizi L, Mehdipour-Ataei S, Yeganeh H. Investigation on the preparation of new sulfonated polyimide fuel cell membranes in organic and ionic liquid media. International Journal of Polymeric Materials and Polymeric Biomaterials. 2014;**63**:149-160. DOI: 10.1080/00914037.2013.812086

[108] Akbarian-Feizi L, Mehdipour-Ataei S, Yeganeh H. Synthesis of new sulfonated copolyimides in organic and ionic liquid media for fuel cell application. Journal of Applied Polymer Science. 2012;**124**:1981-1992. DOI: 10.1002/app.35136

[109] Hartanto Y, Corvilain M, Marien H, Janssesn J, Vankelecom IFJ. Interfacial polymerization of thin-film composite forward osmosis membranes using ionic liquids as organic reagent phase. Journal of Membrane Science. 2020;**601**:117869. DOI: 10.1016/j.memsci.2020.117869

[110] Li P, Coleman MR. Synthesis of room temperature ionic liquids based random copolyimides for gas separation applications. European Polymer Journal. 2013;**49**:482-491. DOI: 10.1016/j.eurpolymj.2012.11.016

[111] O'Harra KE, Kammakakam I, Bara JE, Jackson EM. Understanding the effects of backbone chemistry and anion type on the structure and thermal behaviors of imidazolium polyimide-ionenes. Polymer International. 2019;**68**: 1547-1556. DOI: 10.1002/pi.5825

[112] Shaplov AS, Morozova SM, Lozinskaya EI, Vlasov PS, Gouveia ASL, Tome LC, et al. Turning into poly(ionic liquid)s as a tool for polyimide modification. Synthesis, characterization and CO2 separation

properties. Polym. Chem. 2016;**7**: 580-591. DOI: 10.1039/c5py01553g

[113] O'Harra KE, Kammakakam I, Devriese EM, Doll DM, Bara JE, Jackson EM. Synthesis and performance of 6FDA-based polyimide-ionenes and composites with ionic liquids as gas separation membranes. Membranes. 2019;**9**:79. DOI: 10.3390/membranes9070079

[114] Govindaraj B, Sarojadevi M. Microwave-assisted synthesis and characterization of polyimide/functionalized MWCNT nanocomposites containing quinolyl moiety. Polymer Composites. 2016;**37**: 2417-2424. DOI: 10.1002/pc.23426

[115] Zhou LR, Li YZ, Wang ZQ, Zhang MY, Wang XD, Niu HQ, et al. Preparation of polyimide films via microwave-assisted thermal imidization. RSC Advances. 2019;**9**: 7314-7320. DOI: 10.1039/c9ra00355j

[116] Govindaraj B, Sarojadevi M. Microwave-assisted synthesis of nanocomposites from polyimides chemically cross-linked with functionalized carbon nanotubes for aerospace applications. Polymers for Advanced Technologies. 2016;**29**: 1718-1726. DOI: 10.1002/pat.4275

[117] Wang Y, Huang MY, Lei Y. Improved synthesis and properties of BTDA-TDI/MDI ternary copolyimides via microwave-assisted polycondensation. High Performance Polymers. 2020;**32**: UNSP0954008320908635. DOI: 10.1177/0954008320908635

[118] Cheng K, Hu JP, Wu YC, Shi CQ, Chen ZG, Liu SM, et al. Microwave-assisted synthesis of high thermal stability and colourless polyimides containing pyridine. Royal Society Open Science. 2019;**6**:190196. DOI: 10.1098/rsos.190196

[119] Li TY, Liu JJ, Zhao SS, Chen ZQ, Huang HH, Guo RL, et al. Microporous polyimides containing bulky tetra-o-isopropyl and naphthalene groups for gas separation membranes. Journal of Membrane Science. 2019;**585**:282-288. DOI: 10.1016/j.memsci.2019.05.003

[120] Li TY, Huang HH, Wang L, Chen YM. High performance polyimides with good solubility and optical transparency formed by the introduction of alkyl and naphthalene groups into diamine monomers. RSC Advances. 2017;**7**:40996-41003. DOI: 10.1039/c7ra07142f

[121] Choi JY, Jin SW, Kim DM, Song IH, Nam KN, Park HJ, et al. Enhancement of the mechanical properties of polyimide film by microwave irradiation. Polymers. 2019;**11**:477. DOI: 10.3390/polym11030477

[122] Xiao JD, Qiu LG, Yuan YP, Jiang X, Xie AJ, Shane YH. Ultrafast microwave-enhanced ionothermal synthesis of luminescent crystalline polyimide nanosheets for highly selective sensing of chromium ions. Inorganic Chemistry Communications. 2013;**29**:128-130. DOI: 10.1016/j.inoche.2012.12.028

[123] Seckin T, Kivilcim N, Koytepe S. Synthesis of triazine-based polyimide nanocomposites with flower-like ZnO by microwave-assisted solvothermal technique. Polymer-Plastics Technology and Engineering. 2013;**52**:1368-1375. DOI: 10.1080/03602559.2013.820747

[124] Fujii S, Kawamura S, Mochizuki D, Maitani MM, Suzuki E, Wada Y. Microwave sintering of Ag-nanoparticle thin films on a polyimide substrate. AIP Advances. 2015;**5**:127226. DOI: 10.1063/1.4939095

[125] Dasgupta J, Sikder J, Chakraborty S, Adhikari U, Reddy BVP, Mondal A, et al. Microwave-assisted modified polyimide synthesis. A facile route to the enhancement of visible-

light-induced photocatalytic performance for dye degradation. ACS Sustainable Chemistry & Engineering. 2017;**5**:6817-6826. DOI: 10.1021/acssuschemeng.7b01044

[126] Lin L, Ye P, Cao C, Jin Q, Xu GS, Shen YH, et al. Rapid microwave-assisted green production of a crystalline polyimide for enhanced visible-light-induced photocatalytic hydrogen production. Journal of Materials Chemistry A. 2015;**3**:10205-10208. DOI: 10.1039/c5ta01078k

[127] Wang K, Yuan XJ, Zhan MS. Comparison between microwave and thermal curing of a polyimide adhesive end-caped with phenylethynyl groups. International Journal of Adhesion and Adhesives. 2017;**74**:28-34. DOI: 10.1016/j.ijadhadh.2016.12.008

[128] Imai Y. Recent progress in synthesis of polyimides. Journal of Photopolymer Science and Technology. 1994;**7**:251-256. DOI: 10.2494/photopolymer.7.251

[129] Baumgartner B, Bojdys MJ, Unterlass MM. Geomimetics for green polymer synthesis: Highly ordered polyimides via hydrothermal techniques. Polymer Chemistry. 2014;**5**:3771-3776. DOI: 10.1039/c4py00263f

[130] Baumgartner B, Puchberger M, Unterlass MM. Towards a general understanding of hydrothermal polymerization of polyimides. Polymer Chemistry. 2015;**6**:5773-5781. DOI: 10.1039/c5py00231a

[131] Unterlass MM, Kopetzki D, Antonietti M, Weber J. Mechanistic study of hydrothermal synthesis of aromatic polyimides. Polymer Chemistry. 2011;**2**:1744-1753. DOI: 10.1039/c1py00109d

[132] Baumgartner B, Bojdys MJ, Skrinjar P, Unterlass MM. Design strategies in hydrothermal polymerization of polyimides. Macromoecular Chemistry and Physics. 2016;**217**:485-500. DOI: 10.1002/macp.201500287

[133] Unterlass MM, Emmerling F, Antonietti M, Weber J. From dense monomer salt crystals to CO2 selective microporous polyimides via solid-state polymerization. Chemical Communications. 2014;**50**:430-432. DOI: 10.1039/c3cc47674j

[134] Taublaender MJ, Reiter M, Unterlass MM. Highly crystalline, nanostructured polyimide microparticles via green and tunable solvothermal polymerization. Macromolecules. 2019;**52**:6318-6329. DOI: 10.1021/acs.macromol.9b00985

[135] Cui ZW, Zhou J, Liu T, Wang YC, Hu Y, Wang Y, et al. Porphyrin-containing polyimide with enhanced light absorption and photocatalysis activity. Chemistry, an Asian Journal. 2019;**14**:2138-2148. DOI: 10.1002/asia.201900261

[136] Leimhofer L, Baumgartner B, Puchberger M, Prochaska T, Konegger T, Unterlass MM. Green one-pot synthesis and processing of polyimide–silica hybrid materials. Journal of Materials Chemistry A. 2017;**5**:16326-16335. DOI: 10.1039/c7ta02498c

[137] Zhuo LH, Tang SL, Zhao KY, Xie F, Bai Y. Green facile fabrication of polyimide by microwave-assisted hydrothermal method and its decomposition dynamics. Journal of Applied Polymer Science. 2020;**137**:48484. DOI: 10.1002/app.48484

[138] Jiang SH, Hou HQ, Agarwal S, Greiner A. Polyimide nanofibers by "green" electrospinning via aqueous solution for filtration applications. ACS Sustainable Chemistry & Engineering. 2016;**4**:4797-4804. DOI: 10.1021/acssuschemeng.6b01031

[139] Abadie MJM, Marchal J. Dégradation oxydante du polystyrène à 25°C. II. Étude du mécanisme par amorçage radiochimique en solution chloroformique diluée. Die Makromolekulare Chemie. 1971;**14**: 299-316. DOI: 10.1002/macp.1971.021410123

[140] Ferreira FAS, Battirola LC, Lewicki JP, Worsley MA, Pereira-da-Silva MA, Amaral T, et al. Influence of thermal treatment time on structural and physical properties of polyimide films at beginning of carbonization. Polymer Degradation and Stability. 2016;**129**:399-407. DOI: 10.1016/j.polymdegradstab. 2016.05.001

[141] Hondred PR, Yoon S, Bowler N, Moukhina E, Kessler MR. Degradation kinetics of polyimide film. High Performance Polymers. 2011;**23**: 335-342. DOI: 10.1177/0954008311409262

[142] Hondred PR, Bowler N, Kessler MR. Electrothermal lifetime prediction of polyimide wire insulation with application to aircraft. Journal of Applied Polymer Science. 2013;**130**: 1639-1644. DOI: 10.1002/app.39304

[143] Lin L, Bowler N, Hondred PR, Kessler MR. Influence of thermal degradation and saline exposure on dielectric permittivity of polyimide. Journal of Physics and Chemistry of Solids. 2011;**72**:875-881. DOI: 10.1016/j.jpcs.2011.04.004

[144] Dine-Hart RA, Parker DBV, Wright WW. Oxidative degradation of a polyimide film: I. initial studies. British Polymer Journal. 1971;**3**:222-225. DOI: 10.1002/pi.4980030504

[145] Su X, Xu Y, Li L, Song C. Characterization and thermal degradation kinetics of thermoplastic polyimide based on BAPP. High Performance Polymers. 2017;**30**:787-793. DOI: 10.1177/0954008317729741

[146] Liu Y, Xu X, Mo S, Lan B, Zhu C, Xu J, et al. Kinetic study of the multistep thermo-oxidative degradation of thermosetting siloxane-containing polyimide and unmodified polyimide. Journal of Applied Polymer Science. 2020:e49021. DOI: 10.1002/app.49021

[147] Huertas RM, Tena A, Lozano AE, de Abajo J, de la Campa JG, Maya EM. Thermal degradation of crosslinked copolyimide membranes to obtain productive gas separation membranes. Polymer Degradation and Stability. 2013;**98**:743-750. DOI: 10.1016/j.polymdegradstab. 2012.12.017

[148] Li X, Miyauchi M, Gonzalez C, Nutt S. Thermal oxidation of PEPA-terminated polyimide. High Performance Polymers. 2018;**31**:707-718. DOI: 10.1177/0954008318787852

[149] Hamciuc C, Lisa G, Hamciuc E, Tudorachi N. Thermal decomposition study of some polyimide-polydimethylsiloxane copolymers. Journal of Analytical and Applied Pyrolysis. 2018;**129**:204-214. DOI: 10.1016/j.jaap.2017.11.011

[150] Khazaka R, Locatelli ML, Diaham S, Bidan P. Effects of mechanical stresses, thickness and atmosphere on aging of polyimide thin films at high temperature. Polymer Degradation and Stability. 2013;**98**: 361-367. DOI: 10.1016/j.polymdegradstab. 2012.09.005

[151] Perrota C, Gonona L, Marestinb C, Gebela G. Hydrolytic degradation of sulfonated polyimide membranes for fuel cells. Journal of Membrane Science. 2011;**379**:207-214. DOI: 10.1016/j.memsci.2011.05.063

[152] Xu Y, Zehnder AT. Moisture degradation effects on the mechanical properties of HFPE-II-52 polyimide: Experiments and modeling. Experimental Mechanics. 2017;**57**:857-869. DOI: 10.1007/s11340-017-0281-3

[153] Xu Y, Zehnder AT. Pressure, hydrolytic degradation and plasticization drive high temperature blistering failure in moisture saturated polyimides. Extreme Mechanics Letters. 2017;**16**:49-55. DOI: 10.1016/j.eml.2017.09.002

[154] Sili E, Cambronne JP, Naude N, Khazaka R. Polyimide lifetime under partial discharge aging: Effects of temperature, pressure and humidity. IEEE Transactions on Dielectrics and Electrical Insulation. 2013;**20**:435. DOI: 10.1109/TDEI.2013.6508745

[155] Campbell FJ, Brewer AK, Orr RJ, Janicke TA, Brunining AM. Hydrolytic deterioration of polyimide insulation on naval aircraft wiring. In: Proceeding of the IEEE Conference on Electrical Insulation and Dielectric Phenomena; 16–20 October 1988. Ottawa, New York: IEEE; 1988. pp. 180-188. DOI: 10.1109/CEIDP.1988.26329

[156] Huang X, Pu Y, Zhou Y, Zhang Y, Zhang H. In-situ and ex-situ degradation of sulfonated polyimide membrane for vanadium redox flow battery application. Journal of Membrane Science. 2017;**526**:281-292. DOI: 10.1016/j.memsci.2016.09.053

[157] Harito C, Bavykin DV, Yuliarto B, Dipojono HK, Walsh FC. Inhibition of polyimide photodegradation by incorporation of titanate nanotubes into a composite. Journal of Polymers and the Environment. 2019;**27**:1505-1515. DOI: 10.1007/s10924-019-01443-w

[158] Rusu RD, Constantin CP, Drobota M, Gradinaru LM, Butnaru M, Pislaru M. Polyimide films tailored by UV irradiation: Surface evaluation and structure-properties relationship. Polymer Degradation and Stability. 2020;**177**:109182. DOI: 10.1016/j.polymdegradstab. 2020.109182

[159] Rasmussen K, Grampp G, van Eesbeek M, Rohr T. Thermal and UV degradation of polymer films studied in situ with ESR spectroscopy. ACS Applied Materials & Interfaces. 2010;**2**:1879-1883. DOI: 10.1021/am100219z

[160] Severin D, Balanzat E, Ensinger W, Trautmann C. Outgassing and degradation of polyimide induced by swift heavy ion irradiation at cryogenic temperature. Journal of Applied Physics. 2010;**108**:024901. DOI: 10.1063/1.3457846

[161] Damaceanu MD, Rusu RD, Olaru MA, Timpu D, Bruma M. KrF pulsed laser ablation of thin films made from fluorinated heterocyclic poly (naphthyl-imide)s. Microscopy and Microanalysis. 2012;**18**:545-557. DOI: 10.1017/S1431927612000165

[162] Damaceanu MD, Rusu RD, Olaru M, Stoica I, Bruma M. Nanostructured polyimide films by UV excimer laser irradiation. Romanian Journal of Information Science and Technology. 2010;**13**:368-377

[163] Ennis CP, Kaiser RI. Mechanistical studies on the electron-induced degradation of polymethylmethacrylate and Kapton. Physical Chemistry Chemical Physics. 2010;**12**:14902-14915. DOI: 10.1039/c0cp01130d

[164] Dong SS, Shao WZ, Yang L, Ye HJ, Zhena L. Surface characterization and degradation behavior of polyimide films induced by coupling irradiation treatment. RSC Advances. 2018;**8**:28152. DOI: 10.1039/c8ra05744c

[165] Plis E, Engelhart DP, Barton D, Cooper R, Ferguson D, Hoffmann R. Degradation of polyimide under exposure to 90 keV electrons. Physica Status Solidi B: Basic Solid State Physics. 2017;**254**:1600819. DOI: 10.1002/pssb.201600819

[166] Lin RH, Wang WM, Chen YH, Ho TH. Preparation and characterization of biodegradable